Transnational Agrarian Movements Confronting Globalization

Transnational Agrarian Movements Confronting Globalization

Edited by

Saturnino M. Borras Jr, Marc Edelman
and Cristóbal Kay

A John Wiley & Sons, Ltd., Publication

This edition first published 2008
Originally published as Volume 8, Issue 2/3 of *Journal of Agrarian Change*
Chapters © 2008 The Authors
Book compilation © 2008 Blackwell Publishing Ltd

Blackwell Publishing was acquired by John Wiley & Sons in February 2007. Blackwell's publishing program has been merged with Wiley's global Scientific, Technical, and Medical business to form Wiley-Blackwell.

Registered Office
John Wiley & Sons Ltd, The Atrium, Southern Gate, Chichester, West Sussex, PO19 8SQ, United Kingdom

Editorial Offices
350 Main Street, Malden, MA 02148-5020, USA
9600 Garsington Road, Oxford, OX4 2DQ, UK
The Atrium, Southern Gate, Chichester, West Sussex, PO19 8SQ, UK

For details of our global editorial offices, for customer services, and for information about how to apply for permission to reuse the copyright material in this book please see our website at www.wiley.com/wiley-blackwell.

The right of Saturnino M. Borras Jr, Marc Edelman and Cristóbal Kay to be identified as the authors of the editorial material in this work has been asserted in accordance with the Copyright, Designs and Patents Act 1988.

All rights reserved. No part of this publication may be reproduced, stored in a retrieval system, or transmitted, in any form or by any means, electronic, mechanical, photocopying, recording or otherwise, except as permitted by the UK Copyright, Designs and Patents Act 1988, without the prior permission of the publisher.

Wiley also publishes its books in a variety of electronic formats. Some content that appears in print may not be available in electronic books.

Designations used by companies to distinguish their products are often claimed as trademarks. All brand names and product names used in this book are trade names, service marks, trademarks or registered trademarks of their respective owners. The publisher is not associated with any product or vendor mentioned in this book. This publication is designed to provide accurate and authoritative information in regard to the subject matter covered. It is sold on the understanding that the publisher is not engaged in rendering professional services. If professional advice or other expert assistance is required, the services of a competent professional should be sought.

Library of Congress Cataloging-in-Publication Data is available for this book.

ISBN 978-1-4051-9041-1 (paperback)

A catalogue record for this book is available from the British Library.

Set in 10/12 Bembo by Graphicraft Limited, Hong Kong

01   2008

Contents

	Notes on Contributors	vii
	Henry Bernstein and Terence J. Byres Foreword	ix
1	**Saturnino M. Borras Jr, Marc Edelman and Cristóbal Kay** Transnational Agrarian Movements: Origins and Politics, Campaigns and Impact	1
2	**Philip McMichael** Peasants Make Their Own History, But Not Just as They Please . . .	37
3	**Marc Edelman** Transnational Organizing in Agrarian Central America: Histories, Challenges, Prospects	61
4	**Saturnino M. Borras Jr** La Vía Campesina and its Global Campaign for Agrarian Reform	91
5	**Brenda Baletti, Tamara M. Johnson and Wendy Wolford** 'Late Mobilization': Transnational Peasant Networks and Grassroots Organizing in Brazil and South Africa	123
6	**Ian Scoones** Mobilizing Against GM Crops in India, South Africa and Brazil	147
7	**Peter Newell** Trade and Biotechnology in Latin America: Democratization, Contestation and the Politics of Mobilization	177
8	**Nancy Lee Peluso, Suraya Afiff and Noer Fauzi Rachman** Claiming the Grounds for Reform: Agrarian and Environmental Movements in Indonesia	209
9	**Harriet Friedmann and Amber McNair** Whose Rules Rule? Contested Projects to Certify 'Local Production for Distant Consumers'	239
10	**Jonathan Fox and Xochitl Bada** Migrant Organization and Hometown Impacts in Rural Mexico	267
11	**Kathy Le Mons Walker** From Covert to Overt: Everyday Peasant Politics in China and the Implications for Transnational Agrarian Movements	295
12	**Kevin Malseed** Where There Is No Movement: Local Resistance and the Potential for Solidarity	323
	Index	349

Notes on Contributors

Suraya Afiff is at the Department of Agriculture, University of Indonesia, Jakarta, Indonesia (safiff@inspiritinc.net).

Xochitl Bada is at the Sociology Department, University of Notre Dame, IN, USA (xbada@nd.edu).

Brenda Baletti is at the Department of Geography, University of North Carolina at Chapel Hill, NC, USA (bbaletti@email.unc.edu).

Saturnino M. Borras Jr is Canada Research Chair in International Development Studies at Saint Mary's University, Halifax, Nova Scotia, Canada (sborras@smu.ca).

Marc Edelman is Professor of Anthropology, Department of Anthropology, Hunter College and the Graduate Center, City University of New York, USA (medelman@hunter.cuny.edu).

Jonathan Fox is Professor of Social Sciences, Latin American and Latino Studies, University of California, Santa Cruz, CA, USA (jafox@ucsc.edu).

Harriet Friedmann is Professor of Sociology and Fellow of the Centre for International Studies, University of Toronto, Canada (harriet.friedmann@utoronto.ca).

Tamara M. Johnson is at the Department of Geography, University of North Carolina at Chapel Hill, NC, USA (johnson5@email.unc.edu).

Cristóbal Kay is Professor of Development Studies and Rural Development at the Institute of Social Studies (ISS), The Hague, Netherlands (kay@iss.nl).

Kathy Le Mons Walker is Associate Professor of History, Department of History, Temple University, Philadelphia, PA, USA (kwalker@temple.edu).

Kevin Malseed is Program Fellow in Agrarian Studies, Yale University, New Haven, CT (kevin@khrg.org), and founder and researcher with the Karen Human Rights Group, 1992–2007, Thailand.

Philip McMichael is Professor of Development Sociology, Department of Development Sociology, Cornell University, Ithaca, NY, USA (pdm1@cornell.edu).

Amber McNair is at the Department of Sociology, University of Toronto, Canada (amber.mcnair@utoronto.ca).

Peter Newell is Professor of Development Studies, School of Development Studies, University of East Anglia, Norwich, UK (P.Newell@uea.ac.uk).

Nancy Lee Peluso is Professor of Society and Environment, Department of Environmental Science, Policy and Management, University of California, Berkeley, CA, USA (npeluso@nature.berkeley.edu).

Noer Fauzi Rachman is at the Department of Environmental Science, Policy and Management, University of California, Berkeley, CA, USA (noer@nature.berkeley.edu).

Ian Scoones is Professorial Fellow at the Institute of Development Studies, University of Sussex, Brighton, UK (I.Scoones@ids.ac.uk).

Wendy Wolford is Associate Professor of Geography, Department of Geography, University of North Caroline at Chapel Hill, NC, USA (wwolforfd@email.unc.edu).

Foreword

HENRY BERNSTEIN AND TERENCE J. BYRES

In 2008 we retired from a combined total of 56 years of editorial leadership of the *Journal of Peasant Studies* (*JPS*) and then the *Journal of Agrarian Change* (*JAC*). The *Journal of Peasant Studies* was founded by T.J. Byres in 1973 and we edited it together from 1985 to 2000. We started the *Journal of Agrarian Change* in 2001 to continue the intellectual project of the *Journal of Peasant Studies*. Once the *Journal of Agrarian Change* was firmly established and thriving, we decided that the moment was right to hand over to a new editorial team in 2008. They are Deborah Johnston, Cristóbal Kay, Jens Lerche and Carlos Oya. In the final period of our editorial office we had agreed the preparation of a special double issue on *Transnational Agrarian Movements Confronting Globalization* with Jun Borras, Marc Edelman and Cristóbal Kay as guest editors. It duly appeared as the *Journal of Agrarian Change* Vol 8 Nos 2–3, and is now also published as this book. It was a pleasure to receive the initial proposal for the special issue from its three editors, to discuss it extensively with them and support them in their efforts, and we are privileged by their invitation to write a Preface for this book version.

In using this brief Preface to commend the imagination and illumination evident in the work of the editors and their contributors, we can also locate the importance of their achievement in the changing historical contexts of agrarian issues over the last 35 years since the *Journal of Peasant Studies* started. Its initial agenda manifested the intense intellectual and political currents and concerns of a particular conjuncture, characterized by continuing struggles against imperialism in which rural people were major actors (notably the Vietnamese struggle for national liberation); the prospects for economic and social development in the recently independent, and mostly agrarian, former colonies of Asia and Africa, as well as Latin America; and the ways in which those prospects could be explored through investigating practices, experiences and theories of socialist transformation of the countryside (in which China loomed large). These concerns were informed by a materialist political economy and a strongly historical approach. Thus the *Journal of Peasant Studies* from 1973 to 2000 presented a cornucopia of articles on agrarian social formations before capitalism; transitions to capitalism in its original heartlands and subsequently; agrarian change in colonial conditions and in the revolutionary circumstances of the USSR and China, Vietnam and Cuba; processes of agrarian change in the independent countries of the Third World in the contexts of their various projects of 'national development'; and later the demise of the very different types of state socialism in the USSR, China and elsewhere, and its consequences.

The *Journal* promoted investigation and debate of the social relations and dynamics of production and reproduction, property and power, in agrarian formations and their processes of change, both historical and contemporary. As its name suggested, much of the terrain of its concerns and inquiries was inhabited by rural producers defined in one way or another as 'peasants', who might be considered both victims and active protagonists in the historical dramas of the making of modernity. This also suggests that our understanding of political economy, which guided the *Journal's* approach to 'peasant studies', was an expansive one that incorporated and engaged with many issues of sociology and politics, culture and ideology.

There are, of course, often lags between the rhythms of intellectual inquiry and changes in the wider world, sometimes as rapid as they are contradictory. As the *Journal of Peasant Studies* was establishing its distinctive intellectual identity and international reputation, the conjuncture which inspired its founding and development was starting to undergo profound changes that became more evident over the next two decades. One such change was the accumulation crisis and subsequent restructuring of the international capitalist economy from the 1970s, namely 'globalization', and the political and policy agendas devised by capital to manage and promote it from the 1980s, namely 'neo-liberalism'. International divisions of labour in farming and agribusiness, and markets for agricultural products, were reshaped by the dynamics of globalization and neoliberalism (or 'neoliberal globalization'), albeit subject to the usual unevenness of capitalism. This had widespread effects, both direct and indirect, for 'peasants' and others in the countrysides of 'the South' (as indeed for farmers in 'the North'). One manifestation of these comprehensive changes was the end of the international food regime constructed under US hegemony after the Second World War, as analysed in the pioneering work of Harriet Friedmann and Philip McMichael, both contributors to this volume.

A very different kind of change, albeit no doubt connected in some ways, was that the 1970s arguably marked the end of the long period of 'peasant wars of the twentieth century' over the previous 60 years, famously explored in the work of Eric Wolf. It began to seem that the upheavals of 'classic' peasant revolts and rebellions – a world-historical dynamic central to the making of modernity from the French revolution to the victory of the long Vietnamese struggle for national liberation – were now over. This did not mean, of course, that forms of agrarian and rural politics, of resistance and opposition, other than the epic instances of 'peasant wars', lost their relevance. Indeed, the *Journal of Peasant Studies* presented material across the range of the political sociology of the countryside, including two special issues (also published as books) on 'everyday forms of resistance', highlighted in the work of James Scott, and on the new farmers' movements of India that rose to prominence in the 1980s. In the 1990s the *Journal* also began to publish articles on the new conditions and effects of the internationalisation of agriculture generated by 'globalization'.

This carried over to the *Journal of Agrarian Change* from 2001, which has also published a number of strong articles on the contemporary politics of land, and of land reform, recently revived in the agenda of neoliberal development discourse and policy, for example, by the World Bank, and by agrarian social movements in different parts of the 'global South'. The latter typically articulate their claims in terms of the interests of 'peasants', 'family farmers' or 'people of the land', and sometimes celebrate what they consider processes of 're-peasantization', thereby resonating older traditions of agrarian populism and reasserting their salience to opposition to neoliberal globalization.

One striking manifestation today of the 'new' politics of agrarian resistance is the emergence of 'transnational agrarian movements' (TAMs), the subject of this collection. International organizations of farmers, of very different social composition and political complexion and purpose, are not entirely novel, as the editors usefully show in their introductory chapter. However, current transnational agrarian movements – constituted on a regional and transcontinental level – can plausibly claim new forms of mobilization and action in their opposition to neoliberal globalization and the havoc it wreaks. If research on globalization helps our understanding of the accumulation strategies of contemporary agribusiness, and their effects for farming and farmers, that political economy needs to be matched by a similarly rigorous political sociology. This is a formidable challenge, on which this collection makes its own distinctive and significant mark. As the editors observe in the conclusion to their introduction:

> Transnational agrarian movements are political projects with deep historical roots in diverse national societies, multiple and shifting alliances, varied action repertoires, and complex forms of representation, issue framing and demand making . . . By acknowledging TAMs' contradictions, ambiguities and internal tensions . . . (the authors in this collection) also seek, from the standpoint of engaged intellectuals, to advance a transformative political project by better comprehending its roots, past successes and failures, and current and future challenges.

This intention is formulated with admirable clarity, thoroughness and precision in the editors' own introductory essay, which also provides a valuable contribution to current explorations of 'new' social movements more broadly. The combination of political engagement with intellectual sobriety and responsibility, to which they refer, is a necessary antidote to 'movement' voluntarism and triumphalism, and is expressed in exemplary fashion in their discussion of 'silences in the literature' to date. Those silences concern such issues as class relations, and other bases of social differentiation and inequality, among 'farmers'/'people of the land' – always a central concern of agrarian political economy – as well as complex questions of political representation, legitimacy and leadership, ideological and political differences within transnational agrarian movements and, indeed, their constituent organizations, and the very uneven political geographies of transnational agrarian movements.

The introductory chapter thus provides an excellent springboard for the diverse and rich contributions that follow. These take on board both familiar themes like demands for land and (other) public support for small farmers, and resistance to state oppression and dispossession, and new issues generated by neoliberal globalization like the effects of trade liberalization, corporate control of crop genetic material, and market-led land reform. Moreover, some of the contributions follow the lead given by the introduction in considering the sociological complexities of rural class, gender and ethnic relations, and how they are compounded by movement across, and connections between, countryside and town, and by international labour migration. The various chapters explore the themes noted in many of the major regions and countries of the 'global South', from Central America and Brazil, which provide two of the most emblematic agrarian social movements of our times, to Indonesia and the Philippines, as well as Burma and China which transnational agrarian movements have yet to reach and which demonstrate their own important forms of agrarian and rural resistance. Sub-Saharan Africa, other than South Africa, is a notable gap; while India receives limited attention (in a chapter comparing anti-GM campaigns in India, South Africa and Brazil), although the editors properly note that the KRRS (Karnataka State Farmers' Association) – often heralded as the most important oppositional agrarian movement in India – is dominated by middle and rich farmers.

It is inevitable that there will be some gaps in a collection of this kind. Notwithstanding such gaps, the book provides a highly significant, impressive, and indeed seminal contribution to understanding transnational agrarian movements in the conditions of today's neoliberal globalization. We are confident that its importance will be appreciated and will stimulate much fruitful debate, and further research into the themes and issues it embraces, illuminates and develops.

1 Transnational Agrarian Movements: Origins and Politics, Campaigns and Impact

SATURNINO M. BORRAS JR, MARC EDELMAN
AND CRISTÓBAL KAY

THE RISE OF TRANSNATIONAL AGRARIAN MOVEMENTS

How have recent changes in the global political-economy affected the autonomy and capacity of the 'rural poor'[1] to understand their condition, assess political opportunities and threats, frame their worldviews, forge collective identities and solidarity, build movements and launch interlinked collective actions? What are the emerging forms of local, national and international resistance and how do these impact processes of agrarian change? How and to what extent has the scholarly literature in social movements and agrarian studies been able to catch up with this rapidly changing terrain of ideology, politics and organizations of (trans)national agrarian movements? These are the broad questions addressed in this collection.

Neoliberalism has significantly altered the dynamics of agrarian production and exchange relations within and between countries across the north–south divide. The simultaneous processes of globalization 'from above', partial decentralization 'from below' and privatization 'from the side' of the central state that used to play a key role in the maintenance or development of agrarian systems have shaken rural society to its core (see, for example, Edelman 1999; Gwynne and Kay 2004). These broader societal processes coincided with the most recent wave of agrarian restructuring, providing even greater power to transnational and domestic capital to dictate the terms of agricultural production and exchange (Byres 2003; Friedmann 2004; Bernstein 2006; McMichael 2006; Akram Lodhi and Kay 2008). While there are winners and losers in this global–local restructuring, working people and their livelihoods increasingly face ever more precarious conditions. Diversification of (rural and rural–urban; on-farm, off-farm or non-farm) livelihoods, forced or otherwise, has been widespread (Bryceson et al. 2000; Ellis 2000; Rigg 2006; World Bank 2007). Access to and control over land resources are being redefined and landed property rights restructured to favour private capital (De Soto 2000; World Bank 2003; but see Rosset et al. 2006; Lahiff et al. 2007; Akram Lodhi et al. 2007).

---

[1] The 'rural poor' is obviously a highly heterogeneous category; here, it includes small owner-cultivators, sharecroppers, tenant farmers, rural labourers, migrant labourers, subsistence fisherfolk and fish workers, forest dwellers, indigenous peoples, peasant women and pastoralists. The differentiation among the rural poor has class, gender, race, ethnic and caste dimensions, among others.

These global–local complex processes have affected 'agrarian movements' in a variety of ways, many of which used to (and continue to) operate solely within local communities and national borders. Today, many agrarian movements have 'localized' their struggles in response to partial decentralization, others have focused on 'privatized' activities in a manner akin to 'state-substitution' on development issues such as social service delivery, while others have 'internationalized' their struggles in response to global agrarian restructuring.

In recent years, Transnational Agrarian Movements – or TAMs, for short – (taken here in a loose definition to mean 'movements', 'organizations', 'coalitions', 'networks' and 'solidarity linkages' of the 'rural poor') and some of the national peasants' and farmers' groups directly linked to these transnational movements have gained considerable power and political influence (and in some quarters, perhaps notoriety).[2] La Vía Campesina is the most well known of all the contemporary TAMs, networks or coalitions (Borras 2004; Desmarais 2007). There are, however, numerous other transnational movements, networks and coalitions that are based among rural sectors or advocate for rural people; some are engaged in left-wing politics, while others are less radical (Edelman 2003).

Perhaps the largest agrarian-based and agrarian-oriented transnational network today is the International Planning Committee (IPC) for Food Sovereignty, with more than 500 rural social movements and NGOs, radical and conservative, as members (http://www.foodsovereignty.org). The fisheries sector also has relatively vibrant transnational networks, including the World Forum of Fish Harvesters and Fishworkers (WFF), World Forum of Fisher Peoples (WFFP) and the International Collective in Support of Fish Workers (ICSFW). The agricultural workers' sector is relatively less visible, but the International Union of Food, Agricultural, Hotel, Restaurant, Catering, Tobacco and Allied Workers' Associations (IUF) remains an active and significant global actor. Some agrarian-based networks are relatively new, but others have been around for decades, such as the International Federation of Adult Catholic Farmers' Movements (FIMARC), founded in the 1950s. There are even more numerous agrarian-oriented transnational civil society networks, such as the FoodFirst Information and Action Network (FIAN), the Land Research and Action Network (LRAN), the Erosion, Technology and Concentration Group (ETC Group), the Genetic Resources Action International (GRAIN) and Friends of the Earth (Edelman 2003).[3]

Some of these global movements and networks have ideological and political orientations that are fundamentally different, as is the case with Vía Campesina and the International Federation of Agricultural Producers (IFAP), or Vía Campesina and the Asian Peasant Coalition (APC). Others have closer fraternal relationships, such as Vía Campesina and the International Movement of Catholic

---

[2] For more nuanced academic discussions about the differences and similarities of these categories in the context of cross-border or transnational 'movement' literature, see Fox (2000, 9–12, 45) and Khagram et al. (2002, 9).
[3] The full names of organizations are provided when they are first mentioned. After that most are referred to by their abbreviations.

Agricultural and Rural Youth (MIJARC). Meanwhile, many others are regional in focus, such as the *Coordinadora Latinoamericana de Organizaciones del Campo* (CLOC, Latin American Coordination of Peasant Organizations) and the *Réseau des Organisations Paysannes et de Producteurs de l'Afrique de l'Ouest* (Network of Peasants' and Producers' Organizations of West Africa, ROPPA). Moreover, many of the agrarian movements are differentiated in terms of the class origin of their main mass base. In IFAP, for example, medium and large farmers are dominant, while in Vía Campesina the member organizations consist mainly of poor peasants and small farmers. Nonetheless, the large transnational movements tend to be highly heterogeneous in terms of class as well: IFAP also has members that come from the ranks of poor peasants and small farmers, while Vía Campesina has members that come from the ranks of middle and rich farmers. Finally, many transnational movements, networks and coalitions have overlapping memberships, even those that are rivals, such as Vía Campesina and IFAP.

Many of these groups, and especially Vía Campesina, perhaps the most politically coherent of all contemporary TAMs or networks, have significantly undermined major conferences of important intergovernmental institutions, most notably the World Trade Organization (WTO), International Monetary Fund (IMF) and the World Bank. In Seattle, Washington DC, Cancún, Hong Kong, and elsewhere, theatrical protests with significant peasant and farmer participation contributed to raising public and media awareness of these movements. Following the violence outside the 2001 G-8 summit in Genoa, for example, *Newsweek* singled out the Vía Campesina as one of eight 'kinder, gentler globalist' groups behind the anti-G-8 protests (*Newsweek* 2001, 17). In 2008, the London *Guardian* included Vía Campesina coordinator and Indonesian peasant leader Henry Saragih in its list of 'ultimate green heroes', the '50 people who could save the planet' (*Guardian* 2008).

Many of the national movements that are leading members of transnational networks have engaged in dramatic anti-corporate actions, such as bulldozing a McDonald's fast-food shop in France, burning a Kentucky Fried Chicken outlet in Bangalore, and uprooting a GM soya farm and a eucalyptus nursery in Brazil, among others. The protest suicide of Lee Kyang Hae, a South Korean farmer, during the Cancún WTO negotiations was another form of dramatic action.

Yet, the same agrarian movements also sit in the consultative bodies of some United Nations (UN) agencies, such as the Food and Agriculture Organization (FAO), International Fund for Agricultural Development (IFAD) and the United Nations Commission on Human Rights (UNCHR). They negotiate and bargain with international bureaucrats for a variety of agendas that range from policy reforms to funding support. They negotiate and agree to some compromises with select nongovernmental donor and cooperation agencies about the terms of funding for their activities. Official venues of international representation have witnessed profound changes in recent years when TAMs asserted their own distinct, direct representation in these 'spaces' – challenging the traditional 'occupants' of such representation spaces, particularly intermediary NGOs. In some cases, more radical agrarian movements have been able to gain seats in these official venues, partially undermining the hegemony previously enjoyed by more conservative movements. These challenges to the traditional clout of NGOs and conservative movements such as IFAP can be seen, for example, in Vía Campesina's ability to eventually gain direct representation in the official global interface mechanism (Farmers' Forum) with farmers' groups at the International Fund for Agricultural Development (IFAD), in the Food and Agriculture Organization of the United Nations (FAO), and at the International Conference on Agrarian Reform and Rural Development (ICARRD).

The TAMs' growing international presence, visibility, voice and political influence has inspired a broad range of progressive and radical non-agrarian networks, from environmentalists to human rights groups, resulting at times in new kinds of powerful synergies. The rise of the (new) peasant movements in general and TAMs in particular has also generated a new wave of academic scholarship, from Marxist to postmodern interpretations of such phenomena (see, for example, Harvey 1998; Edelman 1999; Brass 2000). It has provided a much needed concrete justification for some nongovernmental funding agencies' programmes. Indeed, TAMs have achieved a level of reach and influence that can no longer be ignored even by mainstream development and financial institutions, which have intensified their attempts to capture and co-opt, collaborate with or undermine at least some of these transnational movements.

Most observers focus their attention on only the most visible and 'noisy' global movements, such as Vía Campesina. But current TAMs and networks are plural and diverse, as indicated earlier. Some observers tend to assume, as many agrarian activists frequently claim, that contemporary TAMs constitute a *new* phenomenon. But TAMs and networks are not always entirely new. Many comparable groups had existed before. Many of the actually existing movements or networks have been here for decades, as is the case, for example, of *Campesino a Campesino* (Peasant to Peasant) in Central America–Mexico, which started in the 1970s (Holt-Giménez 2006). Moreover, actually existing transnational movements or networks often build directly on older cross-border linkages from well before the neoliberal onslaught that commenced in the early 1980s (see, for example, Edelman 2003, 196–7). Many cross-border and cross-continental links were forged, for example, during the 1970s and 1980s as part of the vast political solidarity networks in Europe and North America that backed national liberation and anti-dictatorship movements in many developing countries, such as Chile, Nicaragua, South Africa and the Philippines. However, the roots of contemporary transnational alliance-building date back much earlier than the most immediate predecessors described above. Understanding the diversity and dynamics of contemporary TAMs partly requires an understanding of past TAMs and networks. It is to this topic that the next section is devoted.

HISTORICAL ANTECEDENTS

Transnational alliance-building among peasant and small farmer organizations accelerated after the late 1980s, but its roots lie as far back as the late nineteenth and early twentieth centuries. This indicates that cross-border organizing is not merely a result of new communications technologies, the growing reach of supra-national governance institutions or a weakening of the contemporary state system under globalization. Early transnational farmers' organizations manifested sometimes eclectic amalgams of agrarian populism, Communism, elite-led reformism and noblesse oblige, pacifism and feminism.

The Green International and the Red Peasant International

In the ten years after World War I, two rival international movements vied for peasant support in central and eastern Europe: the agrarian Green International, eventually headquartered in Prague, and the Moscow-based Peasant International or *Krestintern* (Jackson 1966, 51).[4] Following the War, agrarian or peasant-led political parties came to power in Bulgaria and Yugoslavia and had major influence in Czechoslovakia, Poland, Romania, Hungary, Austria and the Netherlands. The agrarian parties differed in ideology and practice, and each was typically composed of bitterly competing factions, but most sought to shift the terms-of-trade in favour of rural areas, to implement land reforms and to break the power of the traditional landed groups. The latter two goals were, of course, shared by the Communists, with whom the Agrarians had intimate and complex, occasionally collaborative and more usually antagonistic, relations in country after country.

The most formidable agrarian government was in Bulgaria, where in 1919, following a period of violence and instability, Alexander Stamboliski's Agrarian Union won the first postwar elections (Jackson 1966, 161; Bell 1977, 142–3).[5] Stamboliski carried out wide-ranging social reforms, most notably modifying the tax system to favour the rural poor and distributing the few large estates to the peasantry. Over the next four years, the Agrarians won growing electoral support (as did the Communists, the second largest party). Stamboliski – famously hostile to cities and urbanites, whom he repeatedly termed 'parasites' – hoped to turn Bulgaria into a 'model agricultural state' within 20 years (Jackson 1966, 42; Pundeff 1992, 82–3).

Novelist Ernest Hemingway, who met Stamboliski in 1922, wrote in the *Toronto Star* that he was 'chunky, red-brown-faced, has a black mustache that turns up like a sergeant major's, understands not a word of any language except Bulgarian, once made a speech of fifteen hours' duration in that guttural tongue, and is the strongest premier in Europe – bar none' (Hemingway 1987, 149).[6] At home, Stamboliski formed the Agrarian Orange Guard, peasant militias armed with clubs, which were periodically mobilized to meet threats to the government,

---

[4] '*Krestintern*' was a conjunction of the Russian '*Krest'yianskii Internatsional*' or Peasant International.
[5] The full name of Stamboliski's party was the Bulgarian Agrarian National Union (BANU).
[6] The 15-hour speech was to an Agrarian Party meeting. According to Hemingway (1987, 150), 'that speech broke the hearts of the Communists. It is no good opposing a man of few words who can talk for fifteen hours'.

especially from the Communists and right-wing Macedonian nationalists (Pundeff 1992, 82). In foreign policy, he attempted to secure support from agrarian parties in Poland, Czechoslovakia and elsewhere for an international agricultural league that would serve as protection against both the reactionary 'White International' of the royalists and landlords and the 'Red International' of the Bolsheviks (Colby 1922, 108–9; Gianaris 1996, 113).

The Green International first took shape in 1920, when agrarian parties from Bulgaria, Yugoslavia, Austria, Hungary, Romania, the Netherlands and Switzerland began to exchange delegations and set up a loosely organized 'league' under the direction of a monarchist Bavarian physician and peasant leader, Dr Georg Heim (Durantt 1920).[7] The following year, the alliance formally constituted itself as the International Agrarian Bureau and set up a headquarters in Prague (Bell 1977, 143).[8] This effort, due significantly to Stamboliski's initiative, made little headway over the next three years, as the Bulgarian leader was occupied with diverse diplomatic problems and a wide range of domestic opponents, including the Communists, disenchanted urban elites, nationalist and royalist army officers, Russian and Ukrainian 'White' refugees from the civil war in the Soviet Union, and right-wing Macedonian extremists.

In 1923, Stamboliski's enemies toppled him in a bloody right-wing coup that ushered in more than two decades of military and royalist dictatorship.[9] Stamboliski's captors severed his right hand and, after prolonged torture, decapitated him (Bell 1977, 237–8). Sporadic peasant resistance was quickly overcome and dozens of BANU supporters were assassinated in the succeeding weeks. Several months after the coup, a brief, fragile alliance between exiled Bulgarian Agrarians and Communists produced a Communist-led uprising, but this too was rapidly quelled, with an estimated 5,000 rebel fatalities (Pundeff 1992, 85–7; Carr 1964, 209).

The Bulgarian disaster helped pave the way for the 1923 decision of the Communist International (Comintern) to establish a Red Peasant International (Krestintern) and to seek deeper ties with the agrarian parties. Several factors in the Soviet Union and the international Communist movement also contributed to this move. The 1921 introduction of the New Economic Policy (NEP) in the USSR, characterized by greater tolerance of agricultural markets and smallholding property, ushered in a uniquely pro-peasant period in Soviet history that lasted until 1929 when the consolidation of Stalin's rule brought the initial steps toward collectivizing agriculture and 'liquidating the kulaks as a class'. Disappointed by the failure of the 1919 Communist uprisings in Germany and Hungary and by the 1920 defeat of the Soviet invasion of Poland, Moscow increasingly looked to the east as the most likely zone for successful new revolutionary

---

[7] Heim's political views are briefly mentioned in Brown (1923).
[8] In this period, the Green International had a strong Pan-Slav orientation. Its membership in 1921–25 was limited to the Czech, Serbian, Polish and Bulgarian agrarian parties (Bell 1977, 143).
[9] During the coup the Communists declared neutrality in what they saw as a simple quarrel between the urban and rural bourgeoisies (Bell 1977, 231).

movements, but these societies had only tiny industrial proletariats and massive peasantries. At the Krestintern's founding congress in 1923, the group issued an appeal to 'the peasant toilers of the colonial countries' (Carr 1964, 615). The first issue of its journal contained articles by Nguyen Ai-quoc (a pseudonym for Ho Chi Minh) and Sen Katayama, the Japanese Comintern operative whose activities ranged across Asia and as far as Mexico and Central America (Edelman 1987, 12).

The Krestintern only succeeded in attracting non-Communist agrarian movements on a few occasions. In 1924 it briefly recruited as a member Stjepan Radić's Croat Peasant Party, heretofore very much in the agrarian rather than the Communist tradition and, like Moscow, radically opposed to the idea of a Yugoslav federation that might serve as 'a mask for Great Serbian imperialism' (Biondich 2000, 198). Radić, however, who hoped to use the Krestintern affiliation to pressure Belgrade for greater Croatian autonomy, had pacifist leanings and encountered difficulty in collaborating with Yugoslav Communists. He never actually participated in any Krestintern activities and his withdrawal from the Krestintern weakened the broader legitimacy of an already weak organization (Carr 1964, 227–9, 953; Jackson 1966, 139).

China's nationalist Kuomintang (KMT) also flirted with the Krestintern during the mid-1920s as part of its alliance with the Chinese Communist Party (CCP). Several KMT leaders visited Moscow and Krestintern and Comintern operatives, including Ho Chi Minh and a significant group of Vietnamese militants, studied at the CCP's Peasant Movement Training Institute (PMTI), where Mao Tse-tung was an instructor (Quinn-Judge 2003, 82–9). In 1925 the slogan 'Join Krestinern' reportedly appeared on posters in Chinese villages (Carr 1964, 723). But this connection was also severed, in 1927, when the KMT massacred its Communist allies in Shanghai, something which caught Soviet leaders completely by surprise. On the eve of the coup, the Comintern had instructed the CCP to bury its arms (Cohen 1975, 261).

The Krestintern never attained the influence or reach of most of the other 'auxiliary organizations' of the Comintern, such as the Red International of Trade Unions (*Profintern*) or the International Organization for Aid to Revolutionaries (also known as Red Aid or MOPR, its Russian acronym). In early 1925, the Krestintern journal published an apology for the difficulties that had prevented it from holding a second congress. Later that year, Nikolai Bukharin, in an exhaustive report to the Fifth Comintern Congress on efforts around the world to bring the peasantry into revolutionary organizations, failed to even mention the Krestintern. Following the Comintern congress, the Krestintern held a plenum, with 78 delegates from 39 countries. It recommended that its militants participate in existing peasant organizations and try to align them with Communist positions (Carr 1964, 952–7). But this was precisely the approach that two years later led to the Shanghai fiasco and apart from some minor and ephemeral organizing successes, the Krestintern was basically moribund by the end of the 1920s. Pro-peasant figures in the Soviet Party, in particular Bukharin, increasingly found that they had to conform to Stalin's vision of the rural world and most were ultimately eliminated in the purges of the mid to late 1930s (Cohen 1975). The Krestintern's only durable achievement was the founding of International Agrarian Institute in Moscow, which was explicitly intended to serve as a counterweight to the Rome-based International Institute of Agriculture (IIA) (Carr 1964, 956).[10]

From the outside, however, the Red Peasant International did not appear so weak. In 1926–7, in response to the perceived threat of the Krestintern, there were rival efforts to form an international coordinating body for peasant organizations. The first originated with Dr Ernst Laur, general secretary of the Swiss Peasant Union, who sought to unite the Paris-based International Commission of Agriculture (ICA) and the IIA in Rome, which was closely associated with the League of Nations.[11] Laur's plan was to create closer links between national peasant and farmer organizations and the two policy bodies, but it foundered when the ICA and IIA each established competing international coordinating groups of farmer organizations and when the eastern European agrarian parties kept their distance, suspicious of Laur's opposition to state expropriations of large estates and intervention in the agricultural sector (Jackson 1966, 140–50).

By 1926, the Prague International Agrarian Bureau or Green International jettisoned its Pan-Slav orientation and began to reach out more widely to farmer organizations in France, Romania, Finland and elsewhere in Europe. Under the leadership of Karel Mečiř, who had served as Czech Ambassador to Greece, the Green International defined itself as a centre for the exchange of experiences, moral reinforcement and solidarity for peasants and agrarian parties, and as an international adversary to national governments that threatened peasant interests. Its main activities, however, were the publication of a multilingual quarterly bulletin and the holding of annual conventions. At its height in 1929, it included 17 member parties, stretching, in Mečiř's words, 'from the Atlantic Ocean to the Black Sea, from the Arctic Ocean to the Aegean' (Jackson 1966, 149).

The world economic crisis of 1929, the failures of various national agrarian parties, and the rise of fascism all contributed to the demise of the Green International. The Communists, despite occasional flirtations with the agrarian parties, heartily condemned both the Green International and Laur's attempt to unite the Paris ICA and the Rome IIA. In an increasingly polarized central and eastern Europe, with rapidly shrinking political space, the project of a peasant or farmers international did not re-emerge until after World War II, with the founding of the International Federation of Agricultural Producers (IFAP, see below).

---

[10] The IIA, founded in 1905 by an American, David Lubin, with Rockefeller Foundation support, hoped to unify different governments around common approaches to agricultural statistics and research (Jackson 1966, 140–1). The IIA was a remote ancestor of the United Nations Food and Agriculture Organization (FAO), also based in Rome since its founding in 1945.
[11] The ICA was formed in 1889 by French Agriculture Minister Jules Melin. It sought to hold periodic international congresses on technical problems of world agriculture (Jackson 1966, 140–1).

Associated Country Women of the World

Further west, a very different sort of transnational farmer organization began to take shape in the late 1920s, the Associated Country Women of the World (ACWW), which is still in existence.[12] Today ACWW, 'the largest international organisation for rural women', claims a membership of nine million in 365 participating societies in over 70 countries (ACWW 2002). Its proximate origins lie in encounters between leaders of the International Council of Women (ICW) – founded in Washington in 1888 – and the Women's Institute (WI) movement, which began in Canada in the 1890s and spread to the United States, England and many British colonies (Davies n.d.).

The ICW was founded by US activists (and delegates from eight other countries) who had participated in the abolitionist, women's suffrage and temperance movements (Rupp 1997, 15). The Women's Institutes were initiated by leaders of ICW's Canadian affiliate as auxiliaries to the Farmers' Institutes, a provincial extension programme which also existed in the United States (Moss and Lass 1988; McNabb and Neabel 2001). In 1913 Canadian WI activist Madge Watt moved to Britain where she helped found several hundred local Women's Institutes and interested long-time ICW President Ishbel Gordon Aberdeen in starting an international federation. Watt and Lady Aberdeen, an aristocratic feminist whose husband had served as British Governor General of Canada, called a meeting in London in 1929 with women from 23 countries who established an ICW committee on rural women (Drage 1961, 125).[13] The committee published a yearbook (*What the Countrywomen of the World Are Doing*), a journal (*The Countrywoman*) and a newsletter (*Links of Friendship*); it also circulated leaflets in three languages to recruit new national associations (Meier 1958, 5). In 1933, in Stockholm, it became Associated Country Women of the World.

In ACWW's early years, women from the English, Belgian, Romanian, German and Swedish nobility played key roles (Meier 1958, 4–5; Drage 1961, 131–3; *London Times* 1938). By 1936 its first Triennial Conference outside Europe, in Washington, DC, attracted some 7,000 farm women, most of them Americans (Meier 1958, 7). The Association set up speakers' schools for organizers and researched issues such as midwifery services and nutrition. In the pre-war period it worked with the League of Nations. During the War, it moved its headquarters from London to Cornell University in upstate New York. Following World War II, it attained consultative status with several United Nations agencies (Meier 1958). More recently, ACWW has supported development and income-generating programmes and advocated in international fora for women's rights. Despite growing participation by women from less-developed countries and an increasingly sophisticated approach to gender issues, ACWW

---

[12] Parts of the sections on ACWW and IFAP are based on Edelman (2003, 185–8).
[13] At ACWW's founding meeting Lady Aberdeen reportedly 'went fast asleep, enveloped in a large shawl. She woke up however, just at the right moment, and closed the meeting'. 'She was', according to ACWW secretary Dorothy Drage (1961, 134), 'wonderfully talented'.

never transcended its elite British origins. Its conventions are still held in English, without translation services, a practice which limits participation from outside the English-speaking world primarily to educated middle- and upper-class women, most of whom are NGO personnel rather than rural producers (Storey 2002).

International Federation of Agricultural Producers

The International Federation of Agricultural Producers (IFAP) was founded amidst post-World War II optimism about global cooperation and fears of impending food shortages and a recurrence of an agricultural depression like that of the 1930s.[14] In 1946 the British National Farmers' Union convoked a meeting in London of agriculturalists' representatives from 30 countries, with the objective of supporting the newly formed UN Food and Agriculture Organization (FAO) and overcoming differences between commodity-based interest groups – grain farmers and feed-lot livestock producers, for example – within the agricultural sector (*London Times* 1946a, 1946b).

The northern European groups that dominated IFAP already had a decades-long history of international congresses, many involving cooperative societies and Christian farmers' organizations created in the early twentieth century (ICA and IFAP 1967; IFAP 1957, 5). Despite a certain ambivalence about market liberalism, these forces often backed centre-right political parties. They worked with the Rome-based IIA (see above), which engaged in agronomic research, campaigned for uniform systems of statistical reporting and cooperated with the League of Nations in the inter-war period. The FAO, founded in 1945, was explicitly modelled on this earlier experience and IFAP was intended as the FAO's private-sector counterpart or ally.

The post-war food crisis led IFAP to emphasize raising production, even though some delegations, such as the Canadians, called for international marketing mechanisms that 'would distribute abundance efficiently and in such a way that surpluses would not spell disaster to the producers' (*London Times* 1946b). IFAP leaders served in government delegations to FAO conferences, sometimes exercising substantial influence on FAO policies (IFAP 1952). More recently IFAP joined ACWW in publicizing World Rural Women's Day, celebrated each year since the 1995 Beijing Women's Conference on 15 October.

SILENCES IN THE LITERATURE

Studies of social movements and, to some extent, those of agrarian change, have followed the flows of the neoliberal and movement dynamics: some focus on local civil society actions, others emphasize state-substitution initiatives by civil

---

[14] A sense of the critical food situation is provided by the same *Times* article that announced the founding meeting of IFAP. It noted the imminent arrival in London and various other cities of '215,181 boxes of apples from Australia' and 'the first consignment of tomatoes from the Channel Islands', which was to be sent to the north of England and Scotland (*London Times* 1946b).

society organizations, while others concentrate on transnational social movements. These studies have stretched the traditional boundaries in our understanding of contemporary agrarian movements and agrarian change. However, at least three broadly distinct but interrelated gaps remain in the current literature.

First, the rapid expansion of studies on social movements and civil society, especially at the international level, has shed new light on the intermediary organizations, especially NGOs, that usually 'broker' between the rural poor and various institutions of governance, and on issues around human rights, environment and trade negotiations. However, few of these studies provide a full understanding of the internal dynamics of the agrarian movements themselves. Often these receive relatively little systematic examination, despite the TAMs' impressive entry onto the global political scene during the past decade and a half. A glance at the fast-growing transnational social movements literature shows that recent studies focus mainly on networks concerned with human rights, women's and indigenous rights, labour, environment, migrants and international financial institutions (see, for example, Keck and Sikkink 1998; Cohen and Rai 2000; Florini 2000; O'Brien et al. 2000; Seoane and Taddei 2001; Waterman 2001; Smith and Johnston 2002; Pianta and Silva 2003; Taylor 2004; Mayo 2005; Tarrow 2005). Surprisingly, given the dynamism and high profile of transnational agrarian movements, scholars have given them relatively less scrutiny, with a few exceptions that include Edelman (1998, 2003, 2005), Mazoyer and Roudart (2002), Borras (2004), Holt-Giménez (2006) and Desmarais (2007). How movements around each theme (environment, human rights, trade, migrants, agrarian, and so on) ally with or compete against each other, and with what implications, is another question that is generally under-explored. This collection of chapters makes an initial effort to fill in this gap in the transnational social movements literature.

Second, there is a dearth of analysis of the dynamics of interconnectivity, or absence of it, between the international, national and local levels of contemporary agrarian movements. One example of this is the way in which representation and accountability at these different levels tends to be sweepingly assumed rather than systematically problematized and empirically examined. The actually existing local–national–global linkages demand better and fuller understanding. One issue worth deeper examination is the question of 'partial representation': it is very common to hear a global movement claiming to be 'representative' of the 'voice of the rural poor' from a given country, even when that particular organization is far from being representative of the diverse rural poor from its country. Similarly, it is common to hear claims from movements about their 'global' representation; it only takes one to ask whether these movements have representatives from China or the former Soviet Union, or the Central Asian region or the Middle East to discover that their 'global' coverage is quite limited. This collection attempts to initially engage this issue more systematically.

Jonathan Fox (2005) raises a related issue when he distinguishes between 'transnational and translocal memberships' made in the context of cross-border migrant studies or Deborah Yashar's (2005) concept of 'transcommunity' in the context of indigenous peoples' communities. This distinction is also analytically useful for studies of transnational social movements, not just because it calls attention to differences between translocal migrant associations or transcommunity indigenous peoples' linkages, but also it reminds us not to take as unproblematic what is claimed by agrarian movement leaders (for example, 'trans*national* links'), on the one hand, and what may actually exist in reality (for example, 'trans*local*'), on the other.[15]

Furthermore, there is also a question of 'dynamic versus static' views on representation and interconnectivity. Social movements are highly dynamic, undergoing surging and ebbing all the time (see, for example, Edelman 1999 for a Central American case). Movements may come and go, rise and fall, or strengthen but later weaken. However, the official claims of transnational movements about themselves, their constituent organizations and their respective levels of representativeness tend to be static. Failed attempts at sustaining TAMs and other social movements remain significantly under-theorized in the literature, as Edelman notes (this collection).[16]

Third, there tend to be only weak analytical connections between the phenomenal rise of studies in social movements and civil society on the one hand and studies on agrarian change dynamics on the other hand. We are thus generally confronted by two parallel, at times unrelated, sets of studies. Frequently social movement studies describe the political context in which agrarian movements function but lack any analysis of the agrarian structure within which the movements are embedded. Conversely, in many recent studies on global agricultural trade rules, TAMs are completely absent, despite such movements' frequently significant influence on global trade negotiations.

In short, despite increasing scholarly attention to transnational social movements, the agrarian dimension of these has remained significantly under-studied. Meanwhile, the broader field of agrarian studies has been slow to catch up with the phenomenal rise of TAMs. This collection is an initial contribution towards filling in these gaps and towards suggesting some new research agendas. It will, of course, only be able to partially cover the gaps discussed above. But we hope that it will inspire or provoke further studies of the many remaining, underexplored issues.

COMMON ISSUES

The contributions to this collection are diverse in terms of themes, analytic approaches and disciplines, as well as the specific units and levels of inquiry and the contexts, settings and conditions of the agrarian movements examined. However, the contributions share a concern with agrarian structures and

---

[15] Fox's and Yashar's analyses are indications of an infrequently remarked semantic problem. 'Transnational' and 'transnationalism' are used in two separate, albeit related, senses in contemporary social science. One, employed in this chapter and in most of the other contributions to this collection, emphasizes *political* linkages across national borders. The second usage refers mainly to *migration* and *diaspora*-related phenomena. For examples of the latter approach, see Mahler (1999) and Glick Schiller (2004). For an analysis that bridges both usages, see Fox and Bada's chapter in this collection.

[16] See, however, the essays in Anheier (1999).

institutions and they ask some important common questions. These include: (i) What are the characteristics of the agrarian structures from which these movements have emerged (or did not emerge)? (ii) What is the social base of the agrarian movements being examined? What social classes, groups and sectors do they represent (or claim to represent or not represent at all)? (iii) What are the issues and demands put forward by the global, national and local organizations on particular development and policy questions? What are the sources of these demands and the social and political forces that back them? (iv) What are the issues that unite – and divide – agrarian movements, and why? And (v) to what extent have the (discourse and) campaigns and collective actions over time by these movements altered (or not) the very agrarian structures that they sought to change in their favour?

These questions raise issues significant for any rigorous study of TAMs that aims to bring together the social movements and agrarian studies literatures: (i) representation and agendas, (ii) political strategies and forms of actions, (iii) disaggregating and understanding impacts, (iv) TAMs as arenas of action between different (sub)national movements, (v) diverse class origins, (vi) ideological and political differences and (vii) the dynamics of alliance building.

REPRESENTATION CLAIMS AND AGENDAS

The contributions in this collection consider the *diverse* interests of the social classes and groups in whose names TAMs and other global citizens' networks establish policy positions and make demands. The impact of neoliberalism on pre-existing rural production and exchange relations has had varying impacts on different social classes, regions and sectors, and within and between countries in the global north and south. The rural poor have usually suffered adverse effects. Looking upward from below, we see that some have formed organizations and joined social movements, although most of these have remained localized. Some local organizations have forged ties with each other, resulting in the formation of national associations and movements. A few of these national groups were able to link up with movements from other countries, or have been able to forge a certain degree of 'vertical' integration of their movements or networks (Edwards and Gaventa 2001; Fox 2001), and have created transnational networks or movements. Looking downward from above, it is clear that while a few globalized movements have made impressive strides, far more groups and ordinary rural poor people have been left behind or decided, for various reasons, to remain within the boundaries of local and national venues – unlinked to any global agrarian movements. It is in light of this reality that the question of representation within TAMs needs to be discussed more carefully.

Effective representation of the social base's interests within their movements should not be assumed to be automatic or permanent and unproblematic. 'Effective representation' is dynamically (re)negotiated within and between leadership and membership sections of movements over time. It is important to critically examine the question of representation in studying TAMs for at least two interlinked reasons: (i) it is from movements' or networks' claims to represent a group or groups of people that they justify their issue-framing and demand-making initiatives; (ii) movements argue for the importance, urgency, necessity or justness of their cause and make demands based on their claims to represent particular groups of people. Indeed, this is their public raison d'être. In the case of TAMs, it is usually the 'oppressed', 'peoples of the land' or 'the peasants' that are the subjects of these representation claims.

Despite the importance of the issue of representation, there are not many critical efforts to scrutinize this question. It is often the case that the issue of representation is assumed rather than empirically demonstrated. Similarly, the extent of representation is commonly conflated with the degree of visibility and 'loudness' of TAMs; that is, the more visible and the louder the TAMs, the more representative they are assumed to be. For example, a much smaller global network of NGOs may be able to be present in most international conferences and official venues of consultations, while a community-based transnational agrarian movement may be far larger, but may not be able to be present in such global conferences and institutional processes. Uncritical observers, especially among international development agencies, might very well consider the former as the representative network (even when this is not the case in reality). Understanding representation in the context of TAMs requires analysis that goes beyond what is easily observable, i.e. 'visibility and loudness', and into issues that are usually not included in the movements' discourses. This can be accomplished in a number of ways.

First, in most cases, by representation, TAMs are in fact talking about 'partial representation'. Partial representation can be viewed from at least two perspectives: globally and nationally. A quick scan worldwide would reveal that most of the high-profile TAMs today do not have any (significant) presence in large areas of the world, notably Russia, Central Asia, the Middle East and North Africa (MENA) region, and most especially China. Together, these regions host the majority of the world's rural poor. In this collection, Kathy Le Mons Walker examines contemporary rural China, and Kevin Malseed the particular case of the Karen people in Burma, to demonstrate that poor people's demands in these regions may be distinct from, but also often similar to, rural poor people's demands elsewhere. Despite the TAMs' global aspirations and discourse, the cases that Le Mons Walker and Malseed examine are completely beyond the TAMs' radar, notwithstanding the importance, intensity and scale of these and similar struggles.

Nationally, partial representation can be seen by the fact that, no matter what is claimed, no single organization, movement or group of movements can fully represent the vast and diverse groups and interests in an entire country. We can see this by looking at two extreme settings: one where a national member of a TAM is very weak and the other where a national member is very strong. Both South Africa and Brazil have member organizations in Vía Campesina. As Brenda Baletti, Tamara Johnson and Wendy Wolford explain in this collection, the Landless People's Movement (LPM) of South Africa, a 'late mobilizer', started very weak politically and organizationally, and after just a few years, imploded. It is still the only organization that represents the South African rural poor in Vía Campesina despite its extremely limited, currently nearly insignificant, degree of representation. Meanwhile, the Movement of the Landless Rural Workers of Brazil (MST, *Movimento dos Trabalhadores Rurais Sem Terra*) is said to be, and perhaps is, the largest and most politically coherent national movement within Vía Campesina. It certainly represents a great number of poor people in Brazil. However, even in the context of Brazil, MST's representational capacity is, at best, partial. The recent candid admission by its leader, João Pedro Stédile, about its still limited current ability to organize Brazil's landless poor is instructive (Stédile 2007, 195). Moreover, even sympathetic critics of the MST have noted its limited representation among the Afro-Brazilian population and among rural women, some of whom have abandoned the MST to form their own organizations (Stephen 1997, 217–33; Rubin 2002, 45–6). Arguably, all other national movements in Vía Campesina fall somewhere in between these two extreme poles in terms of representation, i.e., varying degrees of *partial* representation.

As will be discussed below, and in the studies in this collection, partial representation has profound implications for the very nature and orientation of issue-framing and demand-making processes and outcomes within TAMs. But one implication of the reality of partial representation is that perhaps it is better seen not from an 'either/or' perspective, i.e., either representative or not. Rather, representation is inherently a matter of degree.

Second, framing representation as a matter of degree suggests that it is not static. Representation is constantly renegotiated within organizations or movements. This means the degree of representation of a particular movement may increase or decrease over time, and in some cases may completely disappear. Marc Edelman's historical analysis of the Central American peasant alliance, ASOCODE, is illustrative of the contested and dynamic character of representation (Edelman, this collection; Edelman 1998).[17] ASOCODE was one of the earlier versions of progressive transnational agrarian movements based among small farmers and landless peasants. Its rapid and phenomenal rise to fame in the early 1990s was an inspiration to many other movements outside of Central America. It was quite vibrant and indeed had solid claims about representation. For various internal and external reasons, ASOCODE later imploded. By the turn of the century, it could no longer claim and could no longer be considered as having a high degree of representation in the region. This is similar to the story told by Saturnino M. Borras Jr (this collection; Borras 2004) in the case of the Democratic Peasant Movement of the Philippines (DKMP, *Demokratikong Kilusang Magbubukid ng Pilipinas*), another member of Vía Campesina that had a significant social base in the 1990s, but rapidly and significantly contracted before the end of that decade. Solidarnosc-Rural in Poland, which was a founding

---

[17] ASOCODE was the *Asociación Centroamericana de Organizaciones Campesinas para la Cooperación y el Desarrollo* (Central American Association of Peasant Organizations for Cooperation and Development).

member of Vía Campesina and sat in its International Coordinating Committee during the latter's formative years, is yet another, similar case. Solidarnosc-Rural went from having a sizable base to being unable to actually demonstrate any significant degree of representation (among small farmers in Poland). It was asked to leave Vía Campesina a few years after Vía Campesina's second world assembly. Agrarian movements go through the natural dynamics of ebb and flow over time. Their capacity to represent the particular groups that they claim to represent is directly affected by such cycles and dynamics, which in turn have some influence on the nature and orientation of the movements' issue-framing and demand-making processes.

Third, a problem is the failure to significantly represent groups of rural people, partly due to restrictive definitions of 'peasant' or 'farmer'.[18] Conventional framing of agrarian movements tend to be 'agriculture-centred', with 'people of the land' as a defining character of the common actors. But the rural sector has long been diverse and plural, as are the livelihoods of the rural poor (see, for example, Bryceson et al. 2000; Ellis 2000; Rigg 2006). To limit organizing effort as well as issue-framing and demand-making to agriculture-oriented themes effectively excludes a significant portion of the world's rural poor. The case of cross-border migrant workers is particularly important because they constitute a large section of the world's rural population (in their areas of origin, but many times also in the receiving countries). Jonathan Fox and Xochitl Bada (this collection) examine how cross-border migrants (Mexico-to-USA) have had a profound impact in the sending areas. Out-migration affects local labour markets in diverse ways and sending communities are, furthermore, increasingly transformed by the huge volume of migrants' remittances. There is a persistent and troubling divide, politically and academically, between the rapidly growing scale of transnational labour migration and the failure of the major TAMs to accord the issue of cross-border migrant workers a central place in their analyses and strategies.[19] Migrant organizations and studies of cross-border migration only rarely embed their political or academic work in agrarian change dynamics (as Fox and

---

[18] Structural factors are also, of course, significant in explaining why it has often been difficult to organize and represent rural as compared to urban people. Geographical dispersion, remoteness, poor communications and transport infrastructure, and the greater differentiation of rural people and variety in social relations in the countryside have historically complicated rural organizing, at least as compared to the efforts to organize the urban industrial working class. The deindustrialization and rise of the informal sector that often accompanies neoliberalism have, however, over the last two decades, contributed to greater difficulties in many parts of the world in organizing the urban poor and working class.

[19] In a number of cases, however, what we might call 'minor TAMs' with a bi-national focus have advocated effectively on behalf of migrant workers. Among notable examples that work in both the United States and Mexico are the Rural Coalition (Rural Coalition 1994) and the Farm Labor Organizing Committee (FLOC), which is affiliated with the US AFL-CIO (American Federation of Labor-Congress of Industrial Organizations). FLOC, which has worked in Mexico to clean up corruption in the recruiting system for farm workers going to the United States, has been a target of violence and in 2007 lost one of its key leaders, Santiago Rafael Cruz, to an assassination (Velásquez 2007). Some of these and other minor TAMs have formal and informal relations with major TAMs, such as Vía Campesina.

Bada observe on the issue of the persistent 'disconnect between migration and development'). The Mexico–US cross-border migration that has direct links to the situation of the Mexican rural sector is not the only significant migratory flow out of a threatened countryside. Others include Nicaragua-to-Costa Rica, Zimbabwe-to-South Africa, Morocco-to-Spain, Poland-to-Britain and the Philippines-to-Malaysia.

Fourth, and finally, as the discussion above suggests, the question of representation is complicated and often contentious. Diverse realities and multiple channels of representation and accountability are often too complex for movement leaders and activists to understand and deal with. For example, the concept of 'class' is frequently invoked in terms that are too abstract and vague, and too divisive. Unstructured, unorganized and covert forms of actions are too fluid and invisible for movement activists and leaders to acknowledge or comprehend. And so, in many ways, movement leaders act like states, inasmuch as they engage in a 'simplification process' to make complex realities legible to and manageable for them, to use the powerful analytic lens offered by James Scott in his book *Seeing Like a State* (1998). But in doing so, a great many important details tend to be taken for granted or missed in the analysis and discourse that the TAMs produce (see Malseed in this collection).

The question of representation is inherently linked to the issue of the agrarian movements' agenda-setting processes and outcomes. A global network of middle to rich farmers that stands to benefit from certain types of free trade will not oppose initiatives for barrier-free trade between countries. A transnational network that does not have a mass base in China or the countries of the former USSR may float the notion of a socialist alternative in an unproblematic manner. It is difficult, if not impossible, to fully understand representation and agendas as separate issues. This points to the importance of not automatically and uncritically accepting as unproblematic claims by transnational movements about their agendas, bases of support and political successes. One way of taking a more nuanced look at this is to identify possible discrepancies between claims about agendas and other matters between national member groups, on the one hand, and the TAMs, on the other. The responses of (trans)national peasant movements and ordinary, local peasants and farmers around the issue of genetically modified (GM) crops illustrates this disconnect. As Peter Newell (this collection) points out in the case of Latin America, and Ian Scoones (this collection) in the cases of India, South Africa and Brazil, GM crops have become a major focus of contention, pitting peasant and small farmer organizations and many environmentalists against agribusiness and large farmer groups. The controversy about GM crops in reality involves a range of interlinked debates: on food safety, contamination of crop land races and threats to genetic diversity, the health of humans and other species, farmers' budgets, intellectual property, farmers' right to produce and save seeds, the economic and ecological vulnerability implied by genetic monocultures, and the concentration of power in the hands of a few giant corporations. Yet as Newell and Scoones suggest, the stakes in this discussion and small farmers' on-the-ground practices are not always as clear cut and straightforward as the contending parties claim or would wish (see also Müller 2006; Herring 2007).

POLITICAL STRATEGIES AND FORMS OF ACTIONS

TAMs generally agree on certain types of common political strategies and forms of action. It is indeed remarkable how highly diverse (sub)national and (sub)regional peasant groups such as the National Farmers' Union (NFU) of Canada, Brazil's MST, the European Peasant Coordination (CPE), KRRS of Karnataka, India or Mozambique's UNAC would be able to agree on common political strategies and forms of action (see Edelman 2003; Desmarais 2007).[20] But while it is critical – politically and academically – to understand such strategic alliances, it is equally important to acknowledge potential and actual differences within and between TAMs. For one, understanding diversity and differences will lead us to appreciate better the accomplishments of TAMs' unity-building efforts. It will also help us grasp other related issues, such as representation and 'inclusion–exclusion' processes within TAMs. In short, it will help us understand TAMs better in all their complexity. We briefly examine this issue along three dimensions.

First, similarities and differences between TAMs and other groups that do not have clear and legible strategies and effective forms of actions. Most transnational agrarian movements look for 'counterpart' organizations in other agrarian societies. When they do not find rural poor peoples' associations that are in their 'image and likeness', they tend to conclude that 'nothing is happening' or that 'there are no social movements' in those societies. This process of TAMs searching for counterparts in practice usually involves looking at the ideological makeup of a potential ally as well as its political strategies, methods and forms of collective action. It is for this reason that TAMs typically characterize organizations in regions and countries such as the former USSR, the Middle East and North Africa (MENA), China and Southeast Asia, and to a significant extent, Sub-Saharan Africa, as having 'no clear political strategies' or being 'politically and organizationally weak'. They may also point to such regions and argue that there is an 'absence of movements' or an 'absence of mobilizations and protests'.

But important works around the 'everyday forms of peasant resistance' (e.g., Scott 1985, 1990; Scott and Kerkvliet 1986; Kerkvliet 2005) and subsequent studies inspired or provoked by this approach (e.g., O'Brien 1996; O'Brien and Li 2006) alert us that such assumptions are problematical. In many settings, peasants engage in everyday forms of resistance that are 'unorganized', 'unstructured' and 'covert' to defend or advance their interests. Le Mons Walker explains that in the case of contemporary China everyday peasant resistance has taken a more overt form, but remains 'unstructured' and 'unorganized' (in the conventional senses

---

[20] KRRS stands for *Karnataka Rajya Ryota Sangha* (Karnataka State Farmers' Association). The European Peasant Coordination is usually abbreviated CPE for *Coordination Paysanne Européenne*, its name in French. UNAC stands for *União Nacional de Camponeses* (National Union of Peasants).

of these terms). Malseed describes how, in the case of the Karen people in Burma, everyday forms of resistance have a lot in common with conventional organized, structured and overt agrarian movements. Le Mons Walker and Malseed each argue (in this collection) that contrary to the dominant assumptions, there may be great potential for transnational linkages between these various peasant initiatives, but that this would require revising many of the established TAMs' conventional assumptions. To a significant extent, the problem of conventional TAMs excluding particular groups because their political strategies and forms of actions are not in their 'image and likeness' is also applicable in the case of cross-border migrant associations, as explained by Fox and Bada (this collection) in the context of Mexican migrant groups. Unless these critical gaps are quickly and seriously addressed by TAMs, their issue-framing, demand-making and overall power on the global scene will be, at best, well below its potential. Critical scholarship on TAMs requires a truly global lens in looking at movements or networks that frame issues and make demands in the name of the *world's* rural poor but that have no presence in key geographic zones.

Second, similarities and differences between TAMs. There are significant differences between various TAMs on political strategies and forms of action. Some favour a combination of confrontation and critical collaboration with (inter)governmental institutions, while others favour uncritical collaboration and formal alliances. More generally, the first type of strategy has guided Vía Campesina's political work, while the latter has characterized the IFAP. Vía Campesina has declared that it does not, and will not, legitimize key international institutions that are among the promoters of neoliberal globalization, such as the International Monetary Fund (IMF), the World Trade Organization (WTO) and the World Bank. It has refused to participate in their institutional processes and attempts to discredit these agencies at every opportunity (Vía Campesina 2000).[21] But Vía Campesina has nonetheless decided to collaborate with particular groups of sympathetic reformers within the Food and Agriculture Organization of the United Nations (FAO) and the International Fund for Agricultural Development (IFAD), a specialized UN agency. This is in contrast to the longstanding harmonious relationship of IFAP with all these agencies. Larger transnational coalitions that include (inter)national members from the more radical and the more conservative groups tend to move back and forth between these two poles, as in the case of IPC for Food Sovereignty.

Some TAMs have identified radical, direct actions as the most appropriate and effective ways of putting their issues onto the official agendas and effecting actual reforms in their favour. These actions include land occupations, physical attacks on GM experimental infrastructures and field testing sites, bulldozing or torching of symbolic icons of transnational corporate greed, occupations of government and corporate offices, and street marches to disrupt important multilateral conferences (see discussions by Scoones, Newell, Edelman, Borras, and Baletti,

---

[21] But see Edelman (2003, 207) on Vía Campesina's brief 1999 attempt to engage in dialogue with the World Bank about agrarian reform.

Johnson and Wolford in this collection). Other TAMs, especially those associated with middle and rich farmers such as IFAP, as well as those dominated by large NGOs such as the ILC, do not engage in dramatic, confrontational forms of collective action. They prefer to send their top executives for official meetings with important international institutions.

Assessing which strategies and forms of actions are more effective depends largely on the goals and targets of a particular campaign. If the goal is to delegitimize specific institutions, then public shaming through confrontational actions may indeed be an appropriate approach. In campaigns where the goal is to secure concessions, as for example, expanding 'invited spaces' (see Gaventa 2002) for civil society participation, then a more critical collaborative interaction would likely be more effective. Overall and in the long run, however, it may not be favourable to a particular TAM to be pigeonholed as having a specific, predictable strategy or action repertoire, such as 'only confrontational actions' or 'only conformist interaction through official channels'. The Vía Campesina's capacity for combining multiple tactics and strategies, as well as a wide range of forms of actions, almost certainly contributes to its effectiveness.

Third, similarities and differences within TAMs. Within large TAMs, different political strategies and forms of actions – sometimes competing and contradictory – can occur simultaneously. Take for instance the strategy of Brazil's MST, which is generally confrontational direct action, as well as 'independence' from the state (Baletti, Johnson and Wolford, as well as Newell, this collection) – and contrast this to the Senegalese group, CNCR (*Conseil National de Concertation et de Coopération des Ruraux*), which works much more closely with government and opts for less confrontational forms of action (McKeon et al. 2004). It is possible to come up with a long list of similar cases that demonstrate the contradictions between TAMs' supposed political strategies and the forms of actions with their (sub)national and (sub)regional members. This diversity and the resulting contradictions are most pronounced and widespread in the larger TAMs, such as the IPC for Food Sovereignty.[22]

Can or should activists in the TAMs and in their national member organizations reach consensus around standards for politically effective strategies or actions? In cases where this appears to have occurred, it usually reflects what the dominant actors within and between TAMs define as standards. This process is illustrative not only of the TAMs' simplification or standardization efforts, but also that TAMs are indeed sites of power. Keck and Sikkink remind us that 'Transnational advocacy networks must also be understood as political spaces, in which differently situated actors negotiate – formally or informally – the social, cultural, and political meanings of their joint enterprise' (1998, 3). They continue, 'Power is exercised within networks, and power often follows from resources . . . Stronger actors in the network do often drown out the weaker ones' (see also Della Porta et al. 2006, 20). It is usually the powerful actors within TAMs

---

[22] See Jordan and van Tuijl (2000) for a relevant discussion on this issue.

that define standards for their movement and, to the extent that they can, for others outside their movement as well.

DISAGGREGATING AND UNDERSTANDING IMPACTS

Disaggregating TAMs' impacts is important in order to assess whether, how and to what extent their actions bring them closer to their goals. This is especially so because, as Tarrow explains, 'advocates of transnational activist networks have highlighted many successful instances of successful intervention on behalf of actors too weak to advance their own claims. In an internationalized world, we are likely to see more of such intervention, so it is important to look at it without illusions. Transnational intervention fails more often than it succeeds' (2005, 200). Keck and Sikkink (1998, 201) propose that transnational networks attempt to realize their objectives through five interlinked processes, namely, (i) by framing debates and getting issues on the agenda, (ii) by encouraging discursive commitments from state and other policy actors, (iii) by causing procedural change at the international and domestic level, (iv) by affecting policy and (v) by influencing behaviour changes in target actors.

Most of the contributions to this collection agree and demonstrate that TAMs' greatest and most palpable impacts have been in (re)framing debates and getting issues on the agenda. This is argued perhaps most forcefully by Phil McMichael, who explains how TAMs, especially Vía Campesina, are reframing debates not only on specific issues, but *the terms of the debates* around the very notion of 'development' and the agency of peasants in this process. McMichael, Scoones, Newell, and Friedmann and McNair (all in this collection) show how transnational linkages have contributed to reframing debates around the political economy of food – from the question of organic food certification to anti-GM crops mobilizations. The notion of 'food sovereignty' as an alternative to the current corporate-controlled and industrial food complex was developed by Vía Campesina and has since then spread to the agendas of many other TAMs, environmental movements, and even governmental and intergovernmental institutions.[23] Meanwhile, Fox and Bada (this collection) demonstrate why out-migration of rural Mexicans to the USA does not always constitute a choice of 'exit' over 'voice', something which is frequently assumed to weaken civil society in the areas of origin of the migrants. They show that in some cases 'exit' can be followed by 'voice', where migrants have reframed the terms of the discussion around local development and their roles in it. Borras (this collection) explains that the Global Campaign for Agrarian Reform (GCAR) led by Vía Campesina has its most significant impact in terms of reframing the debate about land reform and land

---

[23] Vía Campesina and its allies also see 'food sovereignty' as an alternative to the mainly quantitative notion of 'food security' propounded by some intergovernmental and nongovernmental agencies. The regional government of Tuscany (Italy) is one of several sub-national jurisdictions in Europe that has pioneered food sovereignty policies within its boundaries and has lent political support to food sovereignty campaigns in Europe. For a background discussion, see Patel (2006) and Rosset et al. (2006).

policy in reaction to the currently dominant market-led model.[24] Peluso, Afiff and Fauzi (this collection) explain that the convergence of environmental and agrarian movements in Indonesia with links to global networks can produce mutually reinforcing impacts for these movements and lead to a significant reframing of environmental and land issues.[25]

Quite importantly, however, Edelman (this collection) demonstrates that part of the waning of the regional transnational network in Central America was the movements' failure to adapt their agendas sufficiently to the rapidly changing agrarian structures in this region and the massive exodus of migrants from the countryside. He points out, for example, that agrarian movements in the region continue to repackage calls for land reform. But, he argues, without critical and more systematic understanding about how past state-directed land reforms improved or failed to improve the livelihoods of poor peasants, current issue-framing and demand-making around land reform may not be very effective in countering elite and World Bank efforts to implement market-led agrarian reform.

In the other four dimensions that Keck and Sikkink point to as important for evaluating transnational networks' impact, TAMs, as shown in most of the studies in this collection, have so far made little headway. It is possible that one of the most valuable aspects of TAMs is their reframing the terms of relevant policy and political debates internationally, which can in turn help create a favourable context for (sub)national movements to actually make palpable gains.

TAMs AS ARENAS OF INTERACTION BETWEEN (SUB-)NATIONAL MOVEMENTS

Agrarian movement activists tend to represent TAMs as 'single actors', as networks or movements with collective agency. Such representations in turn tend to be uncritically accepted by external observers, including other TAMs, other global civil society networks, and some scholars and development policy experts. These actors then generalize about particular TAMs based on what they perceive as general patterns of behaviour exhibited by particular TAMs over time (e.g., some TAMs are more radical than others, some TAMs are more reformist than others, and so on). The fact that a particular TAM has its own specific characteristics distinct from others, and that it is a group with collective agency, is important for agrarian movement activists in terms of movement-building processes, and is less problematic for academic observers. However, the notion of TAMs as single, unitary actors cannot be detached, empirically and analytically, from another, concomitant feature of TAMs, notably that these are also arenas of interaction between (sub-)national movements. This tends to confirm what Keck and Sikkink have already pointed out – i.e., that transnational networks

---

[24] For a background on the current debates on land reform, see Borras (2007).
[25] Anna Tsing (2005) has studied a similar Indonesian case, and has offered rich and nuanced analysis of the mutually transforming processes in these global connections through what she calls 'friction'.

should be seen as 'network-as-actor' and 'network-as-structure' (1998, 7).[26] These two spheres of TAMs (re)shape each other in a dynamic process similar to Anna Tsing's (2005) notion of 'friction', resulting in the constant transformation of a TAM, its members and the way interactions between members are structured.

Inherent to political spaces where various actors come to interact with one another is the question of power. Different groups enter this space with dissimilar degrees of political power. Therefore, member movements have varying degrees of influence in the process of shaping the character of a TAM, a single actor, as well as the processes that structure the interaction between member movements. Baletti, Johnson and Wolford (this collection), for example, describe the asymmetry in political power between Brazil's MST and South African's LPM, both members of Vía Campesina but with distinct levels of influence within it. These authors even suggest that MST's assistance to LPM in movement building and mobilization, the very type of collaboration that many view as an advantage of TAMs, might have harmed, not helped, the LPM. This highlights the question of 'uneven development of (sub)national movements' as an important issue in studying TAMs. Baletti, Johnson and Wolford employ a categorization of 'early mobilizers' (Brazil's MST) and 'late mobilizers' (South Africa's LPM) and indicate that this dimension of organizational timing shapes the degree of power of component groups within a TAM.

This categorization of 'early' and 'late mobilizers' is indeed useful in the case of MST and LPM. However there are cases for which this analytical approach may not be as useful. Take, for example, the case of the Indonesian Federation of Peasant Unions (FSPI, *Federasi Serikat Petani Indonesia*) examined by Peluso, Afiff and Fauzi (this collection) and compare it with, say, Mexico's UNORCA.[27] The latter was formed in the 1980s, was one of the key founding members of Vía Campesina in 1993 and remains an important movement, especially for the North American regional coordination of Vía Campesina. Meanwhile, FSPI was formed only in late 1996 and gained organizational and mobilization momentum only during the post-authoritarian regime transition in 1998. It is one of the very late mobilizers in Vía Campesina and even within Southeast Asia. However, for various reasons, including some external opportunities (e.g., Vía Campesina wanted to transfer its operational secretariat to Asia), FSPI rapidly increased its global political influence. In 2004, it became the host of the global operational secretariat of Vía Campesina, and its leader, Henry Saragih, becoming Vía Campesina's general coordinator.

In addition to the mode of categorizing movements based on the timeline of organization-building and mobilization as one of the explanatory factors in the differential power positions within a TAM, it may be useful as well to examine the timeline of the act of linking as well as the 'quality of the link' between a

---

[26] See related discussions in Guidry et al. (2000, 3) and Batliwala and Brown (2006).
[27] UNORCA is the *Unión Nacional de Organizaciones Regionales Campesinas Autónomas* (National Union of Autonomous Regional Peasant Organizations).

movement and the TAM. For this purpose, we can use the categories 'early linkers', 'late linkers', 'strong linkers', 'weak linkers', which may also define an organization's position of influence and power within a TAM. For example, KRRS of Karnataka in India is an 'early linker' that contributed to building Vía Campesina. It also has a 'strong link' due to its particular history and capacity to launch dramatic actions against transnational and GM seed companies, something that is central to the global discourse of Vía Campesina. Meanwhile, numerous movements of landless labourers and Dalit workers in India were, in the context of Vía Campesina, 'late linkers'. Many of them attempted to link-up with Vía Campesina only in 1996, but by then the middle and rich farmer-based KRRS was already well-entrenched within the global movement – actively preventing the entry of other movements from India or effectively discouraging many other movements from India from seeking membership in Vía Campesina.[28] Many of these 'late linkers' also have 'weak links' to the global actors for a variety of reasons, which include the relatively low priority to workers' issues within Vía Campesina, whose advocacy caters primarily to surplus producing strata of the peasantry who are engaged with issues of trade and biotechnology. There have been no systematic worker-centred campaigns within Vía Campesina (around wages, for example) except, of course, land reform, which KRRS initially opposed as a focus for a global campaign (see, for example, the discussions by Borras and by Scoones in this collection). Many 'late linkers' are unable to insert themselves effectively into the TAM, especially where rival national movements have already become well-entrenched within the TAM. This is, for example, the case of UNORKA in the Philippines, whose application for membership in Vía Campesina is opposed by KMP (see Borras, this collection).[29] This type of tension is widespread in most of the TAMs.

DIVERSE CLASS ORIGINS

Without a class analysis it is impossible to disaggregate (and fully understand) the processes and outcomes of development. Scholars who have recently argued for bringing class back in to the study of rural development do not claim that a class analytic lens is sufficient for explaining all important dynamics of agrarian change. Other social relations and identities, particularly gender, ethnicity and

---

[28] Despite KRRS's social base among a relatively prosperous sector of the peasantry, the organization's discourse and actions were frequently radical and dramatic, particularly during the period until 2004, when its longtime leader Professor M.D. Nanjundaswamy passed away. KRRS was famous for ransacking the Bangalore offices of Cargill and campaigning against the presence of US fast food chains in India (Gupta 1998). During the 2004 Mumbai World Social Forum (WSF), Nanjundaswamy and KRRS, alleging that the event was conceived and dominated by non-Indian NGOs with little popular backing, boycotted the WSF and organized a parallel 'resistance' forum outside the main event.
[29] UNORKA is *Pambansang Ugnayan ng mga Nagsasariling Lokal na Organisasyon sa Kanayunan* (National Coordination of Autonomous Local Rural People's Organizations). KMP is *Kilusang Magbubukid ng Pilipinas* (Peasant Movement of the Philippines).

religion, also play distinct roles.³⁰ But scholars who point to the explanatory power of non-class identities in recent rural social movements also frequently argue that while ethnicity and other non-class identities are important, class too must be considered in any analysis of movement-building and agrarian change dynamics.³¹ It is this nuanced approach to the relevance of class that we hope to bring to the discussion about TAMs.

In studying today's TAMs there are two main, critical dimensions related to the issue of class: (i) the *extent of domination* of a specific class or classes and (ii) the *quality of insertion* of a particular class or classes, within a TAM. The extent of presence and domination of a particular class or classes within a large transnational movement matters. It influences the general pattern of political behaviour of the movement or network. Take IFAP, for instance, which is dominated by middle and rich farmers in the north, resulting in less radical and generally conservative advocacy positions. However, the quality of insertion into a transnational network by a particular class also matters. Even when a particular class constitutes a minority, under certain conditions it can become a key actor within a movement. This happens when its representatives' positions are adopted by a wider movement that recognizes their political validity and value for organizing and strengthening the TAM. Such is the case of the middle and rich farmer organization of KRRS and its anti-TNC and anti-GM crops agitation and mobilization (see discussions by Scoones and by Borras, this collection).

A class analytic lens is also useful for examining the nature and orientation of various movements. As Peluso, Afiff and Fauzi show (in this collection), environmentalist and land-oriented agrarian movements in Indonesia have links to, and receive influences from, transnational networks. The class composition of these movements is a major factor in how these relations play out. The earlier wave of agrarian movements in the 1950s and 1960s had a strong base among the poor peasantry demanding land. It was violently crushed by the military. What would later emerge was a class-blind environmental movement. While this movement was also constrained by the Suharto regime, even from its beginnings in the early to mid 1980s it was the main presence in rural-oriented civil society in Indonesia and related organizations with international links. Although this movement did not address the political economy of landed property, from its early years it had a strong but necessarily subdued environmental justice orientation. In the early to mid 1990s, environmental justice-oriented activists made alliances with both underground agrarian activists and indigenous peoples' movement groups. The class-oriented issues of land redistribution were later taken up by resurgent agrarian movements during the 1998 national regime

---

³⁰ See, for example, Bernstein's (2007) explanation in the context of a review of current scholarly discourses on rural livelihoods, as well as Herring and Agarwala's (2006) argument for 'restoring agency to class' in the context of, among others, the class basis of the differential positions of various groups 'organized' and 'unorganized' rural-based and rural-oriented groups in India concerned with the issue of GM crops.
³¹ See, for example, Brysk (2000) and Yashar (2007) in the case of indigenous peoples' movements in Latin America.

transition from centralized authoritarian rule. Discussions about class among activists in Indonesia remain generally implicit, but the demands – land to the landless poor peasants and rural labourers, recuperating lands from large corporations, and so on – are clearly class-informed and class-oriented. The conservation movement, meanwhile, has split in part along class lines over land redistribution in national parks and other conservation areas. However, the ambiguous and shifting alliances between different components of these movements make it difficult to define the differences as clearly or solely class-based.

While movement leaders and their allies hardly consider class divisions and struggles among their favourite issues and indeed rarely speak openly about them, actually existing realities within transnational agrarian movements and within their national member organizations point to the significance of class as a critical issue. A few leaders of these transnational networks may, however, occasionally acknowledge the class character of transnational movements. João Pedro Stédile, for example, recently asserted that 'the [Brazilian] MST and Vía Campesina, especially, work with the theory of waves or cycles of class struggle' (Stédile 2007, 194). While this might be a controversial or delicate claim for some Vía Campesina member organizations in other countries (see below), Stédile's main point was that peasants, not only in Brazil and in contrast to organized working classes there and elsewhere, have been in the forefront of anti-neoliberal and anti-imperialist struggles (see relevant discussion by Petras and Veltmeyer 2003). The class issue is also implicit in some Vía Campesina leaders' vision of alliance-building. Paul Nicholson, for example, a farmer from the Basque Country (Spain) and one of Vía Campesina's most important leaders, argues that the most fundamental alliance in the world's countryside today is that between small farmers and rural workers (Vía Campesina 2000).

IDEOLOGICAL AND POLITICAL DIFFERENCES

From time to time some transnational agrarian movement activists admit that ideological and political differences exist and are important issues in movement or network building. But again, as with class, this issue is typically avoided as much as possible and usually addressed only internally. The strategic implications of ideological and political differences within and between TAMs – in movement-, alliance- and coalition-building, representation and accountability, issue framing and demand making – cannot be taken for granted. They do matter. They play important roles in the rise or fall, strengthening or weakening of transnational movements, networks and coalitions. These dynamics can be seen from three perspectives.

(i) Class-based. Some ideological and political differences are clearly informed by class. Groups dominated by particular classes have particular sets of interests and issues different from other groups with different social class compositions. For example, KRRS of India does not support, and even opposes, land reform. It is a movement dominated by middle and rich farmers who own varying sizes of

landholdings. Several movements of landless rural labourers in India have political differences and conflicts with KRRS primarily because of their different class bases. Such a conflict has been internalized in TAMs. Recall our earlier discussion on the contrasting political strategies adopted by Brazil's MST and Senegal's CNCR (see McKeon et al. 2004). On a larger, global scale, we can point to the ideological and political differences that separate IFAP and Vía Campesina, with class as an important, though not the sole underlying reason for such differences.

(ii) Differences in political strategies and calculation. There are also peasant groups that have mass bases in the same social classes in the countryside, for example, poor peasants and rural labourers, but which bitterly oppose each other because of ideological and political differences. The discussion by Borras (this collection) about the conflicts between the three peasant movements in the Philippines – KMP, DKMP and UNORKA – is illustrative of this phenomenon. On a global scale, there is an important difference for example in the political calculation of the International Land Coalition (ILC) and the IPC for Food Sovereignty. The NGOs and IFAP that are within ILC decided that the best strategy to advance pro-poor land policies would be to forge formal and official alliances with international financial institutions (e.g., the World Bank) and intergovernmental institutions (e.g., World Food Programme of the UN). ILC also works to replicate similar approaches within national-level organizations. This is, of course, a strategy rejected by the IPC for Food Sovereignty, which opted to maintain its civil society character and its autonomy from international financial and development institutions. IPC for Food Sovereignty instead tries to build its rural social movement base and NGO alliances in order to push for a more mutually reinforcing interaction with sympathetic international institutions.

(iii) Institutional turf battles and personality differences. Perhaps the most common differences within and between TAMs result from institutional turf battles and personality differences. The great competition for funding from northern development agencies has always been a major source of tension within and between TAMs, though one which is rarely acknowledged. Moreover, personality differences have also played a role in fanning the flames of ever-present political tensions. On many occasions, competition for funding, fights for public recognition and fame, and personality clashes have contributed to deepening faultlines between and within TAMs.

DYNAMICS OF ALLIANCE-BUILDING

TAMs devote considerable resources and efforts to alliance-building activities in order to extend the reach of their collective actions beyond their own ranks. And because this involves at least two different groups trying to find some common ground, there are inherent tensions and fault-lines in alliance building. In the past the most common types of alliances that involved peasant movements were those with political parties and workers' organizations (see, for example, Kay and Silva 1992 on the case of Chile; Salamini 1971 on the case of 1930s Mexico; Herring 1983 for some South Asian cases; Heller 2000 on the case of India). Most of these alliances coincided with the rise to power of left-wing, communist or socialist parties, generally during the first three-quarters of the past century. However, the waning of many of these left-wing political parties in the 1970s–1980s, the neoliberal resurgence since the early 1980s, and the weakening of most workers' movements have increasingly led rural-based movements to eschew these two types of alliance. This strategic shift coincided with the emergence of identity politics among rural social movements, some of which, though certainly not all, rejected class politics altogether (see, for example, the discussions in Alvarez and Escobar (1992) in the context of Latin America). Most of the peasant movements that joined the major TAMs emerged during this period after the decline or disappearance of political party and worker–peasant alliances. This does not mean that these movements do not interact with political parties or trade unions. They usually do, but not in the politically subordinate, 'transmission belt' kind of relationship that marked the previous era. This is, for example, the context in which Brazil's MST emerged in the early 1980s, formally launched in 1985, and now deals with the governing Workers' Party (PT, *Partido dos Trabalhadores*). This is also the historical context of many other movements that would come to play significant roles in contemporary TAMs (the southern European and Mexican movements within Vía Campesina, for example).

In lieu of the political party and the worker–peasant class alliances, contemporary TAMs are confronted by the challenge of at least three broad types of coalitions: the NGO–peasant movement relationship, sectoral alliances, and thematic advocacy alliances.

(i) NGO–peasant movement relations. Perhaps one of the most controversial and contentious alliances in the contemporary terrain of peasant movements is that between the movements and NGOs. In the most general, caricaturized presentation of this issue, some peasant activists have assumed that NGOs do not and cannot represent the rural poor and are undemocratic, but have privileged seats in official international (non)governmental consultation venues, and they have the funds. The same activists see peasant movements as representing the rural poor and as democratic organizations, but lament that they are rarely invited to official international consultative venues and that they do not have funds. It is, in effect, a 'love–hate' relationship between peasant movements and NGOs. Even those peasant movements most critical of NGOs in fact have ongoing dealings with NGOs. The NGOs remain the most significant funders for peasant movement activities. They are also significant facilitators for TAMs in geographic areas where the latter have no previous contacts. The NGOs remain among the most reliable allies of the TAMs. Not all NGOs, of course. For example, Vía Campesina favours working with a select few, including FoodFirst Information and Action Network (FIAN) and the Land Research and Action

Network (LRAN). It receives key economic support from nongovernmental agencies such as the Dutch Oxfam-Novib and the Inter-Church Organization for Development Cooperation (ICCO). Therefore, for Vía Campesina, it is not the NGOs per se that are problematic. Rather, it is the *terms* of the relationship that matter. Sweeping generalizations about, and against, NGOs by some TAM activists and some scholars are usually contradicted by empirical reality. Edelman (this collection), writing about transnational organizing in Central America in the 1990s, notes the irony that the vehemently anti-NGO tone in the political discourse of ASOCODE (which in turn heavily influenced the subsequent position of Vía Campesina on this issue) did not prevent ASOCODE itself from developing many typical features of a 'bad' NGO.

(ii) Sectoral alliances. What makes the IPC for Food Sovereignty an interesting and important global rural-oriented alliance is not only the fact that it is perhaps currently the largest network of movements, but that it also brings in important organizations of various rural sectors: small farmers and poor peasants, middle and rich farmers, fisherfolk, pastoralists, peasant women, indigenous peoples and rural workers, among others. Among the major TAMs today, it is Vía Campesina that has the clearest policy about prioritizing alliances with the movements of other sectors in the countryside, including the generally weak rural workers' movements. To date, however, Vía Campesina has been able to work more closely with peasant women, fisherfolk and indigenous peoples and less with organized rural workers, despite what Nicholson (see above) said about how fundamental the alliance of farmers and rural workers would be in the struggle against neoliberalism. In theory such an alliance might be desirable and easily achieved because these groups have so many common issues and struggles. In reality, however, forging broad rural multi-sectoral alliances has not proven easy. One reason for this is that different sectoral movements (small farmers, indigenous peasants, rural workers, fisherfolk, and so on) often have overlapping constituencies. At times competition between organizations develops in the struggle to recruit and represent these overlapping constituencies. Many of them also compete for the same funds from northern NGOs.

(iii) Thematic advocacy alliances. Perhaps the most common type of actually existing alliances is those that form around thematic advocacy. These are usually multi-class and multi-sectoral alliances, cutting across the rural–urban and global south–north divides. Several are analyzed in this collection: the anti-GM crops campaigns (see the chapters by Scoones and by Newell), the initiatives for organic food production and certification (see Friedmann and McNair), the Global Campaign for Agrarian Reform (see Borras), and environmental-agrarian advocacy (see Peluso, Afiff and Fauzi). It is in this type of alliance that TAMs are currently engaged, giving them greater visibility and the campaigns greater substance. While many of these alliances are tactical, short-lived and oriented to specific campaigns, some have strategic importance to TAMs. This type of alliance will likely remain the most important type of coalition work in the years to come,

and it is in this context that the emerging broad inter-TAM alliance around climate change or agrofuels should be understood.

CONCLUSION

Transnational agrarian movements are political projects with deep historical roots in diverse national societies, multiple and shifting alliances, varied action repertoires, and complex forms of representation, issue framing and demand making. Their leaders and activists, hoping to advance the projects' objectives, frequently paint pictures of organizational coherence, homogeneity and unity. Sympathetic scholars, similarly, have often tended to downplay TAMs' ambiguities and contradictions, just as has occurred in the study of social movements operating in local or national ambits (Rubin 2004). In some cases, these unitary claims result from scholars' heavy methodological reliance on contacts with movement leaders, while in other cases questionable claims about unified movements derive from researchers' own well-intended efforts to advance the political projects which they are studying.

The authors of the chapters in this collection take a different approach. While their methods and subjects vary widely in geographical, temporal and political scope, they share an understanding of TAMs' complexity that grows out of an appreciation of the complicated historical origins and the delicate political balancing acts that necessarily characterize any effort to construct cross-border alliances that link highly heterogeneous organizations, social classes, ethnicities, political viewpoints and regions. The authors in this collection see this complexity as an essential element in understanding TAMs. By acknowledging TAMs' contradictions, ambiguities and internal tensions, they also seek, from the standpoint of engaged intellectuals, to advance a transformative political project by better comprehending its roots, past successes and failures, and current and future challenges.

We have noted above that TAMs have existed since the early twentieth century, but we also argue that a major qualitative shift occurs with the new TAMs that began to emerge in the late 1980s and early 1990s. Indeed, to the surprise of many, peasants and small farmers in diverse world regions have been among the most belligerent forces confronting and critiquing the free-market juggernaut that began in the 1970s with the collapse of the Bretton Woods framework for regulating the world economy and the onset of the globalization era (Helleiner 1994). The rising profile and growing influence of Vía Campesina and regional networks, such as CLOC in Latin America and ROPPA in West Africa, originate in and reflect the broader expansion of transnational civil society that occurred in the same period. The multiple, severe crises that have affected the rural poor, however, have given particular characteristics to these organizations and led them to employ a dazzling range of action repertoires in an ever greater variety of venues, from WTO ministerial meetings to GM crop test sites. Struggles against unfair trade rules, corporate control of crop genetic material and market-led agrarian reform, as well as for innovative approaches to development and to appropriating more of the wealth that peasants and small farmers produce, are among the important areas covered by the chapters in this collection. The issue of alliance-building – with environmentalist, women's and indigenous and minority rights movements, as well as with non-governmental organizations and supra-national governance institutions – has also been a fertile subject for scholarship and for debate among activists. Several of the contributors allude briefly to the complex and at times contentious role of academic scholars and other professional intellectuals in the movements' campaigns and projects of representation, but this and many other significant questions remain for future in-depth research.

REFERENCES

ACWW, 2002. 'Associated Country Women of the World'. http://www.acww.org.uk, Accessed 20 November 2002.

Akram Lodhi, Haroon and Cristóbal Kay, eds., 2008. *Peasants and Globalization: Political Economy, Rural Transformation and the Agrarian Question*. London: Routledge.

Akram-Lodhi, Haroon, Saturnino M. Borras Jr and Cristóbal Kay, eds., 2007. *Land, Poverty and Livelihoods in an Era of Globalization*. London: Routledge.

Alvarez, Sonia and Arturo Escobar, eds., 1992. *The Making of Social Movements in Latin America: Identity, Strategy, and Democracy*. Boulder, CO: Westview Press.

Anheier, Helmut K., ed., 1999. *When Things Go Wrong: Organizational Failures and Breakdowns*. Thousand Oaks, CA: Sage.

Batliwala, S. and D.L. Brown, eds., 2006. *Transnational Civil Society: An Introduction*. Bloomfield, CT: Kumarian Press.

Bell, John D., 1977. *Peasants in Power: Alexander Stamboliski and the Bulgarian Agrarian National Union, 1899–1923*. Princeton: Princeton University Press.

Bernstein, Henry, 2006. 'Once Were/Still Are Peasants? Farming in a Globalising "South"'. *New Political Economy*, 11 (3): 399–406.

Bernstein, Henry, 2007. 'Rural Livelihoods in a Globalizing World: Bringing Class Back In'. Paper for conference on 'Policy Intervention and Rural Transformation: Towards a Comparative Sociology of Development', China Agricultural University, Beijing, 10–16 September 2007.

Biondich, Mark, 2000. *Stjepan Radić, The Croat Peasant Party, and the Politics of Mass Mobilization, 1904–1928*. Toronto: University of Toronto Press.

Borras, Saturnino Jr, 2004. 'La Vía Campesina: An Evolving Transnational Social Movement'. TNI Briefing Series No. 2004/6. Amsterdam: Transnational Institute.

Borras, Saturnino Jr, 2007. *Pro-Poor Land Reform: A Critique*. Ottawa: University of Ottawa Press.

Brass, Tom, 2000. *Peasants, Populism and Postmodernism: The Return of the Agrarian Myth*. London: Frank Cass.

Brown, Cyril, 1923. 'Says Germany also is Ready for Parley'. *The New York Times*, 9 April.

Bryceson, Deborah, Cristóbal Kay and Jos Mooij, eds., 2000. *Disappearing Peasantries? Rural Labour in Africa, Asia and Latin America*. London: Intermediate Technology Publications.

Brysk, Alison, 2003. *From Tribal Village to Global Village: Indian Rights and International Relations in Latin America*. Stanford, CA: Stanford University Press.

Byres, Terence J., 2003. 'Paths of Capitalist Agrarian Transition in the Past and in the Contemporary World'. In *Agrarian Studies: Essays on Agrarian Relations in*

Less-Developed Countries, eds. V.K. Ramachandran and Madhura Swaminathan, 54–83. New Delhi: Tulika; London: Zed.

Carr, Edward Hallett, 1964. *A History of Soviet Russia: Socialism in One Country 1924–1926*, Vol. 3. New York: Macmillan.

Cohen, Robin and Shirin M. Rai, eds., 2000. *Global Social Movements*. London: Athlone Press.

Cohen, Stephen F., 1975. *Bukharin and the Bolshevik Revolution: A Political Biography, 1888–1938*. New York: Vintage.

Colby, Frank Moore, ed., 1922. *The New International Year Book: A Compendium of the World's Progress for the Year 1921*. New York: Dodd, Mead and Company.

Davies, Constance, n.d. 'The Women's Institute: A Modern Voice for Women'. http://www.womens-institute.co.uk/memb-history.shtml, Accessed 7 December 2002.

Della Porta, D., M. Andretta, L. Mosca and H. Reiter, 2006. *Globalization from Below: Transnational Activists and Protest Networks*. Minneapolis, MN: University of Minnesota Press.

De Soto, Hernando, 2000. *The Mystery of Capital: Why Capitalism Triumphs in the West and Fails Everywhere Else*. New York: Basic Books.

Desmarais, Annette, 2007. *La Vía Campesina: Globalization and the Power of Peasants*. Halifax: Fernwood; London: Pluto.

Drage, Dorothy, 1961. *Pennies for Friendship: The Autobiography of an Active Octogenarian, a Pioneer of ACWW*. London: Gwenlyn Evans Caernarvon.

Durantt, Walter, 1920. 'Accord in Balkans Takes Wider Scope'. *The New York Times*, 27 August.

Edelman, Marc, 1987. 'The Other Superpower: The Soviet Union and Latin America, 1927–1987'. *NACLA Report on the Americas*, 21 (1): 10–40.

Edelman, Marc, 1998. 'Transnational Peasant Politics in Central America'. *Latin American Research Review*, 33 (3): 49–86.

Edelman, Marc, 1999. *Peasants against Globalization: Rural Social Movements in Costa Rica*. Stanford, CA: Stanford University Press.

Edelman, Marc, 2003. 'Transnational Peasant and Farmer Movements and Networks'. In *Global Civil Society 2003*, eds. Mary Kaldor, Helmut Anheier and Marlies Glasius, 185–220. London: Oxford University Press.

Edelman, Marc, 2005. 'Bringing the Moral Economy Back In . . . to the Study of Twenty-first Century Transnational Peasant Movements'. *American Anthropologist*, 107 (3): 331–45.

Edwards, Michael and John Gaventa, 2001. *Global Citizen Action*. Boulder, CO: Lynne Rienner.

Ellis, Frank, 2000. *Rural Livelihoods and Diversity in Developing Countries*. Oxford: Oxford University Press.

Florini, Ann M., ed., 2000. *The Third Force: The Rise of Transnational Civil Society*. Washington, DC: Carnegie Endowment for International Peace.

Fox, Jonathan, 2000. 'Assessing Binational Civil Society Coalitions: Lessons from the Mexico–US Experience'. Working Paper No. 26, Chicano/Latino Research Center, University of California at Santa Cruz.

Fox, Jonathan, 2001. 'Vertically Integrated Policy Monitoring: A Tool for Civil Society Policy Advocacy'. *Nonprofit and Voluntary Sector Quarterly*, 30 (3): 616–27.

Fox, Jonathan, 2005. 'Unpacking "Transnational Citizenship"'. *Annual Review of Political Science*, 8: 171–201.

Friedmann, Harriet, 2004. 'Feeding the Empire: The Pathologies of Globalized Agriculture'. In *The Empire Reloaded: Socialist Register 2005*, eds. L. Panitch and C. Leys, 124–43. London: Merlin Press.

Gaventa, John, 2002. 'Exploring Citizenship, Participation and Accountability'. *IDS Bulletin*, 33 (2): 1–11.

Gianaris, Nicholas V., 1996. *Geopolitical and Economic Changes in the Balkan Countries*. Westport, CT: Greenwood Publishing.

Glick Schiller, Nina, 2004. 'Transnationality'. In *A Companion to the Anthropology of Politics*, eds. David Nugent and Joan Vincent, 448–67. Oxford: Blackwell.

Guardian, 2008. '50 People Who Could Save the Planet'. *Guardian*, 5 January, http://www.guardian.co.uk/environment/2008/jan/05/activists.ethicalliving. Accessed 7 January 2008.

Guidry, J., M. Kennedy and M. Zald, 2000. 'Globalizations and Social Movements'. In *Globalizations and Social Movements: Culture, Power, and the Transnational Public Sphere*, eds. J. Guidry et al., 1–34. Ann Arbor, MI: University of Michigan Press.

Gupta, Akhil, 1998. *Postcolonial Developments: Agriculture in the Making of Modern India*. Durham, NC: Duke University Press.

Gwynne, Robert and Cristóbal Kay, 2004. *Latin America Transformed: Globalization and Modernity*. London: Arnold; New York: Oxford University Press (second edition).

Harvey, Neil, 1998. *The Chiapas Rebellion: The Struggle for Land and Democracy*. Durham, NC: Duke University Press.

Helleiner, Eric, 1994. 'From Bretton Woods to Global Finance: A World Turned Upside Down'. In *Political Economy and the Changing Global Order*, eds. Richard Stubbs and Geoffrey R.D. Underhill, 163–75. New York: St. Martin's Press.

Heller, Patrick, 2000. 'Degrees of Democracy: Some Comparative Lessons from India'. *World Politics*, 52 (2): 484–519.

Hemingway, Ernest, 1987. 'Stambouliski of Bulgaria'. In *The Toronto Star: Complete Dispatches 1920–1924*, ed. William White, 155–6. New York: Scribner's [originally in *Toronto Star*, 25 April 1922].

Herring, Ronald, 1983. *Land to the Tiller: The Political Economy of Agrarian Reform in South Asia*. New Haven, CT: Yale University Press.

Herring, Ronald, 2007. 'The Genomics Revolution and Development Studies: Science, Poverty and Politics'. *Journal of Development Studies*, 43 (1): 1–30.

Herring, Ronald and Rina Agarwala, 2006. 'Restoring Agency to Class: Puzzles from the Subcontinent'. *Critical Asian Studies*, 38 (4): 323–56.

Holt-Giménez, Eric, 2006. *Campesino a Campesino: Voices from Latin America's Farmer to Farmer Movement for Sustainable Agriculture*. Oakland, CA: Food First Books.

ICA & IFAP [International Cooperative Alliance & International Federation of Agricultural Producers], 1967. *Cooperation in the European Market Economies*. Bombay: Asia Publishing House.

IFAP, 1952. 'FAO Position on International Commodity Problems'. *IFAP News*, 1 (1) January: 6.

IFAP, 1957. *The First Ten Years of the International Federation of Agricultural Producers*. Paris and Washington: IFAP.

Jackson, George D., Jr, 1966. *Comintern and Peasant in East Europe, 1919–1930*. New York: Columbia University Press.

Jordan, Lisa and Peter van Tuijl, 2000. 'Political Responsibility in Transnational NGO Advocacy'. *World Development*, 28 (12): 2051–65.

Kay, Cristóbal and Patricio Silva, eds., 1992. *Development and Social Change in the Chilean Countryside: From the Pre-Land Reform Period to the Democratic Consolidation*. Amsterdam: CEDLA.

Keck, Margaret E. and Kathryn S. Sikkink, 1998. *Activists Beyond Borders: Advocacy Networks in International Politics*. Ithaca, NY: Cornell University Press.

Kerkvliet, Benedict, 2005. *The Power of Everyday Politics: How Vietnamese Peasants Transformed National Policy*. Ithaca, NY: Cornell University Press.

Khagram, S., J. Riker and K. Sikkink, 2002. 'From Santiago to Seattle: Transnational Advocacy Groups Restructuring World Politics'. In *Restructuring World Politics: Transnational Social Movements, Networks, and Norms*, eds. S. Khagram et al., 3–23. Minnesota: University of Minnesota University Press.

Lahiff, Edward, Saturnino M. Borras Jr and Cristóbal Kay, 2007. 'Market-Led Agrarian Reform: Policies, Performance and Prospects'. *Third World Quarterly*, 28 (8): 1417–36.

London Times, 1938. 'Associated Country Women of the World'. *London Times*, 25 November.

London Times, 1946a. 'Conference of World Farmers: Supporting the FAO'. *London Times*, 20 May: 6.

London Times, 1946b. 'Marketing of Food'. *London Times*, 30 May: 4.

Mahler, Sarah J., 1999. 'Theoretical and Empirical Contributions Toward a Research Agenda for Transnationalism'. In *Transnationalism from Below*, eds. Michael Peter Smith and Luis Eduardo Guarnizo, 64–100. New Brunswick, NJ: Transaction Publishers.

Mayo, M., 2005. *Global Citizens: Social Movements and the Challenge of Globalization*. London: 2 Books.

Mazoyer, Marcel and Laurence Roudart, eds., 2002. *Vía Campesina: Une Alternative Paysanne a la Mondialisation Néolibérale*. Geneva: Centre Europe-Tier Monde.

McKeon, Nora, Michael Watts and Wendy Wolford, 2004. 'Peasant Associations in Theory and Practice'. Civil Society and Social Movements Programme Paper No. 8. Geneva: UNRISD.

McMichael, Philip, 2006. 'Reframing Development: Global Peasant Movements and the New Agrarian Question'. *Canadian Journal of Development Studies*, 27 (4): 471–86.

McNabb, Marion and Lois Neabel, 2001. 'Manitoba Women's Institute Educational Program'. http://www.gov.mb.ca/agriculture/organizations/wi/mwi09s01.html, Accessed 7 December 2002.

Meier, Mariann, 1958. *ACWW 1929–1959*. London: Associated Country Women of the World.

Moss, Jeffrey W. and Cynthia B. Lass, 1988. 'A History of Farmers' Institutes'. *Agricultural History*, 62 (2): 150–63.

Müller, Birgit, 2006. 'Introduction: GMOs – Global Objects of Contention'. *Focaal-European Journal of Anthropology*, 48: 3–16.

Newsweek, 2001. 'The New Face of Protest: A Who's Who'. *Newsweek*, 30 July, 17.

O'Brien, Kevin, 1996, 'Rightful Resistance'. *World Politics*, 49: 31–55, October.

O'Brien, Kevin and Lianjiang Li, 2006. *Rightful Resistance in China*. Cambridge: Cambridge University Press.

O'Brien, Robert, Anne Marie Goetz, Jan Aart Scholte and Marc Williams, 2000. *Contesting Global Governance: Multilateral Economic Institutions and Global Social Movements*. Cambridge: Cambridge University Press.

Patel, Rajeev, 2006. 'International Agrarian Restructuring and the Practical Ethics of Peasant Movement Solidarity'. *Journal of Asian and African Studies*, 41 (1&2): 71–93.

Petras, James and Henry Veltmeyer, 2003. 'The Peasantry and the State in Latin America: A Troubled Past, an Uncertain Future'. In *Latin American Peasants*, ed. Tom Brass, 41–82. London: Frank Cass.

Pianta, Mario and Federico Silva, 2003. *Globalisers from Below: A Survey on Global Civil Society Organisations*. Rome: GLOBI.

Pundeff, Marin, 1992. 'Bulgaria'. In *The Columbia History of Eastern Europe in the Twentieth Century*, ed. Joseph Held, 65–118. New York: Columbia University Press.

Quinn-Judge, Sophie, 2003. *Ho Chi Minh: The Missing Years*. Berkeley: University of California Press.

Rigg, Jonathan, 2006. 'Land, Farming, Livelihoods, and Poverty: Rethinking the Links in the Rural South'. *World Development*, 34 (1): 180–202.

Rosset, Peter et al., 2006. 'Agrarian Reform and Food Sovereignty'. A paper prepared for and presented at the International Conference on Agrarian Reform and Rural Development (ICARRD), March 2006, Porto Alegre, Brazil. http://www.icarrd.org, Accessed 19 February 2008.

Rosset, Peter, Raj Patel and Michael Courville, eds., 2006. *Promised Land: Competing Visions of Agrarian Reform*. Oakland, CA: Food First Books.

Rubin, Jeffrey W., 2002. 'From Che to Marcos'. *Dissent*, 49 (3): 39–47.

Rubin, Jeffrey W., 2004. 'Meanings and Mobilizations: A Cultural Politics Approach to Social Movements and States'. *Latin American Research Review*, 39 (3): 106–42.

Rupp, Leila J., 1997. *Worlds of Women: The Making of an International Women's Movement*. Princeton, NJ: Princeton University Press.

Rural Coalition, 1994. *Building the Movement for Community Based Development: Rural Coalition 1994 Annual Assembly*. Arlington, VA: Rural Coalition.

Salamini, Heather Fowler, 1971. *Agrarian Radicalism in Veracruz, 1920–38*. Lincoln, NE: University of Nebraska Press.

Scott, James, 1985. *Weapons of the Weak*. New Haven, CT: Yale University Press.

Scott, James, 1990. *Domination and the Arts of Resistance: Hidden Transcripts*. New Haven, CT: Yale University Press.

Scott, James, 1998. *Seeing Like a State: How Certain Schemes to Improve the Human Condition Have Failed*. New Haven, CT: Yale University Press.

Scott, James and Benedict Kerkvliet, 1986. 'Everyday Forms of Peasant Resistance in Southeast Asia'. *Journal of Peasant Studies*, 13 (2), special issue.

Seoane, José and Emilio Taddei, eds., 2001. *Resistencias mundiales: De Seattle a Porto Alegre*. Buenos Aires: CLACSO.

Smith, Jackie and Hank Johnston, eds., 2002. *Globalization and Resistance: Transnationalism Dimensions of Social Movements*. Lanham, MD: Rowman & Littlefield.

Stédile, João Pedro, 2007. 'The Class Struggles in Brazil: the Perspective of the MST'. In *Global Flashpoints: Reactions to Imperialism and Neoliberalism, Socialist Register 2008*, eds. Leo Panitch and Colin Leys, 193–216. London: Merlin Press.

Stephen, Lynn, 1997. *Women and Social Movements in Latin America: Power from Below*. Austin, TX: University of Texas Press.

Storey, Shannon, 2002. Marc Edelman interview with Shannon Storey, National Farmers Union (Canada), Saskatoon, 24 November.

Tarrow, Sidney, 2005. *The New Transnational Activism*. Cambridge: Cambridge University Press.

Taylor, Rupert, ed., 2004. *Creating a Better World: Interpreting Global Civil Society*. Bloomfield, CT: Kumarian Press.

Tsing, Anna, 2005. *Friction: An Ethnography of Global Connection*. Princeton, NJ: Princeton University Press.

Velásquez, Baldemar, 2007. 'Alleged Assassin of Santiago Rafael Cruz Detained in the US and Freed by Mexican Authorities'. *News Farm Labor Organizing Committee, AFL-CIO*. Press release 27 November.

Vía Campesina, 2000. 'Vía Campesina Position Paper: International Relations and Strategic Alliances', discussed during the III International Conference in Bangalore.

Waterman, Peter, 2001. *Globalization, Social Movements and the New Internationalisms*. London: Continuum.

World Bank, 2003. *Land Policies for Growth and Poverty Reduction*. Washington, DC: World Bank; Oxford: Oxford University Press.

World Bank, 2007. *World Development Report 2008: Agriculture for Development*. Washington, DC: World Bank.

Yashar, Deborah, 2005. *Contesting Citizenship in Latin America: The Rise of Indigenous Movements and the Post-Liberal Challenge*. Cambridge: Cambridge University Press.

Yashar, Deborah, 2007. 'Resistance and Identity Politics in an Age of Globalization'. *The Annals of the American Academy of Political and Social Science*, 610: 160–81.

2 Peasants Make Their Own History, But Not Just as They Please...

PHILIP McMICHAEL

INTRODUCTION

The narrative of capitalist modernity has overwhelmingly regarded the peasantry as an historical anachronism, or as a receding baseline of development. Neoclassical economic theory and orthodox Marxism alike have reproduced this ontology, on the grounds of scale economies and/or marginality to a revolutionary class politics, respectively. This chapter reflects on the conditions and consequences of such views, arguing that they share a Euro- and state-centric premise, which, in turn, has shaped a developmentalist episteme that is largely responsible for the contemporary rural crisis. Mazoyer and Roudart capture the essence of this crisis when they observe that

> most of the world's hungry people are not urban consumers and purchasers of food but peasant producers and sellers of agricultural products. Further, their high number is not a simple heritage from the past but *the result of an ongoing process* leading to extreme poverty for hundreds of millions of deprived peasants. (2006, 10; emphasis added)

Compare this to Jeffrey Sachs' metaphor of the development ladder, where peasant poverty is an original condition: 'the move from universal poverty to varying degrees of prosperity has happened rapidly in the span of human history. Two hundred years ago . . . [j]ust about everybody was poor, with the exception of a very small minority of rulers and large landowners' (2005, 26).

Walt Rostow (1960) canonized the developmentalist view that peasants inhabit a baseline, traditional stage of human history. Development theory consigns peasants to a prior historical stage, and W. Arthur Lewis (1954) operationalized this episteme by portraying peasants as constituting 'unlimited supplies of labour' to industrializing economies. World-historical treatments notwithstanding (Wolf 1969; Walton 1984), standard social science deviates little from Barrington Moore's 1966 treatment of the political fate of the peasantry in *Social Origins of Dictatorship and Democracy*. While his implication that English liberal democracy is premised on the elimination of the peasantry is emblematic of the modernist perspective, there is an uncomfortable silence in a comparative method that separates England from its overseas empire, populated largely by peasants – those in India producing 20 per cent of England's bread at the turn of the twentieth century (Davis 2001, 299).

There are three key issues here. First, peasant trajectories are conditioned by world, rather than national, history. Second, as an instrument of legitimacy, the development narrative's enabling of an intensified peasant dispossession under a virulent neoliberal regime has become the focal point of a contemporary peasant mobilization. Third, conventional (liberal and Marxist) attempts to schematize modern history in developmentalist terms run aground on the shoals of stage theory – democratic outcomes, nationally imagined, are as partial as representations of peasants as historical relics. An interesting perspective on this imaginary is that of a contemporary Mexican peasant leader, who observes: 'a campesino comes from the countryside. There have always been campesinos. What did not exist before were investors, industrialists, political parties, etc. Campesinos have always existed and they will always exist. They will never be abolished' (cited in Desmarais 2007, 19). The question is what does being a '*campesino*' mean today? In his study of the farmer networks in Mexico and Central America, *Campesino a Campesino*, Food First Director, Eric Holt-Giménez observes:

> Contrary to conventional wisdom, today's campesinos are not culturally static or politically passive. Nor are they disappearing as a social class. Campesino families across Mesoamerica and the Caribbean (and around the world) are constantly adapting to global, regional, and local forces . . . A story of unflagging resistance to decades of a 'development' that sought to eliminate peasants from the countryside and, more recently, to neoliberal economic policies that prioritize corporate profit margins over environment, food security, and rural livelihoods.
>
> [It is] a struggle for cultural resistance because campesino culture has withstood both socialist and capitalist version of progress . . . Even today, campesinos across the Mesoamerican isthmus resist the devastating economic effects of globalization both from their home communities and from the fields, factories, and service sectors of the United States, to which they supply an inexhaustible army of cheap, expendable labour. (2006, xii)

That is, peasant mobilization within and against the neoliberal project,[1] on a world scale, is politically engaged in a way, and for a cause, rendered unthinkable by classical social theory.

The question becomes, is such peasant mobilization 'casting a long shadow of nostalgia and melancholy over modern society' (Bartra 1992, 17)? That is, is this a defensive, and/or reactionary impulse, or something completely different? I argue the latter. While there have always been, and continue to be, peasants, many of whom simply struggle to get by with a range of different livelihood strategies, there is a mobilized segment which is the subject of this chapter. Peasant mobilization, as examined here, reaches beyond the daily round of survival on

---

[1] The 'neoliberal project' is a cultural specification of what I have called the 'globalization project' (McMichael 1996) in order to avoid the economism associated with the colloquial term 'globalization'. The concepts are essentially interchangeable, and refer to political restructuring of the relations of capital on a world scale, legitimized by the normalizing, privatizing creed of neoliberalism.

the land to linking that struggle to a reframing of what is possible on the land in contradistinction to what is being done to the land and its inhabitants by the neoliberal regime. Commentators have noted differences in consciousness, and tensions, between and among peasants, peasant activists and peasant leaders (Wright and Wolford 2003; Caldeira 2004; Wolford 2006; Desmarais 2007). The point here is that the 'peasant mobilization' is transcending conventional peasant politics, reframing its ontological concerns via a critique of neoliberalism, and reformulating the agrarian question in relation to development exigencies today.

Thus, rather than examine this new peasant question through the conventional lens of modern social theory, it is useful to shift epistemological gears and examine the peasant movement[2] through its own discursive practices, as it critically engages with capitalist narratives and their enabling policies. The critique of the neoliberal project by the peasant movement is a response which, invited by the general assault on peasantries in the name of the development narrative, posits future possibilities on the land that transgress the boundaries of conventional modernist theory.

SITUATING THE TRANSNATIONAL PEASANT MOVEMENT

In order to specify the transnational peasant movement historically, both temporal and spatial distinctions are necessary. The temporal distinction situates this movement in relation to the agrarian question, first formulated in the late-nineteenth century. The spatial distinction concerns the ontological relation of the peasant movement to capitalism and its future. As I will suggest, these distinctions are related, and it is in fact the peasant movement itself that underscores the new meaning of the contemporary agrarian question, politicizing agricultural and food relations within and beyond neoliberal capitalism.

Temporal Distinctions

Temporally, the peasant movement today fundamentally transforms the assumptions and projections of the classical agrarian question. Framed in relation to the politics of capitalist transition in agriculture, the original agrarian question concerned how peasants would identify politically within varying processes of 'differentiation' or 'disintegration' of peasant farming, as capital subordinated landed property (cf. Lenin 1972). Karl Kautsky, a 'proletarian exclusivist' (Alavi 1987), who nonetheless argued that centralization of landed property as capital in agriculture

---

[2] By 'peasant movement' I refer to a generic global movement that is nevertheless highly diverse, localized with specific social and ecological projects, and yet with a historic and common politics of resistance to the commodification of land, seed and food, and to a WTO trade regime whose policies systematically disadvantage and dispossess small farmers across the world. Notwithstanding the divisions in and across leading organizations like Vía Campesina, there is a unity in diversity that informs the 'food sovereignty' project, which in turn constitutes (and advocates) a *process* politics (for an example, see Desmarais 2007).

was contingent, rather than path dependent, argued peasant political allegiances with labour depended ultimately on the combined impact of food prices on the viability of peasant agriculture, and on labour's real wages (1988, 317).

As food regime analysis observes, food prices were increasingly governed by international trade relations (Friedmann 1978). Access to empire and cheap foodstuffs ironically blunted a potential contradiction between peasant and proletarian, keeping wages low for capital, but squeezing European peasantries. At the turn of the century, the separate and combined counter-movements of agrarians and workers diverged from the revolutionary projections of Marxists, contributing to the social-democratic resolution described by Polanyi (1957). Resolution acknowledged a dimension overlooked by the classical agrarian question, namely, the world-historical relations contributing to early twentieth-century social transformations. That is, part of the social-democratic compromise, based in the Fordist wage system, included agricultural mercantilism – in effect, publicly-financed protection of First World agriculture, idealized as family farming (Friedmann and McMichael 1989). Third World states have challenged such agricultural mercantilism ever since, culminating in the emergence of the G-20 in 2003 at the WTO's Cancún Ministerial.

The original, classical formulation of the agrarian question, now problematic in the twenty-first century, was governed by a state-centrism, reflecting the nation-building focus of the late-nineteenth century. As Henry Bernstein characterized it, the classical agrarian question concerned the development of a *home* market for capital as 'the *agrarian question of capital*, and specifically *industrial capital*. In the context of transition(s) to capitalism, this was also assumed to be the agrarian question of labour as well as capital, inasmuch as these two definitive classes of an emergent capitalism shared a common interest in the overthrow/transformation of feudalism, and of pre-capitalist social relations and practices more generally' (2003, 209).

The irony of course is that just as the self-referential European project focused on modernization of the nation-state, pre-capitalist social relations and practices more generally were sustained, often in degraded form, in the empire (Davis 2001). Here, the peasantry was an object of exploitation, rather than elimination, as in the modernist scenario. In other words, while the colonial peasantry was a pedestal for metropolitan wage-labour, this role was invisibilized in social theories concerned with modern social forms of accumulation, predicated on a Eurocentric model of development as a national process.[3]

Arguably, developmentalism has institutionalized the trajectory of peasant redundancy across the now complete state system. Certainly, within the terms of the Cold War, under pressure from peasant insurgency, re-peasantization in the model of American family farming was projected through such public initiatives as the Latin American Alliance for Progress, the green revolution, and via foreign policy institutions such as USAID in Egypt (Mitchell 2002). But this

---

[3] Cf. Tomich (2004), who refocuses, and reformulates, a historical understanding of capitalism through the 'prism of slavery', arguing that the relationship between wage and slave labour was critical to this history, rather than presuming wage labour to be the definitive relationship.

phase of developmentalism was essentially an interlude of economic nationalism governed by peasant unrest, postwar mercantilism and the politics of decolonization, legitimized by the UN's 1960 Declaration of Independence.[4] While re-peasantization accompanied what Araghi calls conservative 'first struggle, first served' land reforms in East Asia and Latin America (1995, 346), de-peasantization also proceeded across the Third World, under pressure from the food aid regime (Friedmann 1982), and the inequalities of the green revolution (Gupta 1998).

As the neoliberal project, colloquially known as 'globalization', has replaced the period of economic nationalism, de-peasantization in the global South has intensified under the combined pressures of evaporation of public support of peasant agriculture, the *second* green revolution (privatized biotechnologies and export agricultures to supply global consumer classes), market-led land reform,[5] and WTO trade rules that facilitate targeting southern markets with artificially cheapened food surplus exports from the North. This 'corporate food regime', based in subsidies which reduce farm prices by as much as 57 per cent below actual costs (People's Food Sovereignty 2003), constitutes a 'world price' through trade liberalization, with devastating effects on small farmers everywhere (McMichael 2005).[6] For example, Sharma (2004) reports:

> Indonesia was rated among the top ten exporters of rice before the WTO came into effect. Three years later, in 1998, Indonesia had emerged as the world's largest importer of rice. In India, the biggest producer of vegetables in the world, the import of vegetables has almost doubled in just one year – from Rs92.8 million in 2001–02 to Rs171 million in 2002–03. Far away in Peru, food imports increased dramatically in the wake of liberalization. Food imports now account for 40 per cent of the total national food consumption. Wheat imports doubled in the 1990s, imports of maize overtook domestic production, and milk imports rose three times in the first half of the previous decade, playing havoc with Peruvian farmers.

Under these circumstances, through which food dumping dispossesses millions of peasants (Madeley 2000), the agrarian crisis of the late-nineteenth century, precipitated by cheap foodstuffs from the New World and the colonies, has been generalized – notably through the centralization of capitalist agriculture in the global North, via the mercantilist resolution following the collapse of Britain's free trade regime. That is, while the resolution of the classic agrarian question was mediated politically by GATT protectionism, institutionalizing increasingly

---

[4] For a comprehensive account of cases of agrarian reform during this period, see Rosset et al. (2006).
[5] For a comprehensive analysis of the 'marketing' of land as a neoliberal solution to rural poverty, see Borras (2006).
[6] Elsewhere I have portrayed the 'world price' as the corporate food regime's instrument of dispossession (McMichael 2005). Pressure on food supplies from agrofuels, and from shifting social diets in India and China, in particular, are related to rising food prices – in the past year, maize prices doubled, and wheat prices rose by 50 per cent, bringing the world to a 'post-food-surplus era' (Vidal 2007). Whether and to what extent this new socially-constructed food scarcity will affect the 'world price' mechanism (created through decoupling of northern farm supports and commodity prices), remains to be seen.

corporate farm lobbies across the Northern world, the current agrarian question has now been globalized through the medium of the corporate world market. But rather than play a conservative back-up role in the class politics of capitalist modernity, the peasant movement is transforming the terms of the question. It is no longer about agrarian transition via the path dependence of a theory of accumulation privileging capital, rather it is about agrarian transformation against the accumulation imperative, championed by a transnational coalition of peasants and other social justice movements, busy defetishizing accumulation.

Instead of defending a world lost, transnational movements such as Vía Campesina advocate a world to gain – a world beyond the catastrophe of the corporate market regime, in which agrarianism is revalued as central to social and ecological sustainability. More than a self-protective manoeuvre, the peasant movement proclaiming food sovereignty[7] calls into question the neoliberal 'food security' project, and its trope of feeding the world with food surpluses generated in the North (McMichael 2003). As the Vía Campesina website (2003) observes:

> Neo-liberal policies prioritize international trade, and not food for the people. They haven't contributed at all to hunger eradication in the world. On the contrary, they have increased the peoples' dependence on agricultural imports, and have strengthened the industrialization of agriculture, thus jeopardizing the genetic, cultural and environmental heritage of our planet, as well as our health. They have forced hundreds of millions of farmers to give up their traditional agricultural practices, to rural exodus or to emigration. International institutions such as IMF (International Monetary Fund), the World Bank, and WTO (World Trade Organization) have implemented those policies dictated by the interests of large transnational companies and superpowers . . . WTO is a completely inadequate institution to deal with food and agriculture-related issues. Therefore Vía Campesina wants WTO out of agriculture.[8]
>
> All over the world, low priced agricultural imports are destroying the local agricultural economy; take for instance European milk imported in India,

---

[7] First articulated at the 1996 World Food Summit, by the transnational peasant movement Vía Campesina, the concept of food sovereignty initially proclaimed 'the right of each nation to maintain and develop its own capacity to produce its basic foods, respecting cultural and productive diversity . . . and the right to produce our own food in our own territory' (cited in Desmarais 2007, 34). In elaborating 'the right of peoples, communities and countries to define their own agricultural, labour, fishing, food and land policies which are ecologically, socially, economically and culturally appropriate to their unique circumstances' (cited in Ainger 2003, 11), the concept of food sovereignty particularizes the socio-ecological function of agriculture to territorial coordinates (as a political tactic).
[8] Desmarais notes the slogan 'WTO out of agriculture' was a compromise among Vía Campesina chapters. While India's Karnataka Rajya Ryota Sangha (KRRS) advocated dismantling the WTO, Canada's National Farmer's Union (NFU) and Mexico's *Unión Nacional de Organizaciones Regionales Campesinas Autónomas* (UNORCA) favoured reforming the power relations of the international trade regulatory system, and the French *Confédération Paysanne* proposed human rights conventions for the WTO. Thus, 'the Vía Campesina demanded a reduction in the organization's powers by taking agriculture out of its jurisdiction, as well as the building of new structures within a transformed, more democratic, and transparent UN system' (2007, 111).

American pork in the Caribbean, European Union meat and cereals in Africa, animal food in Europe, etc. Those products are exported at low prices thanks to dumping practices. The United States and the European Union had a new dumping practice ratified by WTO, which replaces export subsidies by a strong reduction of their agricultural prices combined with direct payments made by the State. To achieve food sovereignty, dumping must be stopped.

The critique of dumping is not simply about unfair trading practices. It illuminates the politics of circulation – in effect the institutional construction of a corporate market premised on 'naturalizing' peasant redundancy, through political means. The contemporary rural crisis supersedes the original agrarian question concerning the political implications of agrarian class transformation for national political alliances. Home markets are no longer coherent, and political regulation of the global market essentially has been privatized, with states playing a 'clean-up' role, where they are involved at all. In the classical version, food registered only through the impact of its price on class identity and/or accumulation patterns. In the current agrarian question posed by the food sovereignty movement, food embodies a broader set of relations, becoming a window on the social, demographic and ecological catastrophe of neoliberalism.

As Vía Campesina put it: 'the massive movement of food around the world is forcing the increased movement of people' (2000). Trade in food surpluses contributes to cycles of dispossession, which in turn make land and agricultural labour available for corporate and export agriculture as Northern labour costs, land rents and environmental regulations rise and encourage the spread of agribusiness (including agrofuel)[9] estates and food processing plants across the South – a foundation on which the recent retailing revolution is being built (McMichael and Friedmann 2007). Further, de-peasantization contributes to the swelling ranks of casual labour in the world labour market at large. While on a national scale there may be a radical decoupling of industrialization from urbanization (Davis 2006, 17), on a global scale the accumulation of capital depends on these cycles of dispossession (McMichael 2005). Temporally, the scale of the agrarian question has shifted from the problematic of nation-building to that of global political-economy.

Through capitalist transition on a global scale, based in a corporate-led process of agricultural commodity production on a 'least-cost' supply principle,[10] recursive

---

[9] See, for example, the special issue on Agrofuels in *Seedling*, July 2007, available at http://www.grain.org.
[10] As noted, farm subsidies in the global North overwhelmingly support agribusiness, by decoupling prices from farm support, allowing market prices to fall to a fictitious low cost (to traders and processors) which in turn depresses prices throughout the world market. The food sovereignty movement – whether in Europe or the global South, favours social payments for producing public goods such as staple/cultural foods, environmental services, and so on. Thus the *Coordination Paysanne Européene* (CPE) claims: 'Public support to agriculture may well be legitimate, for instance for sustainable family farming to exist in every region, provided that this support is not used for low-price exports' (cited in Madeley 2006).

world price dynamics confound the path-dependent national model of the classical agrarian question. Overproduction and a generalized regime of agro-exporting artificially depress agricultural commodity prices and undercut farming everywhere. Thus:

> Thailand is known as one of the top food exporters in the world, particularly in rice. But a study on Thailand shows that while the country experiences an increase in its rice exports, farmers do not benefit from this success. The farm gate prices have not increased over the last decade. The stagnation in real income has been accompanied with a sharp rise in the debt burden of rural households. In short, more exports do not lead to an increase in farmers' welfare . . . (Jacques-Chai Chomthongdi, cited in NGOs 2004)

By denaturalizing the phenomenon of a 'world price', drawing attention to the corporate subsidy system as a foundation of the WTO trade regime, the food sovereignty movement has transformed the agrarian question. It accomplishes this by revealing a capital/state nexus (in the multilateral institutions) as a *global* force, generating a labour reserve of dispossessed peasants for a corporate development project, expressed in the 'mass production of slums' (Davis 2006, 13).

Spacial Distinctions

The 'spatial distinction' of the transnational peasant movement refers to its ontological relation to the neoliberal project. As suggested above, the food sovereignty movement engages critically with the political infrastructure of neoliberal capitalism, denaturalizing the market narrative as a precondition for elaborating an alternative narrative. That is, the challenge to capital occurs within its relations, but not its terms, of subjection (cf. Beverley 2004), positing a different agrarian ontology. The spatial dimension concerns the re-centring of the agrarian question – not as a political sideshow in an industrializing political culture, but as a political solution to the catastrophe of neoliberal industrialization.

This spatial distinction is framed usefully through Harvey's juxtaposition of contemporary ('anti-globalization') resistances as 'movements against accumulation by dispossession' and movements around 'expanded reproduction' (2005, 203). The former refers to peasant movements, insofar as they oppose displacement, withdrawal of public support for small farming, and the appropriation of environments, knowledges and cultures, amplifying values and cultural practices outside of the capital accumulation relation. The latter includes movements where 'the exploitation of wage labour and conditions defining the social wage' are central (2005, 203). Harvey claims finding the organic link between these different movements is 'an urgent theoretical and practical task' (2005, 203).

Arguably, Vía Campesina's politics unites these resistances, organically, in linking the accelerated movement of food with the accelerated dispossession of the peasantry. Neoliberal industrialization of agriculture on a world scale simultaneously generates casual labour and reduces capital's wage costs. Combining a politics of circulation with a politics of production and reproduction offers a world-historical critique of the conditions and consequences of the corporate food regime (cf. McMichael 2005). Vía Campesina de-reifies the euphemism of 'free trade', revealing its corporate/state origins and its unequal and devastating consequences. Further, to show that the expanded reproduction of capital depends upon the generation of an expendable global wage-labour force, also shows that corporate agriculture, as such, is not simply about producing cheap food, it is also about securing new conditions for accumulation by lowering the cost of labour worldwide. It is in this sense that agriculture is central to the solution.

In building an organic link between movements against dispossession and against the relations of expanded reproduction, Vía Campesina importunes us to recognize that accumulation is not simply about the concentration and centralization of the power of capital, but also is about dispossessing alternative practices and foreclosing options for alternative futures. In particular, the ontology of capitalist modernity, rooted in economism, rules out a place for peasants, physically expelling them from the land, and epistemologically removing them from history. Conversely, the ontology of the food sovereignty movement critiques the reductionism and false promises of neoliberalism, positing a practice and a future beyond the liberal development subject, and the science of profit. This emerging ontology is grounded in a process of revaluing agriculture, rurality and food as essential to general social and ecological sustainability, beginning with a recharged peasantry.

Economic theory posits the disappearance of peasants as a consequence of the law of rising productivity, reinforced by the low-income elasticity of demand for food, so that farm populations decline in relative and absolute terms. Unexamined here are assumptions about the conditions and consequences of rising agricultural productivity. Economic logic fetishizes growth in quantitative terms, standardizing agriculture in input–output terms. In externalizing ecological effects such as chemical pollution, soil and genetic erosion, carbon emissions, and discounting energy costs and subsidy structures for agribusiness, this logic seriously undervalues the economic costs of agro-industrialization (Martinez-Alier 2002, 146–7). In so doing, small-scale agriculture is presumed to be inefficient – as evidently confirmed by de-peasantization trends. While such abstract economic valuation is artificial, it nevertheless has real, and violent, consequences.

Alternatively, a grounded ecological perspective offers a range of values concerning the multifunctional and epistemic contributions of agriculture to humans and nature alike. This is the ontological break that informs the food sovereignty movement and its advocacy of revaluing small farming. Even where neoliberalism attempts to overcome the limits of its economism, as Martinez-Alier reminds us:

> The monetary values given by economists to negative externalities or to environmental services are a consequence of political decisions, patterns of property ownership and the distribution of income and power. There is thus no reliable common unit of measurement, but this does not mean that we cannot compare alternatives on a rational basis through multi-criteria

evaluation. Or, in other terms, imposing the logic of monetary valuation . . . is nothing more than an exercise in political power. Eliminating the spurious logic of monetary valuation, or rather relegating it to its proper place as just one more point of view, opens up a broad political space for environmental movements. (2002, 150)

Building such a political space is a component of the food sovereignty movement's consolidation of an ontological detour that rejects the grand narrative of modernity, industrialism and proletarianization as an unfulfilled dream/palpable nightmare, and affirms an alternative, historically-grounded narrative sensitive to place and value incommensurability, as concrete universals, rather than the abstract logic that justifies accumulation by dispossession. This is an ontology that offers a politics of voice and struggle *on* the land, in addition to struggle *for* land (Flavio and Sanchez 2000), thereby politicizing the social-ecology of property relations. Thus Paul Nicholson, European representative to the International Coordinating Committee of Vía Campesina, notes: 'to date, in all the global debates on agrarian policy, the peasant movement has been absent: we have not had a voice. The main reason for the very existence of the Vía Campesina is to be that voice and to speak out for the creation of a more just society' (cited in Desmarais 2002, 96). Further, the International Planning Committee (IPC) for Food Sovereignty (a 500-strong coalition of heterogeneous organizations, to which Vía Campesina belongs) states:

> No agrarian reform is acceptable that is based only on land distribution. We believe that the new agrarian reform must include a cosmic vision of the territories of communities of peasants, the landless, indigenous peoples, rural workers, fisherfolk, nomadic pastoralists, tribes, afro-descendents, ethnic minorities, and displaced peoples, who base their work on the production of food and who maintain a relationship of respect and harmony with Mother Earth and the oceans. (2006, n.p.)

In these senses, the food sovereignty movement is constituting an increasingly significant political economy of representation (Patel 2006) that combines politicization of neoliberal policy, claiming rights beyond market rights, with an agrarian identity based in a value complex weaving together ecological subjectivity and stewardship as a condition for social and environmental sustainability. Defending the peasant way is not just about preserving a 'culture', but strengthening cultural practices committed to transcending the subordination of food and agriculture to the price form. In so doing, the food sovereignty movement asserts the incommensurability of diverse agri- and food-cultures with a monocultural exchange-value regime that objectifies food, incorporating its production, and consumption, into the process of capital accumulation in general. At the same time, it is a politics of 'agrarian citizenship' (Wittman 2005), based in coalitions with other social justice movements on the margins, or in the centre, of the expanded reproduction of capital. Thus, the Movimento dos Trabalhadores Rurais Sem Terra (MST) reconstitutes the 'rural' as a civic base through which to confront Brazilian class and neoliberal politics, by developing cooperative forms of rural labour, producing staple foods for the working poor, and building alliances with, and offering livelihood security to, the urban unemployed (Wright and Wolford 2003). And the IPC for Food Sovereignty maintains that 'food sovereignty is not just a vision but is also a common platform of struggle that allows us to keep building unity in our diversity', 'agrarian reform and food sovereignty commit us to a larger struggle to change the dominant neoliberal model' and 'we will carry these conclusions back to debate with our social bases, and will use these ideas to confront the policies of international bodies like the FAO, and our governments' (2006, n.p.).

Arguably, in creating a space for an alternative ontology, the food sovereignty movement not only occupies a pivotal perspective challenging neoliberal capitalism (cf. Starr 2001, 224), but also reasserts the 'centrality of agriculture' in a post-capitalist modernity (Duncan 1996). The re-centring of agriculture within this political vision constitutes, then, an 'agrarian question of food' (McMichael forthcoming).

THE AGRARIAN QUESTION OF FOOD

Reformulating the agrarian question as a question of food shifts epistemological gears, switching focus from production to social reproduction. In the classical agrarian question, the politics of agricultural transition presumed a base/superstructure model in which emergent capitalist production relations would develop possibilities for a proletarian politics. The terms of reference were governed by a narrative of expanded reproduction of (industrial) capital, which in turn *enclosed* the meaning of 'social reproduction', limiting it to the reproduction of labour-power through the wage relation. Notwithstanding the marginalization of household labour, Polanyi understood capitalist ontology as 'fictitious', insofar as it at one and the same time limited and stimulated the 'discovery of society' – institutionalized in the resolution of the 'double movement' via the social contract associated with the mid-twentieth century citizen-state. While Polanyi's vision was framed in the state-centric terms of the time, it underlined the question of social reproduction discounted by the logic of capital accumulation. Furthermore, the appeal of Robert Owen's cooperatives, within a general conception of socially-embedded material relations, counterposed an alternative subjectivity to the liberal one of 'economic self-interest'.

Arguably, the food sovereignty movement is making an analogous claim, namely, that the neoliberal project of installing a 'self-regulating market' on a world scale *encloses* questions of social reproduction within a legitimating rhetoric of 'feeding the world'. In other words, in an era in which the market, not the state, is the organizing principle, social reproduction is fetishized as a market function. The WTO plays midwife to this project, managing a trade regime dedicated to 'food security' through market access. The food sovereignty movement argues, in contrast, that 'family-farm and peasant-based production for domestic purposes is responsible for approximately 90% of the world's food production, much of which does not even pass through markets' (People's Food Sovereignty 2003).[11] Accordingly, social reproduction through the world market is both ineffectual and licenses an ongoing violence against *extant* forms of social reproduction, as well as enlarging spaces of 'social exclusion' in the countryside and planet-wide urban slums (Davis 2006; Cameron and Palan 2004). The IPC's response to this consequence is both programmatic and subjective.

Programmatically, from a global perspective, it argues that 'in the context of food sovereignty, agrarian reform benefits all of society, providing healthy, accessible and culturally appropriate food, and social justice. Agrarian reform can put an end to the massive and forced rural exodus from the countryside to the city, which has made cities grow at unsustainable rates and under inhuman conditions' (IPC for Food Sovereignty 2006).

And its confederation within the European context, the *Coordinacion Paysanne Européene* (CPE), proposes:

> The European Union would benefit a lot by maintaining sustainable family farming, not only for guaranteeing food supply (food security), but also as regards the social and multi-functional role of agriculture. The present trend must be reversed: instead of concentrating the farms, an important fabric of small and medium-sized farms should be maintained, since they play an irreplaceable role in the following fields: a quality and diversified food production, landscape upkeep, wood and forest clearing, human territory occupation, etc. Maintaining the number of people working in agriculture is not a sign of economic backwardness but an added value. (2003, n.p.)

Subjectively, Vía Campesina echoes Polanyi's 'discovery of society' through the catastrophe of disembedded markets, qualifying a state-centric protectionism with a programme of substantive rights:

> The government should introduce policies to restore the economic condition of small farmers by providing fair allocation of these production [water, forest, local genetic and coastal] resources to farmers, recognizing their rights as producers of society, and recognizing community rights in managing local resources. (2005, 25, 31)

The right to produce society, and manage local resources, is a claim underscoring the ontological distinction with which Vía Campesina works. This is the broader meaning of Wittman's term, 'agrarian citizenship'. It involves the re-territorialization of states through the revitalization of local food ecologies under small-farmer stewardship, in the interests of society at large. Polanyi's claim that the social impulse to protect against, and re-embed, the market, resulted, for him, in the 'discovery of society', realized through the citizen-state. The food

---

[11] According to McCalla (1999), about 90 per cent of the world's food consumption occurs where it is produced; while urbanites depend on the market for almost all their food consumption, rural populations consume 60 per cent of the food they produce.

sovereignty movement, by contrast, recognizing the complicity of states in the neoliberal market project,[12] 'rediscovers' society through a substantive, rather than formal, set of rights, to be exercised as a means to the end of social reproduction, rather than an end in themselves (McMichael 2005). As Raj Patel puts it, the food sovereignty movement views rights as a 'means to mobilizing social relations', in turn 'a call for a mass re-politicization of food politics, through a call for people to figure out for themselves what they want the right to food to mean in their communities, bearing in mind the community's needs, climate, geography, food preferences, social mix and history' (2007, 88, 91).

What is at issue here is a departure from the problematic that the agrarian question is for capital, or labour, to resolve. Rather than view the peasantry as an enabler or spoiler of a revolutionary class politics, Vía Campesina ruptures the social time–space assumptions of classical social theory, politicizing the meaning of the 'expanded reproduction' of capital. To view peasant dispossession through the capital/labour prism is to discount agrarian cultures and to characterize the dispossessed as unemployed labour. Arguably, peasant mobilization articulates a more complex perspective on the crisis of neoliberal capitalism. It rejects teleological assumptions about class and accumulation deriving from a productivist understanding of the movement of capital, and views capital as a relation of production *and circulation*. In this way, it politicizes the privatized organization, and fetishized representation, of the global market, proposing a restoration of public, rights-based international institutions. In thereby reformulating questions of rights, social reproduction and sustainability, the peasant movement poses an 'agrarian question of food', where food embodies social, cultural and ecological values over and above its material value. Thus a farmer participant in an E-Forum for farmers' views, from movements including Vía Campesina, claimed:

> people must re-establish the link with their mother earth – which nourishes them – by practicing farming in balance with natural elements, farming that must be sparing and respectful of natural resources . . . Above all, what we eat must meet our physiological needs (not too much, not too little). Still today, 843 million people in the world, of whom three-quarters are small farmers, suffer from hunger (malnutrition or under-nutrition). The first challenge is therefore to meet the physiological needs of the world population through access to food of sufficient quality and quantity, sharing natural resources and practicing sustainable farming based on fair market rules. What we eat is also shaped by cultural, agricultural and culinary factors. These cultural aspects need to be retained by respecting other people's different beliefs and dietary habits. This is what contributes to the wealth of our planet. (Jean-Baptiste Pertriaux, cited in Pimbert et al. 2005, 15)

---

[12] Contrary to the International Federation of Agricultural Producers (IFAP), which accepts neoliberal principles, and works within the WTO, seeking to enhance farmers' opportunities (Desmarais 2007, 105).

50 *Philip McMichael*

From here it is but a short step to reasserting the foundational role of agriculture in civilization – in the epistemological, rather than simply the chronological, sense. Thus another farmer proclaimed:

> Anyone who speaks of life must speak of water and land, elements as vital as air to living. Farming came into being by combining these elements to make life last longer while constantly improving it. From subsistence farming, necessarily more self-contained and sparing of resources because of space and quantity restrictions, from that pure function of nourishment, we went over to 'commodisation' . . . Then we found globalization, a big word that could have meant discoveries and exchanges, but instead became a vector of slavery, competition, expropriation and exploitation (not that exploitation has not always existed . . .). And delusions of grandeur, constantly wanting more, took over the world. As a result water and land ceased to be vital elements for life, being turned into accessories in the pursuit of profit and market shares. The very notion of food no longer counts, as small farmers themselves have lost the notions of rights and duties, self-respect, respect for their labour, for others, for water and land. (Chantal Jacovetti, quoted in Pimbert et al. 2005, 10–11)

MAKING HISTORY

The agrarian question of food inverts the original focus of the agrarian question, on agrarian transition. Rather than raising questions about the trajectory of a given narrative, the food sovereignty movement questions the narrative itself. In a sense, a mobilized peasantry is making its own history. It is 'mobilized' precisely because it cannot do this just as it pleases – its political intervention is conditioned by the historical political-economic conjuncture through which it is emboldened to act. And it is emboldened precisely because neoliberal capitalism's violent imposition of market relations, with severe social and ecological consequences across the world, is catastrophic. Capitalism is evidently deepening its internal contradictions, but this process is complicated by a politics of dispossession that complicates and/or transcends class analysis. The commodification of natural and intellectual (*qua* social labour) relations crystallizes material and cultural values distinct from those of the dominant economic discourse. Such values are fundamentally ecological, and concern how humans construct, understand and experience their relations of social reproduction.

In addition to undermining social reproduction on the land, in generating a 'planet of slums' neoliberal capitalism reveals the social and ecological limits of the development narrative. The so-called 'unlimited supplies of labour' from the countryside metamorphose into a seemingly unlimited supply of unemployed slum-dwellers, exiting increasingly degraded habitats. Certainly, as Marx (1967) argues, capital precedes landed property as the proper methodological point of departure, for analysis, but this does not mean that its subordination of landed property is ecologically, or even socially, appropriate, even in advancing the socialist ideal. As Duncan claims:

> The proper scaffolding for a well-founded, hence potentially permanent, socialism would be a complex pattern of federated interacting elements of the world's dispersed population, each of which would be aware that they severally and collectively live within a living environment into whose local cycles they must insert their agricultural and industrial activities. There would be a complex division of labour but it would involve several scales of social entity – neighbourhood, municipality, region, and so on up to global levels – many tied to places. . . . This conception, which accords centrality to agriculture, must explicitly deal with the obvious, albeit misconceived, charge that it tends to counsel an antimodern, even 'peasantist,' approach to social questions. . . . The key point is denial of the claim that all cases of modernity depend(ed) necessarily on marginalizing agriculture. (1996, 48–9)

The significance of the food sovereignty movement is that, in the narrative of capitalist modernity, its project is virtually unthinkable (cf. Trouillot 1995). Social scientific categories, including the market episteme, render a 'peasantist' approach to social questions almost incomprehensible, since agriculture is ultimately viewed as a branch of industry divorced from natural cycles of regeneration. Thus analysis of the peasant question through the capital/labour lens posits an 'agrarian question of labour' (Bernstein 2004), with peasants recast as semi-proletarian (Kay 2006; Moyo and Yeros 2005). However, while the peasant movement may frame its struggle in conventional terms (sovereignty, agrarian reform, citizenship, rights), these terms assume new meanings in an alternative 'peasantist' ontology (cf. Mitchell 2002; McMichael 2006). Arguably, food sovereignty discourse offers a method of developing an alternative modernity, re-centred on agriculture and food.

'Food sovereignty' itself is a problematic term, as it evokes protectionism. However, a reflexive understanding of this concept situates it in relation to the politicization of 'food security' as the neoliberal design on feeding the world through the market. Judit Bodnar argues that Bové's activism against McDonalds was about resistance to agro-industrialism, rather than a territorialist response to a symbol of US/market culture threatening French cheese producers. It elevated democratic economy and fair trade principles over the reactionary 'link between land and nation' (Bodnar 2003, 143). The identification with a global civil society in formation is confirmed by Vía Campesina's uncompromising opposition to state complicity in the neoliberal project (including its opposition to G-20 *'free trade'* politics at Cancún), and political goal of 're-territorializing' states from within ('agrarian citizenship'), and from without, through multilateral institutions dedicated to fair trade and global justice (Bové and Dufour 2001). Essentially, 'food sovereignty' serves to appropriate and reframe dominant discourse, as a mobilizing slogan, and as a political tactic to gain traction in the international political-economy *en route* to a global moral economy organized around 'cooperative advantage' – as a counterpoint to 'comparative advantage' and its licensing of corporate manipulation of the state system and world economy as a chessboard for accumulation.

In world-historical terms, the peasant movement stands on the shoulders of previous movements for rights of self-determination. Of course, *what* is to be self-determined is the question, which can only be posed, and answered, historically. My argument is that it is not a question of peasant rights, *per se*. Rather, the rights discourse also concerns questions of social and ecological sustainability, as undermined by the neoliberal project. Reversing dispossession and reclaiming the right to farm, as a general act of social and ecological reproduction, is also critical to provisioning the 2.5–3 billion rural poor immiserated by the corporate food regime.

The food sovereignty vision, contesting the subjection of food to an unequal and unsustainable trade regime, calls into question the subdivision of the world into competing states, beholden to stabilizing their national accounts and currencies by authoring rules of economic liberalization. Such rules undermine principles of human solidarity, and decimate rural populations. Thus José Bové's comrade, François Dufour claims: 'the market has abolished frontiers, and seeks to impose uniformity on the planet. It's up to us, as citizens of the world, to raise the question of rights for everyone. Human rights don't stop at frontiers; we must globalize them' (Bové and Dufour 2001, 190). While demanding a *formal* guarantee of food sovereignty rights (including a certain measure of protection), the movement maintains that the content of these rights (access to land, credit and fair trade, and to decisions about what food to grow and how) is to be determined by the communities and countries themselves – thereby asserting a *substantive* reformulation of sovereignty through context-specific rights, situated in particular, historical subjectivities (cf. Patel 2007).

How to realize and sustain new subjectivities is a key question. This is a long-term process, and various constituents of the food sovereignty movement are so engaged – not only with struggles for land, but also with the 'struggle on the land', informed, or overdetermined by the collective struggle against neoliberalism (Desmarais 2007; Wright and Wolford 2003). Across space, politicization of subjectivity is enacted differently, since, in the first place, chapters of the food sovereignty movement bear different relations to the state system. Bové notes a basic division: 'for the people of the South, food sovereignty means the right to protect themselves against imports. For us [in Europe], it means fighting against export aid and against intensive farming' (Bové and Dufour 2001, 96). And, 'the strength of this global movement is precisely that it differs from place to place . . . The world is a complex place, and it would be a mistake to look for a single answer to complex and different phenomena. We have to provide answers at different levels – not just the international level, but local and national levels too' (Bové and Dufour 2001, 168). Within Europe, the struggle over subsidized farming takes the form of revealing how the rhetoric of 'multi-functionality' of agriculture (rural employment, environmental services, food provision, landscape) conceals a form of European protectionism (via the infamous WTO 'green box'), which directly favours agribusiness and industry, with detrimental consequences for small farmers in Europe and across the world. The food sovereignty movement advocates alternative farm subsidies, delinked from trade liberalization, especially in the global South, as necessary to survival of small farmers and the elaboration of a 'development box' (McCarthy 2005; McMichael 2007).

Ultimately, the realization of multi-functionality via the principles of food sovereignty involves an 'uneven and combined' approach within the movement at large. Solidarity, based on mutual recognition of different struggles, is the ideal:

> The difficulty for us, as farming people, is that we are rooted in the places where we live and grow our food. The other side, the corporate world, is globally mobile . . . The way in which we've approached this is to recognize there are people like us everywhere in the world who are farming people, who are rooted, culturally rooted, in their places. And what we need to do is build bridges of solidarity with each other which respect that unique place each of us has in our own community, in our own country. These bridges will unite us on those issues or in those places where we have to meet at a global level. (Vía Campesina founding member, Nettie Wiebe, cited in Ainger 2003, 10–11)

Thus, Laura Carlsen (2007) reported from a 2006 Vía Campesina international forum in Mexico City:

> For most peasant farmers in Mexico, Asia has always seemed literally and figuratively a world apart. But when Uthai Sa Artchop of Thailand described how transnational corporations sought to patent and control their varieties of rice seed, Mexican peasants realized that the Thais' rice was their corn. When Indonesian farmer Tejo Pramono spoke of how remittances from sons and daughters working in Hong Kong and the Middle East subsidize a dying countryside, Mexican farmers thought of their own relatives forced to migrate to the United States.
>
> Both sides nodded knowingly at the other's descriptions of the loss of markets to imports, the drop in producer prices due to unfair competition, and government cutbacks to producers except the large exporters. The January tortilla crisis in Mexico found its counterpart in the May palm oil crisis in Indonesia, when the price of both staple foods soared due to diversion to agrofuels and transnational control of markets.

The transformation of subjectivity operates simultaneously across and within particular chapters of the food sovereignty movement. Carlsen's report rehearses the co-production of peasant subjectivity within the neoliberal conjuncture, and its various negative impacts on farmers across the world. As she asks, 'who would have thought that in the age of globalization, small farmers' weaknesses would prove to be their strength?' – another way of observing the transnational convergence of a peasant politics via the shared experiences of neoliberal rule. This represents the unifying dimension in the 'unity of diversity' politics of the food sovereignty movement.

Within movement organizations, the transformation of consciousness is neither an overnight accomplishment nor necessarily pre-ordained, and its trajectory varies by spatio-economic and cultural history. Because of distinct experiences, the transformation project involves interpretive ambiguities, as shown in Wolford's research on the Brazilian landless-workers' movement. For example, in the sugarcane region of northeastern Brazil, where the MST's organizational difficulties are interpreted as foundering on the shoals of 'individualism', Wolford notes that settlers' tenuous relationships to the movement are filtered through their experience of land rights on the margins of the plantation system, where land embodies cultural values beyond being simply a means of production. In this case, contrary to the leadership's ideological investment in the collective value of land settlement, movement settlers value land access via individual forms of '"localized" common sense', deriving from an historic desire for privacy from the 'captivity' of planter patronage (Wolford 2005, 2006). Peasant subjectivity in southern Brazil differs, where 'small farmers who decided to join the MST were tied into a spatially expansive form of production that they valued as a part of a broader community. Family and community ties that were forged and re-forged through everyday practices working on the land helped to lower the threshold for participation in MST' (Wolford 2003, 202). Thus subjective conditions can neither be presumed, nor separated from their spatial and temporal coordinates.

In making this observation, Wolford poses important methodological and interpretive questions. She refocuses the conventional, structural question regarding the origin and impact of social movements, to the ethnographic question of why people join (and leave) the movement. Comparative ethnography across space reveals the fallacy of base-superstructure predictions or interpretations of the transformation of consciousness. It also contextualizes why it is that settlers do not automatically embrace the vision of the leadership (Caldeira 2004; Wolford 2006), and underscores the importance of continuous struggle on the land, once land has been occupied by the landless movement. Struggle on the land of course means different things in different places, lending complexity (and a multi-perspectival politics) to a movement that is represented on the world stage via the 'single-point perspective' of 'food sovereignty' – as a political intervention in the 'food security' discourse (cf. Ruggie 1993).

What this means is that while the food sovereignty movement is dedicated to mobilizing peasants and landless peoples around peasant rights versus the depredations of neoliberal rule, it must of necessity model subjective consciousness-raising to reveal and dispel the ways in which neoliberal subjectivity articulates with long-standing undemocratic cultural values in particular locales. Movement learning networks attempt to internalize and concretize the democratic principles that animate 'food sovereignty', via workshops and programmes on addressing unequal gender, class and ethnic relations (Swords 2007; Desmarais 2007; Eber 1999).

Micro-politics articulate with macro-politics in the sense that the strength of these individual movements draws on the ability of members to recognize and connect their particular conditions and political projects. Immediate goals of access to land as property or as a condition for exercising labour on the land (cf. Wolford 2007), or of protecting an *extant*/tenuous peasant culture, are distinctive. Nevertheless, in all cases, the struggle on the land (e.g. MST cooperatives, cross-border peasant learning networks, local wisdom networks,[13] small farming unions like Brazil's *Movimento de Pequenos Agricultores* against the agribusiness juggernaut) endures through the cultivation of reflexive subjects with the capacity to refract their struggle through questions of development, sociality, citizenship and co-production of sustainable living patterns. While micro-politics are the substance of movement, macro-politics constitute the social and world-historical frame, through which to situate, and develop, new subjectivities. By the same token, macro-politics are filtered through particular, or localized, experiences.

Thus, in representing the 'environmentalism of the poor', Martinez-Alier claims environmental conflicts are often expressed through the dialectic of 'macro/micro' value disputes, rooted in 'a clash in the standards of value to be applied, as when losses of biodiversity, or in cultural patrimony, or damage to human livelihoods, or infringements on human rights or loss of sacred values, are compared in non-commensurable terms to economic gains from a new dam, from a mining project or from oil extraction' (2002, 47–8).

Further, Vía Campesina's ecological critique 'localizes' universal themes:

> Issues of global environmentalism such as biodiversity conservation, threats from pesticides and energy saving, are transformed into local arguments for improvements in the conditions of life and for cultural survival of peasants, who are learning to see themselves no longer as an occupation doomed to extinction . . . This is not a phenomenon of post-modernity, in which some live (or try to make a living) by buying Monsanto shares, others eagerly eat hogs grown with transgenic soybeans, others are macrobiotic, and still others do organic farming. It is rather a new route of modernity, away from Norman Borlaug, a modernity based on scientific discussion with, and respect for, indigenous knowledge, improved ecological-economic accounting, awareness of uncertainties, ignorance and complexity, and, nevertheless, trust in the power of reason. (Martinez-Alier 2002, 147)

Ultimately, this is a politics that, in rejecting the uniform vision of capitalist modernity and the singular liberal subject, articulates distinct social, cultural and ecological realities as part of a complex movement in process. Constructing another world, with diversity as its durable theme, depends on the articulation of values concretizing a global macro-politics in micro-settings.

---

[13] In Thailand, for example, farmers in the semi-arid Northeast organize 'local wisdom networks' using the concept of 'learning alliances' to rehabilitate local ecological relations and promote health before wealth in agricultural practices. Since the 1997 financial crisis, these alliances have supported partnerships between farmer networks and government, dedicated to improved water conservation, participatory technologies, community forest management and biodiversity promotion. The goal is to convert monoculture to integrated, diversified farming and community development, and to convert state agencies to a rural sustainability paradigm (Ruaysoongnern and de Vries 2005). See also Holt-Giménez (2006).

CONCLUSION

Interpretations of contemporary peasant movements that process them through the lens of capitalist modernity and/or view them as a romantic phenomenon may render them historically redundant, even as they collectively reveal the crisis of neoliberalism (cf. Bernstein 2004; Petras 1997; Otero 1999). In arguing that Vía Campesina 'does not entail a rejection of modernity, technology and trade accompanied by a romanticized return to an archaic past steeped in rustic traditions [but is based on] ethics and values where culture and social justice count for something and concrete mechanisms are put in place to ensure a future without hunger', Desmarais emphasizes its critique of capitalist modernity through its engagement 'in building different concepts of modernity from their own, alternative and deeply rooted, traditions' (2002, 110). But this alternative modernity, including the politicization of subjectivity, is fully engaged with addressing, rather than simply revealing, the crisis of neoliberalism.

When the food sovereignty movement claims 'hunger is not a problem of means, but of rights' (cited in Starr 2005, 57), it reveals not just the crisis, but the evident limits, of the neoliberal project. That is, it focuses attention on the reductionism of the market paradigm, by which the achievement of 'food security' is reduced to a question of quantity and market 'supply'. As the food sovereignty movement demonstrates, market supply meets corporate, rather than human, needs – corporate food production does not address or generate demand so much as generate hunger. Market control in the name of development systematically violates the rights of people of the land to co-exist and secure the social reproduction of the majority of the world's people, and practice ecological sustainability.

As above, the food sovereignty movement is not without its tensions and contradictions, and its world-historic impact is yet to be determined. What appear to be small victories in the larger scheme of things (e.g. land settlements, reinforcement of the 'development box' discourse at the WTO, incorporation of 'food sovereignty' into FAO debate, agro-ecological learning alliances, witness to violence, etc.) represent the combined faces of a world movement, with an agenda dedicated to denaturalizing the neoliberal order. But more than a question of restoring the state, over the market, as a just organizing principle, the food sovereignty movement's strategic intervention not only problematizes the 'market' as a corporate project, but also problematizes the 'state' as complicit in this project. While Vía Campesina recognizes the jurisdictional authority of the state, it also seeks to transform that authority, by challenging the state *system* to enable states 'to have the right and the obligation to sovereignty, to define, without external conditions, their own agrarian, agricultural, fishing and food policies in such a way as to guarantee the right to food and the other economic, social and cultural rights of the entire population' (IPC for Food Sovereignty 2006, n.p.).

At the same time, the obligation to sovereignty entails securing the 'laws, traditions, customs, tenure systems, and institutions, as well as the recognition of territorial borders and the cultures of peoples' (IPC for Food Sovereignty 2006, n.p.) – repositioning citizenship as a vehicle for minority, as well as human, rights. In advocating an alternative modernity, including 're-territorialization' of, *and among*, states, the food sovereignty movement fundamentally challenges the institutional relations of neoliberal capitalism that contribute to mass dispossession – paradoxically reproducing the peasantry as an 'unthinkable' social force, as a condition for its emergence as a radical world-historical subject.

REFERENCES

Ainger, K., 2003. 'The New Peasants' Revolt'. *New Internationalist*, 353: 9–13.

Alavi, H., 1987. 'Peasantry and Capitalism: A Marxist Discourse'. In *Peasants & Peasant Societies*, ed. T. Shanin. Oxford: Basil Blackwell.

Araghi, F., 1995. 'Global De-Peasantization, 1945–1990'. *The Sociological Quarterly*, 36 (2): 337–68.

Bartra, R., 1992. *The Cage of Melancholy*. New Brunswick, NJ: Rutgers University Press.

Bernstein, H., 2003. 'Land Reform in Southern Africa in World-Historical Perspective'. *Review of African Political Economy*, 30 (96): 203–26.

Bernstein, H., 2004. '"Changing Before Our Very Eyes": Agrarian Questions and the Politics of Land in Capitalism Today'. *Journal of Agrarian Change*, 4 (1/2): 190–225.

Beverley, J., 2004. 'Subaltern Resistance in Latin America: A Reply to Tom Brass'. *The Journal of Peasant Studies*, 31 (2): 261–75.

Bodnar, J., 2003. 'Roquefort vs Big Mac: Globalization and Its Others'. *European Journal of Sociology*, 44: 133–44.

Borras, S. Jr, 2006. 'The Underlying Assumptions, Theory, and Practice of Neoliberal Land Policies'. In *Promised Land: Competing Visions of Agrarian Reform*, eds P. Rosset, R. Patel and M. Courville, 99–128. Oakland, CA: Food First Books.

Bové, J. and F. Dufour, 2001. *The World Is Not For Sale*. London: Verso.

Caldeira, R., 2004. 'A House and a Farm or Social Transformation? Different Modes of Action and One Dominant Utopia'. Paper presented at Poverty, Inequality and Livelihoods workshop, University of Reading, June.

Cameron, A. and R. Palan, 2004. *The Imagined Economies of Globalization*. London: Sage.

Carlsen, L., 2007. 'Vía Campesina Sets an International Agenda'. http://www.mstbrazil/org/?=laviacampesinasetsagenda, Accessed 15 July 2007.

CPE (Coordination Paysanne Européene), 2003. 'For a Legitimate, Sustainable, and Supportive Common Agricultural Policy'. 15 November. http://www.cpefarmers.org/w3/article.php3?id_article=50, Accessed 31 January 2008.

Davis, M., 2001. *Late Victorian Holocausts: El Nino Famines and the Making of the Third World*. London: Verso.

Davis, M., 2006. *Planet of Slums*. London: Verso.

Desmarais, A.A., 2002. 'The Vía Campesina: Consolidating an International Peasant and Farm Movement'. *The Journal of Peasant Studies*, 29 (2): 91–124.

Desmarais, A.A., 2007. *La Vía Campesina: Globalization and the Power of Peasants*. Point Black, NS and London: Fernwood Books & Pluto Press.

Duncan, C., 1996. *The Centrality of Agriculture: Between Humankind and the Rest of Nature*. Montreal: McGill Queen's University Press.

Eber, C.E., 1999. 'Seeking Our Own Food: Indigenous Women's Power and Autonomy in San Pedro Chenalhó, Chiapas (1980–1998)'. *Latin American Perspectives*, 26 (3): 6–36.

Flavio de Almeida, L. and F. Ruiz Sanchez, 2000. 'The Landless Workers' Movement and Social Struggles Against Neoliberalism'. *Latin American Perspectives*, 27 (5): 11–32.

Friedmann, H., 1978. 'World Market, State, and Family Farm: Social Bases of Household Production in an Era of Wage Labor'. *Comparative Studies in Society and History*, 20 (4): 545–86.

Friedmann, H., 1982. 'The Political Economy of Food: The Rise and Fall of the Postwar International Food Order'. *American Journal of Sociology*, 88: 248–86.

Friedmann, H. and P. McMichael, 1989. 'Agriculture and the State System: the Rise and Fall of National Agricultures, 1870 to the Present'. *Sociologia Ruralis*, 29: 93–117.

Gupta, A., 1998. *Postcolonial Developments: Agriculture in the Making of Modern India*. Durham, NC: Duke University Press.

Harvey, D., 2005. *A Brief History of Neoliberalism*. Oxford: Oxford University Press.

Holt-Giménez, E., 2006. *Campesino a Campesino: Voices from Latin America's Farmer to Farmer Movement for Sustainable Agriculture*. San Francisco, CA: Food First Books.

IPC for Food Sovereignty, 2006. *Sovranita Alimentare*. Final Declaration: For a New Agrarian Reform Based on Food Sovereignty. 9 March. http://movimientos.org/cloc/fororeformagraria/show_text.ph, Accessed 31 January 2008.

Kautsky, K., 1988/1899. *The Agrarian Question*. Volume 2. London: Zwan Publications.

Kay, C., 2006. 'Rural Poverty and Development Strategies in Latin America'. *Journal of Agrarian Change*, 6 (4): 455–508.

Lenin, V.I., 1972. *The Development of Capitalism in Russia: Collected Works*, Volume 3. Moscow: Progress Publishers.

Lewis, W.A., 1954. 'Economic Development with Unlimited Supplies of Labour'. *Manchester School of Economic and Social Studies*, 22: 139–91.

Madeley, J., 2000. *Hungry for Trade*. London: Zed Books.

Madeley, J., 2006. 'The Enduring Racket: Why the Rich Won't Budge on Farm Subsidies'. *Panos Online*, 28 July. http://www.globalpolicy.org/socecon/trade/subsidies/2006/0728panos.htm, Accessed 31 January 2008.

Martinez-Alier, J., 2002. *The Environmentalism of the Poor: A Study of Ecological Conflicts and Valuation*. Cheltenham: Edward Elgar.

Marx, K., 1967. *Capital*. Vol. 1. Moscow: Progress Publishers.

Mazoyer, M. and L. Roudart, 2006. *A History of World Agriculture: From the Neolithic Age to the Current Crisis*. New York: Monthly Review Press.

McCalla, A.F., 1999. 'World Agricultural Directions: What Do They Mean For Food Security?'. Presentation to Cornell Institute for International Food and Development, Cornell University, 30 March.

McCarthy, J., 2005. 'Rural Geography: Multifunctional Rural Geographies – Reactionary or Radical?'. *Progress in Human Geography*, 29 (6): 773–82.

McMichael, P., 1996. *Development and Social Change. A Global Perspective*. First edition. Thousand Oaks, CA: Pine Forge Press.

McMichael, P., 2003. 'Food Security and Social Reproduction'. In *Power, Production and Social Reproduction*, eds S. Gill and I. Bakker, 169–89. New York: Palgrave Macmillan.

McMichael, P., 2005. 'Global Development and the Corporate Food Regime'. In *New Directions in the Sociology of Global Development*, eds F.H. Buttel and P. McMichael, 265–99. Oxford: Elsevier.

McMichael, P., 2006. 'Peasant Prospects in a Neo-Liberal Age'. *New Political Economy*, 11 (3): 407–18.

McMichael, P., 2007. 'Sustainability and the Agrarian Question of Food'. Plenary Presentation, *European Rural Sociology Congress*, Wageningen, August.

McMichael, P., forthcoming. 'Food Sovereignty, Social Reproduction and the Agrarian Question'. In *Peasant Livelihoods, Rural Transformation and the Agrarian Question*, eds A.H. Akram-Lodhi and C. Kay. London: Routledge.

McMichael, P. and H. Friedmann, 2007. 'Situating the Retailing Revolution'. In *Supermarkets and Agri-Food Supply Chains*, eds G. Lawrence and D. Burch, 291–319. London: Edward Elgar.

Mitchell, T., 2002. *Rule of Experts: Egypt, Techno-Politics, Modernity*. Berkeley, CA: University of California Press.

Moore, B. Jr, 1966. *Social Origins of Dictatorship and Democracy. Lord and Peasant in the Making of the Modern World*. Boston, MA: Beacon Press.

Moyo, S. and P. Yeros, eds, 2005. *Reclaiming the Land: The Resurgence of Rural Movements in Africa, Asia and Latin America*. London: Zed Books.

NGOs, 2004. 'WTO Rules Must Respect Food Sovereignty'. http://www.southcentre. org/info/southbulletin/bulletin80/bulletin80-09.htm Accessed 31 January 2008.

Otero, G., 1999. *Farewell to the Peasantry? Political Class Formation in Rural Mexico*. Boulder, CO: Westview Press.

Patel, R., 2006. 'International Agrarian Restructuring and the Practical Ethics of Peasant Movement Solidarity'. *Journal of Asian and African Studies*, 41 (1/2): 71–93.

Patel, R., 2007. 'Transgressing Rights: La Vía Campesina's Call for Food Sovereignty'. *Feminist Economics*, 13 (1): 87–93.

People's Food Sovereignty, 2003. 'Peasants, Family Farmers, Fisherfolk and Their Supporters Propose People's Food Sovereignty as Alternative to US/EU and G20 Positions'. 15 December. http://www.ukfg.orguk/docs/Peoples%20Food%20Sovereignty %20declaration.doc, Accessed 31 January 2008.

Petras, J., 1997. 'Latin America: the Resurgence of the Left'. *New Left Review*, 223: 17–47.

Pimbert, M., K. Tran-Thanh, E. Deléage, M. Reinart, C. Trehet and E. Bennett, eds, 2005. *Farmer's Views on the Future of Food and Small Scale Producers*. London: The International Institute for Environment and Development.

Polanyi, K., 1957. *The Great Transformation: The Political and Economic Origins of Our Times*. Boston, MA: Beacon.

Rosset, P., R. Patel and M. Courville, 2006. *Promised Land: Competing Visions of Agrarian Reform*. Oakland, CA: Food First Books.

Rostow, W.W., 1960. *The Stages of Economic Growth: A Non-Communist Manifesto*. Cambridge: Cambridge University Press.

Ruaysoongnern, S. and F.P. de Vries, 2005. 'Learning Alliances Development for Scaling Up of Multi-Purpose Farm Ponds in a Semi-Arid Region of the Mekong Basin'. Paper presented at Learning Alliances Conference, Delft, 6–10 June 2007.

Ruggie, J.G., 1993. 'Territoriality and Beyond: Problematizing Modernity in International Relations'. *International Organization*, 47 (1): 139–74.

Sachs, J., 2005. *The End of Poverty*. New York: Penguin.

Sharma, D., 2004. 'Entitled to Subsidies!'. *India Together*, October. http://indiatogether. org/opinions/dsharma, Accessed 31 January 2008.

Starr, A., 2001. *Naming the Enemy: Anti-Corporate Movements Confront Globalization*. London: Zed Books.

Starr, A., 2005. *Global Revolt: A Guide to the Movements Against Globalization*. London: Zed Books.

Swords, A., 2007. 'Neo-Zapatista Network Politics: Transforming Democracy and Development'. *Latin American Perspectives*, 34 (2): 78–93.

Tomich, D.W., 2004. *Through the Prism of Slavery: Labor, Capital and World Economy*. New York: Rowman and Littlefield.

Trouillot, M., 1995. *Silencing the Past: Power and the Production of History*. Boston, MA: Beacon.

Vía Campesina, 2000. 'Declaration of the International Meeting of the Landless in San Pedro Sula'. Honduras, July.

Vía Campesina, 2003. 'What is Food Sovereignty?'. January 15. Available at www.viacampesina.org.

Vía Campesina, 2005. *Impact of the WTO on Peasants in South East Asia and East Asia*. Jl. Mampang Prapatan, XIV (5).

Vidal, J., 2007. 'The Looming Food Crisis'. *The Guardian*, 29 August.

Walton, J., 1984. *Reluctant Rebels: Comparative Studies of Revolution and Underdevelopment*. New York: Columbia University Press.

Wittman, H., 2005. 'The Social Ecology of Agrarian Reform: The Landless Rural Workers' Movement and Agrarian Citizenship in Mato Grosso, Brazil'. Unpublished PhD Dissertation, Development Sociology, Cornell University.

Wolf, E., 1969. *Peasant Wars of the Twentieth Century*. New York: Harper.

Wolford, W., 2003. 'Families, Fields and Fighting for Land: The Spatial Dynamics of Contention in Rural Brazil'. *Mobilization: An International Journal*, 8 (2): 201–15.

Wolford, W., 2005. '"Every Monkey has its Own Head": Rural Sugarcane Workers and the Politics of Becoming a Peasant in Northeastern Brazil'. Paper prepared for The Colloquium in Agrarian Studies, Yale University.

Wolford, W., 2006. 'The Difference Ethnography Can Make: Understanding Social Mobilization and Development in the Brazilian Northeast'. *Qualitative Sociology*, 29 (3): 335–52.

Wolford, W., 2007. 'Land Reform in the Time of Neo-Liberalism: A Many Splendored Thing'. *Antipode*, 39 (3): 550–70.

Wright, A. and W. Wolford, 2003. *To Inherit the Earth: The Landless Movement and the Struggle for a New Brazil*. San Francisco, CA: Food First Books.

3 Transnational Organizing in Agrarian Central America: Histories, Challenges, Prospects

MARC EDELMAN

INTRODUCTION

In the recent past Central America had some of the most powerful and successful agrarian movements and peasant-based guerrilla armies in the western hemisphere and some of the most extensive agrarian reforms.[1] At its height in the 1970s, for example, after decades of intense conflict, nearly a quarter of the landless and land-poor population of Honduras benefited from the agrarian reform (Isaula 1996, 13; Ruhl 1984, 53; Posas 1985). In El Salvador, about 25 per cent of rural households received land under either the 1980s agrarian reform, which pro-US regimes implemented in hopes of heading off more radical transformations, or as part of the distribution programme that followed the end of the 12-year-long civil war in 1992 (Hecht and Saatchi 2007, 669). In the 1980s in Nicaragua the Sandinista government carried out far-reaching agrarian reforms, granting access to land to an estimated 37 per cent of poor peasants and rural workers eligible to receive grants; under the first post-Sandinista government an almost equivalent proportion of claimants, many of them ex-combatants in the civil war that ended in 1989, received land (Jonakin 1997, 98, 101; Ruben and Masset 2003, 486). In Guatemala as well, ex-combatants and other landless and land-poor *campesinos* received land under the 1996 peace accords.[2]

Central America was also one of the key zones where contemporary transnational peasant organizations emerged. The Vía Campesina, arguably the pre-eminent transnational peasant and small farmer coalition since the early 1990s, has deep historical roots in Central America. It drew heavily on Central American experiences with cross-border organizing, included among its small group of founders several peasant leaders from the region, had its proximate origins in a congress of Nicaraguan and other farmers in Managua, and was headquartered

---

[1] 'Central America', for the purposes of this analysis, will generally be considered to include the five countries that made up Spanish Central America (Guatemala, El Salvador, Honduras, Nicaragua and Costa Rica). Organizations from Panama and Belize, and later Mexico, have also participated in transnational peasant organizing in the region, as noted below.
[2] Frequently, however, post-conflict land distribution programmes had a strong pro-market orientation and a socially demobilizing impact (Gauster and Isakson 2007; De Bremond 2007). They coincided with significant processes of 'counter-reform', especially in Nicaragua (Fiallos 2002, 5; Jonakin 1997; Ruben and Masset 2003); and Honduras (Ruben and Fúnez 1993).

during 1996–2004 in Tegucigalpa, Honduras (Desmarais 2007, 32, 75).[3] Peasants constituted the vast majority of combatants on both sides of the civil wars in Guatemala, El Salvador and Nicaragua. While this had tragic human costs, it contributed to a perception across the left–right spectrum that the rural poor were an important political actor, to be courted and nurtured by those seeking radical social change or to be neutralized and controlled by those seeking to block it.

In the early twenty-first century, however, the situation is far less encouraging for those who once viewed the peasantry as a major political protagonist and motor of social transformation. In every country of the isthmus, the peasant struggles that, from the 1970s to the 1990s, sustained pressure for agrarian reform and better production conditions have diminished in size and frequency. Rural zones are losing population and political clout, as thousands of peasants migrate within or outside the region. The ambitious, high-profile supranational coalitions of small producers that pioneered transnational organizing in the 1990s and that sought to put the brakes on neoliberal 'reform' are largely in disarray, have disappeared completely, or exist primarily as occasional meeting 'spaces' or even 'paper' organizations or 'dot causes'.[4] The consolidation of neoliberal regimes, while more complete in some countries than in others, continues to meet with significant resistance from organized civil society. But in the countryside, in particular, growing numbers of small producers and landless are opting for 'exit' instead of 'voice', eschewing non-market, political approaches to resolving their difficulties and choosing instead to abandon agriculture, rural zones and the organizations that once led the struggle to defend peasant interests and livelihood.[5]

This chapter attempts to link the study of organizations – their antecedents, formation, internal tensions and forms of representation – with the socioeconomic

---

[3] See also the contribution by Borras in this collection. Similarly, the CLOC (Coordinadora Latinoamericana de Organizaciones del Campo), a network that has overlapping membership with Vía Campesina and that works closely with it, had its headquarters in Guatemala City during 2001–2005. CLOC's decision (at its Third Congress in August 2001, which the author attended) to shift its office to Guatemala was contentious. Even though the Congress was held in Mexico City, the only flag on the stage was Cuba's. Several small Mexican organizations initiated a campaign to move CLOC's headquarters to Havana as a gesture of solidarity with Cuba. Proponents of moving instead to Guatemala argued that a Central American location would insert CLOC in a zone with a long history of agrarian struggles and would also limit the isolation that could occur from being headquartered in Cuba. After an acrimonious debate, a small majority of delegates voted for the Guatemala City option.

[4] The phenomenon of 'social movements' that have a presence on the Internet but only limited or indeterminate on-the-ground backing has profound methodological, political, ethical and representational dimensions that scholars and activists only occasionally acknowledge. More than two decades ago Tilly (1984, 311) nonetheless raised the problem of 'fictitious' organizations; more recently, Anheier and Themudo (2002, 209–10) analyzed 'dot causes' or Internet-based advocacy organizations with minuscule constituencies, Tarrow (2005, 165) described the formation of 'paper coalitions' and – perhaps most cynically – some commentators now refer to the practice of forming fake 'grassroots' organizations as 'astroturfing'.

The idea of activist 'spaces', as distinguished from 'social movements', comes from recent Brazilian debates over the character and possibilities of the World Social Forums. At issue is the extent to which periodic get-togethers of activists genuinely facilitate radical social action or are, alternatively, merely 'idea factories' or 'meeting grounds' (Whitaker 2006, 68; Santos 2006, 75–6). This debate echoes in some respects the discussions over Vía Campesina as an 'arena of action' (Borras 2004 and in this collection).

[5] On the notions of 'exit' and 'voice' see Hirschman (1970).

analysis of the contexts in which they operate. It thus moves beyond the study of 'political opportunity' as a conjunctural opening or closing of political space to examine medium- and long-term structural changes that affect the possibilities for collective struggles.[6] The chapter first outlines the meteoric rise and subsequent decline of a regional coalition of *campesino* groups, the Association of Central American Peasant Organizations for Cooperation and Development (Asociación de Organizaciones Campesinas de Centroamérica para la Cooperación y el Desarrollo, ASOCODE), which during 1991–1998 enjoyed an extraordinarily high profile in isthmian politics. It briefly considers several more recent attempts to reconstitute a Central America-wide peasant 'platform'. It then analyzes several major, interrelated changes in Central American economies and societies that impact the rural sector and that have negatively impacted both national and transnational peasant organizations. These include the declining importance of agriculture in the region's economies; the worsening situation of small basic grains and coffee producers; the growing influence of urban culture and the intensification of pluriactive livelihoods and rural–urban and extra-regional migration; and the reconfiguration of rural organizations from broad-based struggles to narrow sectoral demands.[7] Finally, the chapter raises questions about the implications of these changes and of the region's new development models for peasant struggles for agrarian reform and related demands.

TRANSNATIONAL PEASANT ORGANIZING IN CENTRAL AMERICA

The period from the late 1980s to the mid-1990s saw an efflorescence of transnational peasant organizing in Central America.[8] These campaigns built on contacts that grew out of longstanding international labour migrations, decades of intensive local- and national-level organizing, and more recent cross-border refugee flows in a region of small states and porous frontiers. The international solidarity that backed the Sandinistas in their struggle against the Somoza dictatorship and, later, during their decade in power in Nicaragua also provided inspiration for rural activists from different countries, who after July 1979 often met each other at regional events sponsored by Nicaraguan government agencies or pro-Sandinista organizations. Throughout the region, as civil wars ended or ebbed in the 1990s, political space opened and debate about the shape of post-conflict society intensified. At the same time, plummeting coffee prices, the liberalization of intra-regional trade in basic grains and the imposition of structural adjustment programmes in all the countries of the isthmus contributed to intensifying the economic pressures on smallholding agriculture. In much of the region, most notably in Honduras and Costa Rica, peasant organizations were in the forefront of national struggles against economic structural adjustment programmes.

---

[6] For critiques of conventional notions of 'political opportunity', see Bevington and Dixon (2005) and Brockett (2005).
[7] Gaete et al. (2005, 22) discuss this dynamic in Costa Rica.
[8] The roots of these efforts are detailed in Edelman (1998). See also Biekart (1999).

The Rise and Demise of ASOCODE

In 1991, rural activists who had attended a European Community-funded food security education course in Panama founded the Association of Central American Peasant Organizations for Cooperation and Development (Edelman 1998). Led initially by a charismatic young Costa Rican, Wilson Campos, ASOCODE rapidly garnered growing legitimacy with its elite interlocutors and foreign funders, as well as with *campesino* activists from different countries, who increasingly recognized that their counterparts elsewhere confronted similar problems and shared the same concerns.[9] Widely perceived as a significant actor in regional politics, over the next five years ASOCODE managed to have representatives attend numerous presidential and ministerial summits, sent frequent delegations to Europe and North America, published diverse position papers and a bi-monthly newsletter in Spanish and English, and sponsored meetings and courses for peasant activists.[10] It also became one of the first and most vocal 'civil society' members in the Consultative Council of the SICA (Sistema de Integración Centroamericana), which was created as part of the 1991 Tegucigalpa Protocol that reformed, systematized and rejuvenated Central America's main supra-national regional governance institutions as an adjunct to the peace accords already signed in Nicaragua and that were taking shape in El Salvador and Guatemala.

ASOCODE's formation clearly responded to a shift in European cooperation strategies in the late 1980s and early 1990s when multilateral, bilateral and NGO donors let it be known that they preferred to support projects that had a Central American regional, as opposed to a national or local, focus. Not surprisingly, given these changing funding priorities, the new regional peasant coalition attracted a copious flow of cooperation funds, which reached an annual peak in 1996 at US$1.5 million, largely from the Dutch agency HIVOS, Ibis-Denmark, various branches of Oxfam, and other NGOs in the Copenhagen Initiative for Central America (Biekart 1999, 204–6, 280). By 1995 ASOCODE was providing a monthly subvention of US$4,000–5,000 to each of the seven participating national coalitions, most of which was spent as salary for the two representatives that each country assigned to work in ASOCODE's coordinating council

---

[9] Campos epitomized the figures whom Tarrow (2005, 29) termed 'rooted cosmopolitans', 'people . . . who are rooted in specific national contexts, but who engage in contentious political activities that involve them in transnational networks of contacts and conflicts'. Scholars of transnational agrarian movements have largely failed to theorize systematically this social type or even to acknowledge its existence. Leaders of peasant and farm organizations who have university educations (as was the case with Campos), who speak two or more languages, or who have otherwise been upwardly mobile or accumulated significant cultural capital have been prominent in transnational peasant and farmer organizing from its inception. Not all 'rooted cosmopolitan' peasant leaders in Central America shared these characteristics (certainly most did not), but especially in the movements' early years the importance of a few key figures as cultural and intellectual 'brokers', between the organizations and funding agencies for example, cannot be overstated.

[10] Two collections of panegyrical essays (Tangermann and Valdés 1994; Biekart and Jelsma 1994) contributed to the impression, especially among foreign funding agencies and visiting intellectuals, that ASOCODE was an ascendant, potent and durable organization. Some ASOCODE leaders claimed that the English edition of the newsletter was for a Belizean audience, although others conceded that it was primarily for foreign consumption.

(Edelman 1998). Headquartered in a spacious house in an upper-middle-class neighbourhood of Managua, the Association had all of the typical trappings of a developing-country NGO: computers, photocopy machines, faxes, secretaries, maids, a driver, technicians who generated a never-ending stream of project proposals and 'strategic plans', and foreign 'cooperators', first from Denmark and later Canada.[11]

In the early to mid-1990s ASOCODE also initiated, encouraged or participated in the formation of several new networks: the Indigenous and Peasant Community Agroforestry Coordinating Group (Coordinadora Indígena y Campesina de Agroforestería Comunitaria, CICAFOC); the Central American Indigenous Council (Consejo Indígena Centroamericano, CICA); the Civil Initiative for Central American Integration (Iniciativa Civil para la Integración Centroamericana, ICIC); the Latin American Coordination of Rural Organizations (Coordinadora Latinoamericana de Organizaciones del Campo, CLOC); and the Vía Campesina, which eventually included representatives of farmers' organizations from some 60 countries (Desmarais 2007; Edelman 2003).[12] Following the massive devastation of Hurricane Mitch in October 1998, ASOCODE, together with CICAFOC and a regional network of small coffee producers, set up a new Central American Rural Coordinating Group (Coordinadora Centroamericana del Campo) to join a broader Central America Solidarity Coordinating Group (Coordinación Centroamérica Solidaria) at the Stockholm meeting of European cooperation organizations involved in the reconstruction effort. The proliferation of networks meant a sharp rise in the number of regional and extra-regional meetings, many of which were attended by the same individuals. Some of the more enterprising activists reportedly received full-time salaries simultaneously from more than one network.[13]

Success in fundraising and in the formation of new networks, together with an intense round of activities, could easily be mistaken for political impact and influence. These processes, however, also accentuated the top-down character of decision-making within ASOCODE, lessening its accountability to its national components and simultaneously creating 'new needs' that essentially responded to donor offers and priorities (Biekart 1999, 286–93). In 1994–97, according to an internal ASOCODE report, 'cooperation resources were so abundant that they exceeded the capacity of the headquarters' to administer them (ASOCODE 1999, 24). 'Overfunding' contributed to struggles for resources that exacerbated already existing factionalism and ultimately led funding agencies to cut off support.

---

[11] A 1999 internal retrospective evaluation indicated candidly that by 1994 the Association had institutionalized its 'function as a cooperation agency' (ASOCODE 1999).
[12] ASOCODE's member organizations also coordinated activities with the Campesino a Campesino Movement, a peasant-led extension programme that began in the early 1970s in Guatemala and subsequently spread to Mexico, the rest of Central America and Cuba. For a poignant insider history of Campesino a Campesino, see Holt-Giménez (2006).
[13] This 'double dipping' is rather like what Riles (2001, 47) reports among women's civil society networks in the Pacific. Indeed, her analysis of the weak points of such networks, discussed further below, has great resonance in the Central American context.

In 1997 and 1998, ASOCODE entered into a period that its own internal evaluation characterized as 'crisis and rupture' (ASOCODE 1999, 25). First the English and then the Spanish edition of the newsletter ceased publication. Organizational divisions involved more and more energy, and the Association diminished its lobbying at regional and international meetings. Much of the discord manifested itself in a factional split – present from the Association's beginnings – between the Panamanian and Salvadoran 'verticalists' (with orthodox Leninist proclivities), on the one hand, and the five other countries' representatives on the other. Additional controversies divided the coordinating council along different lines, for example, whether the Association should be a regional *campesino* lobby or attempt to resolve immediate on-the-ground problems in the member countries. The diversity of the constituent organizations and their social bases – agricultural workers, indigenous groups, independent peasants, cooperative members – once seen as a strength, became a further source of polarization as some countries' representatives (Nicaraguans and Costa Ricans, in particular) argued for a narrower orientation toward smallholding producers. Worsening disputes over resources between the headquarters and the different national coalitions also led some significant organizations to withdraw and others to be expelled from the national coalitions. Sometimes this occurred because national organizations saw the coalitions which represented them at the regional Central American level as too involved in ASOCODE to attend to pressing issues in their home countries.

In 1999, the Association abandoned all efforts at lobbying international, regional and national institutions. When donor organizations became aware of the turmoil and withdrew their support, the downward spiral accelerated. Some agencies indicated that henceforth they would reverse their previous practice of funding regional initiatives and channel support earmarked for ASOCODE to its constituent national organizations. CONAMPRO, the coalition that had represented Guatemala in ASOCODE, dissolved, decimated by the loss of cooperation funds and eclipsed by the rise of other more dynamic alliances of Guatemalan peasant organizations (in which it had briefly participated).[14] ASOCODE's lavish headquarters in Managua – an example of what one prescient critique termed network 'macrocephaly' (Morales and Cranshaw 1997, 55) – closed its doors. 'Since March 2000,' an internal report commented in April 2001, 'we have not had any financial support from any cooperation agency or organization . . . the different activities have been carried out with the support

---

[14] CONAMPRO (Coordinadora Nacional de Pequeños y Medianos Productores de Guatemala) was the last of ASOCODE's national 'mesas' or coalitions to form and, in contrast to the other countries' 'mesas', some of which had existed for many years, it actually took shape following the founding of ASOCODE itself. CONIC (Coordinadora Nacional Indígena y Campesina) and CNOC (Coordinadora Nacional de Organizaciones Campesinas), also both formed in the early 1990s, with CONIC initially forming part of CNOC. Both continue to carry out some joint activities, as they did earlier with CONAMPRO. CONIC tends to emphasize indigenous and human rights issues and CNOC agrarian reform and rural development. On the contemporary Guatemalan peasant movement, see Leanza (2004) and Santa Cruz (2006).

provided by their organizers. Operating expenses until December 2000 . . . were obtained through the sale of equipment from the headquarters' (ASOCODE 2001, 1).

In April 2001, 25 delegates (15 of them women) from five countries met in Tegucigalpa (representatives from Belize and Guatemala were invited and confirmed their participation, but never arrived). Their agenda was to consolidate what some described as a 'transition'. Instead of a costly headquarters and a regional coordinating committee, ASOCODE decided to divide into issue-specific working groups that would communicate virtually or meet physically on an ad hoc basis. COCOCH (Consejo Coordinador de Organizaciones Campesinas de Honduras), the Honduran member coalition, already host to the global Vía Campesina network, agreed to serve as ASOCODE's office as well and to begin the paperwork needed to establish its legal personality in Honduras. Despite the near demise of the Association, the transition commission's report pointed to a wide range of activities over the previous year: regional 'encounters' on agrarian problems and the landless in Honduras, El Salvador and Nicaragua, on rural women in Nicaragua, and on preparing proposals for the civil society Consultative Council of the Central American Integration System (SICA) in El Salvador, as well as participation in forums in Montreal, Madrid, Bangalore and Nairobi.

Nor did the Tegucigalpa 'transition' meeting neglect public relations. The ASOCODE brochure was updated, a press conference scheduled at the end of the event, and plans were made to announce the new organizational structure on the Association's web page. The Panamanians urged the rest of the delegates 'to reaffirm the presence of ASOCODE in Central America, taking advantage of all the documents on the letterhead of the different national 'mesas', adding the name of ASOCODE [to each, for] example APEMEP-ASOCODE, ADC-ASOCODE, COCOCH-ASOCODE' (ASOCODE 2001, 8).[15]

New Networks in a Reconfigured Space

In 2001 two new transnational civil society networks emerged in the Central American region, one with significant peasant participation and the other identified as solely a peasant effort. A decade earlier the most salient elite-led regional free-market project was the Central American Integration System (SICA) and peasant efforts to 'globalize from below' took place within the Central American region and in explicit opposition to the vision of the dominant groups. Now, however, anxieties about the SICA had shifted and political space was reconfigured as a result of Mexican President Vicente Fox's proposed Plan Puebla-Panamá (PPP), a gigantic new regional integration and infrastructure project, funded primarily by the Inter-American Development Bank, that sought to link southern

---

[15] These abbreviations refer to APEMEP (Asociación de Pequeños y Medianos Productores de Panamá), ADC (Alianza Democrática Campesina [El Salvador]), and COCOCH (Consejo Coordinador de Organizaciones Campesinas de Honduras).

Mexico and the Central American isthmus in a single free-trade, development and – perhaps contradictorily – environmental conservation zone.

The first new network, the Mesoamerican Initiative for Trade, Integration and Sustainable Development (Iniciativa CID), took a stance of critical, cautious engagement in the discussions about Plan Puebla-Panamá, the proposed Free Trade Treaty of the Americas, the US-Central American Free Trade Treaty and – most recently – the proposed free trade agreement between Central America and the European Union.[16] Supported in part by various branches of Oxfam, Catholic Relief Services and the US AFL-CIO (American Federation of Labor-Congress of Industrial Organizations), Iniciativa CID contended that the proposed accords might present opportunities and that it was important to lobby during the negotiations for measures to compensate peasants for low market prices caused in part by the subsidies given to US farmers (Iniciativa CID 2002). Initially, Iniciativa CID included some of the region's most important peasant organizations, such as the Guatemalan CNOC, the Honduran COCOCH, the Costa Rican UPANACIONAL (Unión Nacional de Pequeños y Medianos Productores Agropecuarios), and the Nicaraguan UNAG (Unión Nacional de Agricultores y Ganaderos), UNAPA (Unión Nacional de Productores Asociados) and ATC (Asociación de Trabajadores del Campo).[17] By 2006, however, these organizations had ceased or minimized their participation in the network, leaving a small group of research-oriented NGOs to develop critiques of the proposed free trade agreement with the European Union.

The second new network – the Mesoamerican Peasant Platform or Meeting (Plataforma Campesina Mesoamericana or Encuentro Campesino Mesoamericano, ECM) – arose to oppose the PPP, fuelled in part by Guatemala's CONIC and by the CLOC, the Latin America-wide network that had recently moved its headquarters to Guatemala. In 2002 the new inclusion of Mexico in a reconfigured political space led the president of an almost moribund ASOCODE to remark that the new ECM network 'was betting on Mesoamerica as a space of convergence' for opponents of Plan Puebla-Panamá (CCS-Chiapas 2002). The new group's Action Plan called for gaining it 'public recognition as Regional Coordinator' of the organized peasantry in Mexico and Central America, a status previously claimed, in the latter zone at least, by ASOCODE. At the network's third meeting in Tegucigalpa in 2003, it agreed to 'institutionalize' itself as the MOICAM (Movimiento Indígena y Campesino Mesoamericano). The

---

[16] The Free Trade Treaty of the Americas, heavily promoted by the US Administration of George W. Bush, foundered as a result of popular opposition and the misgivings of several South American governments, notably Brazil. The US-Central American Free Trade Agreement became the Dominican Republic-US-Central American Free Trade Agreement (DR-CAFTA).

[17] UNAG was a pro-Sandinista organization that included large- and medium-size producers, as well as some small-scale peasant farmers and cooperatives. It played a key role in founding ASOCODE and Vía Campesina, although it early on withdrew from the latter network (see Borras in this collection). UNAPA was made up primarily of former combatants from the National Resistance or contras. ATC is a pro-Sandinista organization which advocates for the interests of small producers and rural labourers and which now participates in several transnational networks.

Tegucigalpa meeting also proposed that MOICAM 'link itself to already existing networks, ASOCODE in Central America, CLOC at the continental level, and Vía Campesina at the world level' (ECM 2003).

In the case of ASOCODE, at least, 'already existing' was an overly optimistic description. Despite ASOCODE's continued Internet presence and its 'office' inside the COCOCH office in Honduras, its irrelevance and collapse were sufficiently well known by 2004 that the participating mesas decided to dissolve it and reorganize themselves in 'a new regional entity, Vía Campesina Centroamericana' (Desmarais 2007, 6).

The new MOICAM, despite its significant Mexican participation and generous funding from Oxfam Great Britain, was hardly immune to the types of problems that had made ASOCODE founder. The closing statement of the 2003 Tegucigalpa meeting lamented that

> the action plans agreed to in these meetings have been fulfilled only minimally and the Liaison Commission which we agreed to form in the [2002] Managua meeting has functioned very little. In other words, these meetings [acercamientos] between Mesoamerican peasants still have not turned into a true regional organization. (ECM 2003, np)

Three years later, Juan Tiney, general coordinator of the Guatemalan CONIC and a CLOC leader as well, candidly noted that the MOICAM was still focused primarily on periodic regional meetings of activists and had achieved little else:

> We have to improve our communication. The MOICAM cannot continue to be only a space and time for meeting each other, analyzing the situation with which we are already familiar, and committing ourselves to a series of activities which later we don't follow up. Because of all this we must dynamize the coordinating commission in order to dynamize our MOICAM. The indigenous peoples, while they are participating, need to concretize their own platform, which might permit them to have a more active role in the MOICAM. We must commit ourselves to reactivating the participation of organizations from Belize and Panama and continue strengthening the integration of the organizations from the other countries. (Tiney 2006, np)

Armando Bartra, one of Mexico's most prominent long-time 'campesinista' scholar-activists, has been a tireless promoter of the MOICAM (Bartra 2005, n.d.).[18] He highlights, for example, the march on 20 November 2006 of some 'ten thousand' largely indigenous protestors 'convoked by the MOICAM' to coincide with the conclusion of a MOICAM meeting in San Cristóbal de las

---

[18] Much of this promotional work appears on the website of the Observatorio Social del Agro Mesoamericano, sponsored by the Instituto de Estudios para el Desarrollo Rural 'Maya' A.C (IEDRM), which Bartra directs. He is also a professor at the Universidad Autónoma Metropolitana-Xochimilco in Mexico City. On the campesinista-descampesinista debate and Bartra's contribution to it, see Edelman (1999, 203–7) and Kay (2001, 377–86).

Casas, Chiapas (Bartra 2006). Nevertheless, an electronic bulletin by a correspondent from the Centro de Estudios para el Cambio en el Campo Mexicano (CECCAM), an organization generally sympathetic to peasant interests, put the number of protestors at 4,000 and pointed out that their main concerns were not so much those raised by the MOICAM but rather solidarity with the protest movement in Oaxaca and the possibility of viewing on a giant screen the mock inauguration in Mexico City of presidential candidate Andrés Manuel López Obrador, who was widely believed to have been denied electoral victory as a result of fraud (Enríquez 2006).

A DIFFICULT CONTEXT

The chronic difficulties that have affected transnational peasant movements in Central America may be explained in part by internal organizational weaknesses, about which more will be said in the conclusion of this chapter. It is also the case that the peasant organizations' external environment has become much less favourable in many respects. Political opportunities have diminished, as has the importance of agriculture in the region's economies; conditions for smallholders, particularly basic grains and coffee producers, have deteriorated; and massive migration from the countryside has transformed rural households, cultural expectations and political aspirations in ways that do not bode well for the peasant organizations' work.

The Declining Importance of Agriculture

In all the Central American countries the importance of the agricultural sector in the economy has diminished in recent years and nowhere does it now represent even one-quarter of GDP (see Table 1). The proportion of the population

Table 1. Central America: agriculture as a percentage of GDP, 1995–2005 (in constant dollars, year 2000)

	1995	*2000*	*2005*
Costa Rica	9.3	8.6	7.7
El Salvador	10.8	9.8	9.4
Guatemala	24.1	22.8	22.8
Honduras	15.2	14.0	13.6
Nicaragua	18.4	18.5	17.9

Source: CEPAL (2006, 87, 94). 'Agriculture' includes hunting, forestry and fishing. Because shrimp farming and export-oriented fishing have expanded greatly during this period and are included in the agricultural sector data, the declines in crop agriculture are almost certainly more significant than indicated by the figures in this table.

Table 2. Central America: agricultural employment as a percentage of total employment, 1990–2005

	1990	1995	2000	2005
Costa Rica	25.4	21.0	16.9	15.0
El Salvador	–	25.6	20.7	18.4
Guatemala	48.0	37.6	36.5	36.2
Honduras	42.0	38.2	34.0	36.3
Nicaragua	30.6	33.7	32.4	29.0

Source: CEPAL (2006, 42).

Table 3. Central America: agricultural exports as a percentage of total exports, 1995–2005

	1995	2000	2005
Costa Rica	43.6	23.3	21.8
El Salvador	27.6	12.0	6.2
Guatemala	42.3	37.6	28.0
Honduras	60.0	37.5	29.5
Nicaragua	48.6	44.1	23.1

Source: CEPAL (2006, 162, 166, 171, 174, 218). 'Agricultural exports' include those from hunting, forestry and fishing.

working in agriculture is now a minority, even in the more agrarian societies of Guatemala and Honduras (see Table 2).[19] Despite the huge push since the mid-1980s to stimulate non-traditional agricultural exports (Barham et al. 1992), the proportion of export earnings derived from agriculture has dropped very rapidly and dramatically in all countries of the region (see Table 3).[20]

In the early 1990s, Salvador Arias and Eduardo Stein, both key figures in the emergence of transnational peasant organizing in the region, maintained in an important book, *Democracia sin pobreza*, that agroindustry could be the engine of growth in Central America and that new forms of ownership could democratize

---

[19] As Segovia (2004, 19) points out, the discrepancy between the proportion of GDP from agriculture, which is now quite low (Table 1) and the somewhat higher proportion of the population employed in agriculture (Table 2), very likely reflects both the increasing involvement of rural labour in non-agricultural and non-farm work and the differential labour productivities of agricultural and non-agricultural activities.

[20] 'Traditional agricultural exports' usually refers to coffee, bananas, sugar and beef. 'Non-traditional agricultural exports' include ornamental plants, winter vegetables, fresh fruits and tubers for the US Hispanic market. Exports of cultivated shrimp and fresh fish have also grown rapidly, especially in Honduras and Nicaragua, and are included in CEPAL's data on 'agricultural exports'.

the benefits that this might generate.[21] Arias, whose views are echoed by other contributors to the volume, argued that

> one of the few remaining options for positive prospects for the regional economies is to develop the industrialization of agriculture and a regional agro-food system, temporarily protected, as part of a broader Food Security policy . . . Domestic demand must become an element that dynamizes supply alongside demand from the international economy (Arias Peñate 1992, 64).

In addition to calling for protection for infant industries, Arias advocated the 'construction of productive linkages' between the region's heterogeneous agriculture and other sectors, which would in turn 'generate new comparative advantages' (Arias Peñate 1992, 80).

Today, it is more difficult to make this argument. Instead of an agriculture-based development model with high levels of protection, the region's traditional agro-export economies have weakened or even collapsed and governments have opted since the late 1980s for what some scholars term the 'New Economic Model' (or 'Export-Led Development'), based on opening to the global economy and, with the exception of Costa Rica, a greatly reduced state role in the economy and social welfare. New activities – particularly textile and electronics assembly factories, non-traditional agricultural exports, financial and other services, tourism and remittances from abroad – have become the main sources of foreign exchange, key targets for regional and extra-regional capital, and important sources of employment and macroeconomic stability (Díaz Porras and Pelupessy 2004, 25; Segovia 2004, 7–8; Robinson 2003, 156–213). The huge out-migration from both rural and urban areas has also generated new and often lucrative employment opportunities for 'coyotes' (smugglers of migrants), couriers who carry funds and goods between Central America and the United States, and providers of related services. The specific features of this neoliberal model and the mix of activities of course vary from country to country, with El Salvador and Costa Rica furthest along the road to having primarily non-agricultural economies. The implementation of the Dominican Republic-United States-Central America Free Trade Agreement (DR-CAFTA), which all governments in the region have now endorsed, is likely to accelerate and deepen the application of the new economic model.

---

[21] Arias and Stein, both economists, were affiliated with the European Community-funded and Panama City-based Comité de Apoyo al Desarrollo Económico y Social de Centroamérica. CADESCA sponsored a food security training programme in the late 1980s and early 1990s, taught largely by Arias, which became one of the first venues in which Central American peasant activists from different countries met and exchanged experiences (Edelman 1998, 57–61). Arias went on to become a member of the Political Commission of the Frente Farabundo Martí para la Liberación Nacional and since 2003 has been an elected FMLN deputy in El Salvador's Legislative Assembly. Stein headed the Organization of American States electoral observer team that in 2000 contributed to the breakdown of the authoritarian regime of Peruvian President Alberto Fujimori. He has been Foreign Minister (1996–2000) and Vice President (2004–present) of Guatemala.

Basic Grains and Coffee, Erstwhile Pillars of the Smallholding Sector

The two activities historically of greatest importance to Central America's small farmers – basic grains and coffee – are those that have been hardest hit over the past two decades. In the former case, this represents a cruel irony, since increased demand for maize for ethanol production in the United States produced an unprecedented price spike in 2006–2007. This price rise, however, occurred at a time when, measured since the early 1990s, the smallholding maize sector had virtually disappeared in Costa Rica, stagnated in El Salvador, Honduras and Guatemala, and expanded very significantly only in Nicaragua, most likely in the latter case as a result of a return to subsistence production by the most impoverished sectors of the population.[22] While the impact of the post-2006 maize price increase on Central American peasants is still undetermined, it is probable that the smallholding sector is poorly positioned (in terms of its access to working capital, land, transportation, machinery services and storage facilities) to take advantage of the boom. Moreover, in all the Central American countries, population growth has far outpaced the production of basic foods and the modest expansion of maize harvests and area.[23] Rising maize prices have become a major threat to poor households, including many peasants who are net consumers of corn or who rely significantly on purchased food. Policymakers and market analysts are also concerned that smallholders may sell their maize reserves to take advantage of high prices and then not be able later to purchase seed or maize for consumption (USAID and MFEWS 2007).

In all of the Central American countries, basic grains producers have been the group most impacted by structural adjustment reforms and regional market integration and liberalization in the 1980s and 1990s. For approximately two decades, until the price spike of 2006–2007, most kinds of basic grains production have been unprofitable. Even small farmers who cultivated maize for household consumption were likely to think twice about continuing to do so if they purchased any manufactured inputs, hired labour or machinery services, rented land, or considered the opportunity cost of their labour. Producing for the household and selling a small surplus on the market ceased to be a sensible strategy, except perhaps for those who minimize their interactions with the market economy. Despite Central American peasants' millennial attachment to maize culture, for many it was a better option to simply buy the maize they needed. While rising maize prices in the United States after 2006 have already reduced exports to Central America, two decades of dumping and liberalized intra-regional trade in

---

[22] This supposition is supported by data that show major declines since the 1980s in Nicaragua in overall fertilizer use, tractors per hectare of arable land and percentage of cropland irrigated (Acevedo Vogl 2003, 54–5). While low-input maize cultivation may sometimes be an adequate survival strategy for some of the rural poor, it is unlikely to give rise to a sustained process of accumulation that might assure them improved levels of well-being.

[23] This is based on data on per capita food production, maize production and area harvested from the FAOSTAT (2008) on-line data base. See also Proyecto Estado de la Región-PNUD (2003, 138), which reports that the area in basic grains in Central America and Panama rose 4 per cent between 1978 and 2001, while the rural population grew 55 per cent in the same period.

basic grains had a devastating effect on smallholder livelihoods. When DR-CAFTA is fully implemented, few smallholders are likely to benefit and many face probable ruin from increased competition, as occurred in Mexico with NAFTA (Acevedo Vogl 2003; Carazo Vargas 2004; Galián 2004; Moreno 2003).

The precarious position of smallholding basic grains producers has been exacerbated over the past decade by other elements of Central America's new economic model and, more broadly, by neoliberal globalization. These include the privatization of state development banks, extension agencies and commodities boards; the de-funding of public-sector agronomic research, particularly on small-scale, hillside and sustainable agriculture; the ever-growing concentration and vertical integration among the giant corporations that supply inputs and control the lion's share of agricultural trade; the reduction or elimination of tariff protections; and the disappearance of subsidies (for credit, crop insurance, support prices, inputs and technical assistance) once provided by now debilitated public-sector institutions. Small producers are ever more at the mercy of local loan sharks, intermediaries and increasingly capricious and distant invisible forces. The evisceration and reversal of the agrarian reforms of the 1960s, 1970s and 1980s, which elites abhor as obstacles to the free play of market forces, have contributed to reducing the peasant land base and shattering hope. The costs of these processes have been substantial in terms of social decomposition in the countryside, migration to other rural areas, urban slums and abroad, and social pathologies of all kinds (see below).

Until a rebound that began in 2006, coffee prices were at historic lows since the virtual collapse in 1989 of the International Coffee Agreement, which assigned quotas to exporting countries.[24] This nearly two-decade long depression in the coffee sector was exacerbated by the rise of new producing countries, notably Vietnam and Indonesia, and the advent of new roasting technologies that made it possible to incorporate lower quality, robusta beans into premium blends, something which disadvantaged Central American producers of high quality arabicas. Central American coffee producers in some areas simply abandoned farms that were no longer profitable to harvest. Others attempted, with modest success, to cater to niche markets for organic and fair trade coffee. Slightly more than half of Costa Rica's coffee exports, for example, now go to 'high quality' markets that pay premium prices (Barquero 2007). As with the recent rise in maize prices, however, it is too early to tell whether improved perspectives in the coffee market will redound to the advantage of Central American smallholders. It takes a longer time to respond to positive price signals in the coffee market than is the case with annual crops and reversing nearly two decades of institutional abandonment of the sector (and of their crops by former coffee producers) is likely to be challenging at best. Moreover, coffee has ceased to be a major political priority for Central American governments. While coffee yields and area have increased in the past 15 years, the contribution of coffee to

---

[24] The price jump between January 2005 and December 2007 was 80.8 per cent for robusta varieties, but only 19.8 per cent for the 'other mild arabicas' produced in Central America (ICO 2008).

Table 4. Central America: coffee exports as a percentage of total foreign exchange earnings, 1980–2003

	1980–4	*1990–4*	*2000–3*
Costa Rica	2.2	8.1	2.6
El Salvador	43.1	11.2	2.7
Guatemala	25.4	13.2	6.4
Honduras	19.2	12.1	6.3
Nicaragua	24.5	12.9	8.0

Source: CEPAL data cited in Segovia (2004, 14).

export earnings has dropped rapidly in all the Central American countries and especially in Costa Rica and El Salvador, formerly the region's pre-eminent coffee producers (see Table 4).[25]

Beyond the Campesinista-Descampesinista Debate

In Central America, as elsewhere, scholars and 'modernizing' policymakers have predicted the partial or complete disappearance of the peasantry at several key transitional moments: first, with the shift to mechanized agriculture; then, with the advent of green revolution technological packages; later, with imposition of structural adjustment programmes, economic liberalization and the reversal of state-led agrarian reforms (often termed 'counter-reform'); and, most recently, with the signing of free trade agreements and the advent of market-led agrarian reform. In most cases, the debate over the fate of the peasantry has centred on the interlinked problems of the economic viability of peasant farms and the social reproduction of peasant households. These are not inappropriate foci of discussion, as the analyses of coffee and basic grains production above or of migration and pluriactivity below suggest. Nonetheless, from the standpoint of understanding the possibilities of peasant movements, and especially transnational ones, several additional, largely political considerations need to be taken into account.

In Central America, the momentous upheavals of the past two to three decades have drastically altered the balance of class forces and contributed to a recomposition of elite and non-elite forces in every country. The elites, despite their historical ties to and, in some cases, continuing involvement with agro-export sectors, have diversified. In part, this resulted from the insecurity of many rural zones for investors during the civil wars in Guatemala, El Salvador and Nicaragua;

---

[25] Table 4, based on CEPAL data that Segovia (2004) aggregated in multi-year periods, is suggestive of the broad trend. 'Total foreign exchange earnings' includes remittances and exports, so the proportion provided by coffee, as indicated in the table, is smaller than would be the case if the data were coffee exports as a percentage of total exports alone. CEPAL data which use the latter formula suggest slightly higher percentages (CEPAL 2006).

the application of radical agrarian reforms in the 1980s in the latter two cases also led many large landowners to withdraw from the countryside, if not from agriculture. The growing flows of remittances (see below) have become a major factor in the emergence of a powerful banking elite that charges migrants and their relatives for funds transfers and engages in regional and extra-regional portfolio investments and acquisitions and expansion of financial institutions.[26]

The intensifying export orientation of Central American agriculture in the 1980s and after has brought a proliferation of medium-size, highly capitalized enterprises. The most powerful elite sectors in several Central American countries are now fundamentally industrial and financial in orientation. Their interest in the new kinds of medium-size agro-export firms is not so much as direct investors, but rather as providers of inputs, marketers of outputs or financial backers. At the risk of generalizing perhaps too much, it is evident that throughout the region the shift away from a model of accumulation centred around traditional agro-exports (coffee, bananas, sugar, beef) to one based on services and on more diverse sources of foreign exchange, including non-traditional agro-exports, has changed the balance of power within the agrarian elites and also weakened them vis-à-vis non-agrarian elites (Robinson 2003; Segovia 2004). The possibilities for cross-class sectoral alliances between, for example, large and small basic grains producers have diminished as well, since all but the best-endowed large grain producers have moved into more lucrative activities and the small grain producers have been decimated by the two-decades-long drop in prices and by public-sector retrenchment. The recomposition of the elites has thus intensified the political marginalization and social exclusion that affect the poorer sectors of the peasantry.[27]

As the importance of agriculture in the region's economies declines, the economic and political centre of gravity has moved away from the rural sector, making it more difficult to wage struggles around either the terms of trade between agriculture and industry or around the class-based demands of the smallholding and landless peasantry for genuine agrarian reform and improved production conditions.[28] As rural areas throughout the region lose population,

---

[26] This tendency is epitomized by the Salvadoran-owned Banco Cuscatlán, which has operations in El Salvador, Guatemala, Honduras, Costa Rica, Panama and the United States, and which has acquired banks in the United Kingdom and elsewhere. In 2007, Banco Cuscatlán was acquired by the New York-based Citi group.

[27] The term 'social exclusion' became widely used in the mid-1990s, when European Union and International Labour Organization analysts began to employ it to refer to what they saw as new kinds of social and economic problems resulting from globalization (Gacitúa and Davis 2000, 13). Popular organizations adopted the language, notably in the formation of a transnational network based in Brazil called the 'Cry of the Excluded' (Grito dos Excluídos/Grito de los Excluidos), which overlaps with and has close links to the CLOC. More recently, Bebbington (2007) and others have argued that a more analytically useful concept is 'adverse incorporation', since it suggests that the poorest of the poor are frequently thoroughly integrated, albeit in highly disadvantageous ways, into the dominant economic and social system.

[28] Advocacy of renewed state-led agrarian reform, a key plank in the platforms of Vía Campesina and its member organizations, is further complicated in a context in which large numbers of beneficiaries of the reforms of the 1960s and after have given up and abandoned their lands, often selling them to large agribusinesses (Ruben and Fúnez 1993; Ruben and Masset 2003).

Table 5. Central America: percentage of urban population in total population, 1995–2010

	1995	2000	2005	2010
Costa Rica	54.2	58.7	62.6	66.0
El Salvador	52.5	55.2	57.8	60.3
Guatemala	36.5	43.0	50.0	57.2
Honduras	42.9	45.3	47.9	50.6
Nicaragua	54.0	55.4	56.9	58.2

Source: CEPAL (2006, 33). The definition of 'urban' corresponds to that used in each country. Data for 2010 are projections.

they lose not only political representation in formal institutions, but also political clout in interest-group politics and lobbying and in struggles that arise outside of those institutions. Urbanization also tends to reduce the concern of urban citizens and consumers with addressing rural problems. With the exception of those small producers who have achieved a successful integration into the new economic model – winter vegetable growers in Guatemala and Costa Rica, for example – it has simply become exceedingly difficult for peasant voices to make themselves heard. And in the absence of 'voice', rapidly growing numbers are opting for 'exit'.

URBANIZATION, MIGRATION AND PLURIACTIVITY

In Central America, as elsewhere, non-agricultural employment and entrepreneurial activity almost everywhere have a growing role in the rural poor's survival. This is reflected in indicators that chart the changing composition of the economically active population in rural areas, as well as at the level of household strategies.[29] This diversification of individual and family strategies has contributed to the expansion of a semi- and unskilled rural workforce that often maintains some land base and/or links to urban areas and that depends fundamentally on a complicated mix of wage labour, self-employment and involvement in 'micro-enterprises', small-scale agriculture, artisanal or industrial production, and other activities. In some major urban areas, large numbers of city residents travel, often periodically or seasonally, to nearby rural zones to work in agriculture and agroindustry.

In all the Central American countries the proportion of the population living in urban areas continues to grow. Even in Guatemala and Honduras, the most rural countries overall, half of the population is now or will soon be living in cities (see Table 5). This demographic and geographic shift has profound implications not just for the sectoral composition of the Central American economies (the expansion of the service sector, of urban 'informality', and so on), but for the reconfiguration of labour markets and of cultural and political imaginaries.

[29] In all the Central American countries more than 80 per cent of economically active rural women in 1998 were involved in non-agricultural activities; the proportion for men ranged from 21.5 per cent in Honduras to 57.3 per cent in Costa Rica (Proyecto Estado de la Región-PNUD 2003, 135).

In recent years, urban and rural livelihoods and culture have converged in so many ways that it is possible to speak of the 'urbanization' of the rural labour force and the 'ruralization' of the urban one, as well as of the creation of a single labour market (Viales Hurtado 2002, 161–2). An urban imaginary and urban consumption expectations now have a rapidly growing influence in rural zones. Even in remote rural areas, improved access to schooling, growing reliance on off-farm employment and declining average fertility reduce the numbers of available family labourers and intensify pressures on those who are still farming and/or on those working elsewhere to provide an ever higher level of consumption for the entire household. This may strain extended family ties and limit possibilities of participating in diverse kinds of collective endeavours.

The convergence of urban and rural culture and expectations has not signified a narrowing of differences in living standards, however. The rural–urban gap – in income, in consumption, in life chances – remains pronounced in most of the region. Poverty rates are still higher in rural areas, although in some countries (El Salvador, Guatemala, Nicaragua) the rural–urban poverty gap has narrowed slightly in recent years. In this sense the situation in Central America differs dramatically from the 'new rurality' that some European scholars celebrate. If rural areas once provided labour, food and raw materials for the rise of industry, such scholars suggest, under the 'new rurality' the countryside experiences growth in the service and manufacturing sectors, which no longer require the same degree of spatial concentration as they did in a Fordist economy and which can serve in combination or in lieu of agriculture as pillars of farm families' domestic economy (Marsden 1995).[30] Except for the parts of rural Central America that have experienced major growth in tourism (primarily in Costa Rica), the limited applicability of this paradigm should be obvious. The subordination of agriculture to industry remains important outside of Europe and particularly in Central America. The combination of converging expectations and diverging life chances has potentially explosive consequences. The outcomes, though, can be quite variable, ranging from involvement in crime ('maras' or gangs, drug mafias, and the 'ruralization' of urban-type crime), to collective action for improved conditions, to individualism and political demobilization.

Social pathologies previously little known in the countryside, such as gang violence and drug trafficking and consumption, are increasingly present in rural areas (Andrade-Eekhoff and Silva Ávalos 2004, 72–4). The reasons are complex and interrelated, ranging from the United States' deportation of California gang members to their countries of origin to the continuing importance of remote Central American rural airstrips and agro-export packing facilities in the drug cartels' shipping routes, the high unemployment among demobilized combatants from the region's civil wars, the easy availability of weapons in the post-conflict period, the lack of adequate programmes of social rehabilitation, and ongoing

---

[30] For more critical views, see Rubio (2002) and Kay (2005, 11–15).

Table 6. Central America: remittances from abroad, 1995–2005 (millions of current US dollars)

	1995	2000	2005
Costa Rica	133.9	93.4	270.4
El Salvador	1,388.6	1,797.1	2,864.6
Guatemala	491.2	868.2	3,557.5
Honduras	242.5	746.9	2,002.4
Nicaragua	138.0	406.2	750.4

Source: CEPAL (2006, 140). The data refer to net current transfers from the rest of the world.

processes of social exclusion.[31] Most notable, for the purposes of this analysis, is that in the late 1980s when social scientists began to research Central American gangs or '*maras*', their participants were overwhelmingly 'young people born and raised in the city' (Levenson 1989, 7). Within a decade, however, *maras* had spread to small towns and villages (especially in Guatemala, El Salvador and Honduras), had expanded into the business of human trafficking (north into Mexico and the United States), and had developed such a formidable transnational dimension that US military and foreign policy officials began to view them as dangerous 'non-state actors' operating in 'failed states' and as major threats to national security (Arana 2005; Manwaring 2005).

Rural households increasingly depend not just on off-farm employment, but also on extra-regional employment and remittances, a pattern that has long been pronounced in zones such as eastern El Salvador, but which is now characteristic also of many other areas – including urban ones – in every country in the region apart from Costa Rica.[32] The latter has become a major receiving country for Nicaraguan migrants, who now constitute approximately 10 per cent of the population (Sandoval García 2002, 264–8). Remittances have not only grown extremely rapidly, but they have also become increasingly essential to social reproduction and to the macroeconomic stability of the Central American economies. In El Salvador, Guatemala, Honduras and Nicaragua, funds sent from abroad are now equivalent to well over half of all export earnings (and in Guatemala close to three-quarters) (see Table 6 and Table 7).

Migration abroad keeps unemployment low, serves as a safety valve for reducing political pressure, and is, in effect, a kind of redistributive measure in a context in which state-led redistributional programmes have weakened or

---

[31] The use of agro-export packing plants to facilitate drug smuggling is one of the ironical unintended consequences of the push to stimulate non-traditional agricultural exports. Such methods have become so common that US Drug Enforcement Administration officials reportedly term them 'the broccoli routes' (Fischer and Benson 2006, 27).
[32] In the eastern Salvadoran department of La Unión, an estimated 41 per cent of households receive remittances from abroad (Hecht and Saatchi 2007, 667).

Table 7. Central America: remittances from abroad, 1995–2005 (as a percentage of total export earnings)

	1995	2000	2005
Costa Rica	3.0	1.2	2.8
El Salvador	67.5	49.9	62.6
Guatemala	17.4	22.3	71.4
Honduras	14.0	30.0	58.5
Nicaragua	22.7	43.2	54.6

Source: CEPAL (2006, 135, 140). Net current transfers from rest of world as percentage of exports of goods and services.

disappeared. Migration has also contributed to the ageing and feminization of the peasant population, which tends to undermine the smallholder economy, particularly in zones of unmechanized hillside agriculture where hard manual labour is required (Morales and Castro 2002, 117). Large remittance flows have also given rise to a leisure effect, discouraging labour and investment and encouraging consumerism (Segovia 2004). This, in turn, has brought a configuration of new, transnational economic spaces in high-remittance zones, border areas where the lowest-wage jobs in agriculture and other sectors are now held mainly by non-citizens, such as the growing number of Nicaraguan and Honduran migrants in eastern El Salvador or of Nicaraguans in Costa Rica (Morales and Castro 2002, 12). Perhaps most importantly, for the purposes of this chapter, migration frequently undermines the capacity for political action. Referring to the western Salvadoran municipality of San Antonio Pajonal, for example, Katharine Andrade and Claudia Silva ask, 'What does it mean in terms of effective citizenship that 80 per cent of the registered voters may live in Los Angeles [California]?' (Andrade-Eekhoff and Silva Ávalos 2004, 58). San Antonio Pajonal may be an extreme case of out-migration, but the weakening of civic participation and political commitment as a result of transnational migratory flows (and of *aspirations* to migrate) is nonetheless significant in other communities throughout the region.

CONCLUDING QUESTIONS

Central America, with a long tradition of cross-border solidarity and flows of people, was in the 1990s one of the first and most dynamic centres of transnational peasant organizing. Yet scarcely more than a decade after the emergence of the first regional peasant networks, the prospects for consolidating strong transnational peasant organizations are far from promising.[33] Understanding this changed

---

[33] These difficulties are not always acknowledged by peasant movement activists in their internal strategy discussions or by the intellectuals who accompany them (but see Gaete et al. 2005); indeed, as indicated above at various points, there is frequently a problematic denial or euphemization regarding internal organizational or external obstacles.

scenario requires attention to the interrelated problems of the organizations' internal politics and styles of work and to their external environment.

Increasingly over the course of the 1990s, efforts to theorize 'global' or 'transnational' civil society employed the notion of 'network' as an analytical category, a metaphor for a social condition and a description of emerging organizational and institutional forms, communications technologies and knowledge practices. Typically 'network' analyses generated great enthusiasm among proponents of 'global civil society', even as they produced trepidation among counter-insurgency strategists.[34] As Jonathan Fox (2000, 9–12, 45) has pointed out, however, there is often a conceptual slippage between both scholars' and activists' use of 'network', 'coalition' and 'movement'. While it is clear that 'transnational civil society exchanges *can* produce networks, which *can* produce coalitions, which *can* produce movements' (Fox 2000, 10, italics in original), it is also apparent from the Central American cases discussed above that this does not occur automatically and indeed it may not occur at all. As in the broader collective action literature, where the study of unsuccessful social movements is distinctly under-theorized (Edelman 1999, 2001), only a few lone voices – notably Annelise Riles in her ethnographic tour-de-force on women's organizations in the Pacific – question the prevailing view of the 'network' as a durable organizational morphology and call attention to the way 'networks' frequently point to their own networking activities as measures of success. Riles goes so far as to suggest that the 'network' may be 'a form that supersedes analysis and reality' and that its '"failure" is endemic, indeed ... [an] effect of the Network form' (2001, 174, 6).[35] While perhaps not applicable to all networks in all times and places (as Riles seems to claim), this latter perspective deserves consideration when analyzing the rise and decline of ASOCODE in Central America in the 1990s or the difficulties faced by its successors, such as the MOICAM.

If peasant activists and the scholars who accompany them have been loath to acknowledge the slippage between 'networks', 'coalitions' and 'movements', discussing the overlap between NGOs and popular organizations has been even more taboo. In the competition for European funding that occurred in Central America in the 1990s, claiming to be a 'popular organization' that 'represented' a historically marginalized or excluded sector of the population came to be of the utmost importance. Given the historical antagonism in the region between 'popular organizations' and NGOs, it is striking how much the two forms converged, at least in the case of ASOCODE. Analyzing Latin American feminist movements, Sonia Alvarez (1998) calls this a process of 'NGO-ization' of popular organizations. ASOCODE, headed by some of the most belligerent anti-NGO

---

[34] The two poles of this discussion are epitomized by Riechmann and Fernández Buey's *Redes que dan libertad* (1994), on the one hand, and Arquilla and Ronfeldt's *Networks and Netwars* (2001), on the other.

[35] Comaroff and Comaroff (1999, 33) also call attention to how 'Euro-modernist forms' of civil society 'may be emptied of substance ... turned into a hollow fetish ... [or] a dangerous burlesque'. At the same time, they acknowledge, as Riles rarely does, that 'civil society' nonetheless serves as a vessel for utopian visions and for opening up democratic spaces.

activists in the region, resembled nothing so much as a large NGO. In addition to its top-heavy, expensive infrastructure, headquarters and organizational form, its leaders and technical staff became disturbingly fluent in the banal and repetitive 'NGO speak' ('sustainability', 'transparency', 'participation', 'accountability', etc.) devastatingly lampooned in Argentine human rights activist Gino Lofredo's (1991) parody, 'Get rich in the 1990s. You still don't have an NGO?'[36]

The demise of ASOCODE did generate lessons that some of its associated networks and coalitions learned. Groups such as Vía Campesina and CLOC, for example, have largely eschewed the heavy reliance on cooperation funding, the maintenance of expensive headquarters, and leadership styles characterized by personalistic ties and a lack of generational succession, all of which contributed to undermining ASOCODE (and some of the region's major national organizations, as well). At the same time, having learned such lessons is not, in and of itself, sufficient to build a dynamic movement, as the example of the MOICAM surely suggests. The conception of this and other organizations as 'networks' and 'spaces' defines limits to their political efficacy, especially at the transnational level.

Much of the political potency of transnational civil society derives from what Keck and Sikkink (1998, 12–13) famously termed 'the boomerang pattern' and others have variously called 'venue shifting' (van Rooy 2004, 20) or 'leap-frogging' (O'Brien et al. 2000, 61). Essentially, movements that are unable to attain their objectives in domestic politics seek out international allies in order to pressure governments to conform to international norms. The empirical referents for Keck and Sikkink's model were human rights and environmental campaigns yet, as Paul Nelson (2004) has pointed out, other issue areas, notably trade and financial policies, have been less vulnerable to 'boomerang' strategies. This is because trade and financial policies are more central to powerful G8 governments than, say, World Bank project lending, human rights or debt relief for poor, mostly small countries. International financial and trade organizations, such as the World Bank, International Monetary Fund and World Trade Organization, are less vulnerable to civil society pressure and persuasion than is usually the case with national governments. In part this is because NGOs that attained considerable legitimacy as critics of development and cooperation policies have yet to achieve similar levels of credibility as critics of macroeconomic and financial policies and institutional reforms. Finally, civil society responses to the latter sort of issues frequently call for strengthening – rather than 'boomeranging' – national governments in areas as diverse as labour and environmental regulations, trade protection and intellectual property.

Certainly state-led agrarian reform, better production conditions for small agricultural producers, food sovereignty and other peasant organization demands are, like trade and financial issues, also less amenable to 'boomerang' international pressures on national governments. The efforts of Vía Campesina and associated organizations, such as FIAN, to employ a rights discourse could begin to give

---

[36] For a less jocular approach to the same problem, see Stirrat (1996).

the 'boomerang' greater impact (this is clearly their intention) (Vía Campesina 2002, 2005; FIAN 2006). But the expansion of the domain of 'rights' to include economic and social rights faces daunting obstacles in an age when the hegemonic conceptions are limited to narrow notions of individual expression and the 'rights' of economic actors in the market.

This difficulty in expanding norms about rights is part of a broader problem of making peasant voices heard in societies where thousands are abandoning the countryside every day and where powerful elites and policymakers no longer view agriculture as the motor force of economic development. The changes described above have shifted the balance of power in Central American politics away from rural zones and toward urban ones, away from broad-based *campesino* movements and toward a variety of sectoral and product-specific groups with varied demands, and away from the traditional agro-export elites towards entrepreneurial groups involved in non-traditional export activities and financial and other services. The massive migration to the United States (or, in the case of Nicaragua, to Costa Rica) has reduced political pressure, fortified the financial-sector elites and demobilized what was earlier a highly politicized rural population through the leisure effect of family remittances and the generalized aspiration to 'exit' rather than to stay in the countryside and struggle.

Like the migration to which it is related, the growing 'pluriactivity' of rural households and the increasing inter-penetration of city and countryside complicate the question of *campesino* identity in ways that have ramifications for how people view their struggles and their participation in collective efforts for change. The first thing to acknowledge is that the *campesino* of today is usually not the *campesino* of even 15 years ago. The individual who farms during the rainy season, works on urban construction sites during the dry season, rents out his pickup and provides mechanic services to neighbours, sells imported Chinese toys and pirated CDs in town on weekends, and receives a monthly wire of 50 dollars from a son living in New York is less likely to fully identify with the agenda of the historic *campesino* organizations than was his more fully 'peasantized' parent or grandparent. Can or should a *campesino* organization attempt to convince this individual that obtaining more or better land or better production conditions might resolve his economic difficulties? Obviously there are profound ethical, cultural and other reasons for trying to do so. But can this argument be made persuasive under current conditions? Is the argument likely to be more persuasive when the organization making it is sectoral (broccoli or ornamental plant producers, for example) or when it is national or even transnational and all-inclusive?

The near collapse of the two main pillars of peasant agriculture – basic grains and coffee – suggests that the *campesino* organizations have to propose immediate alternatives even as they also struggle for longer range goals such as the food sovereignty that might guarantee the survival of small maize producers. Is it possible that part of the rise of sectoral organizations in the countryside and the decline of the all-inclusive, broad-based peasant coalitions has to do precisely with this dynamic? Honduran maize producers will probably never sell maize in Iowa, but they might, in a different macroeconomic environment (with a guarantee of food sovereignty), at least be able to sell it in Tegucigalpa or San Pedro Sula. In the short term, however, if they are to survive, they need other options. The niche markets for coffee, while relatively attractive, are unlikely to resolve the problems of all coffee producers. Do the *campesino* organizations have something practical to offer the small-scale agriculturalist that can no longer produce maize or coffee?

Elsewhere I have argued that one of the most notable successes of transnational peasant and small farmer organizing has been to put the issue of agrarian reform back on the international development agenda in the mid to late 1990s (Edelman 2003, 206–7). FIAN and Vía Campesina played critical roles in this. It is not easy, however, to make convincing arguments for new state-led agrarian reforms in contexts, such as Central America, where so many beneficiaries of previous waves of reform have abandoned the land and when so many reform cooperatives have gone belly up. The World Bank, of course, has vast resources for promoting its 'market-assisted' or 'community-managed' version of land reform. The credibility of the Bank's proposals, however, is enhanced not only by its huge promotional apparatus and its connections to the centres of power, but by the manifest failure of so many state-led agrarian reform projects. In calling for a new wave of state-directed agrarian reform, peasant and small farmer organizations have strong ethical arguments on their side and powerful environmental and development arguments as well, since redistribution can be crucial in conserving soils and biodiversity, reducing poverty and stimulating demand. But do the organizations and the researchers who accompany them have a solid analysis of what went wrong with previous agrarian reforms? Most importantly, in opposing the World Bank's market-oriented approach, do they have persuasive arguments about how such problems might be avoided or minimized in the new era of state-led agrarian reforms that they hope to bring about?

Some of Central America's historic *campesino* organizations continue to back campaigns for land and related resources and for environmental justice (such as the anti-mining campaigns in Guatemala), and some have participated in coalitions that oppose DR-CAFTA, Plan Puebla-Panamá, the proposed EU-Central America free trade accord and other measures to liberalize the region's economies. In several countries, organized land recuperations have been targets of violent repression. In contrast to the historic *campesino* organizations involved in these struggles, a range of other, generally much smaller groups has concentrated on making either sectoral, product-specific or local demands, usually through conventional political channels. Both types of organization are attempting to again make the peasantry a protagonist in the region's history, albeit in very different ways. Even though the prospects are not encouraging, strategies for confronting the many obstacles and questions outlined above ought to be on the organizations' agenda.

REFERENCES

Acevedo Vogl, Adolfo José, 2003. *Impactos potenciales del Tratado de Libre Comercio Centroamérica-Estados Unidos en el sector agrícola y la pobreza rural de Nicaragua*. Managua: Ediciones Educativas.

Alvarez, Sonia E., 1998. 'Latin American Feminisms "Go Global": Trends of the 1990s and Challenges for the New Millennium'. In *Cultures of Politics/Politics of Cultures: Re-Visioning Latin American Social Movements*, eds Evelina Dagnino, Sonia E. Alvarez and Arturo Escobar, 293–324. Boulder, CO: Westview.

Andrade-Eekhoff, Katharine and Claudia Marina Silva Ávalos, 2004. 'La globalización de la periferia: flujos transnacionales migratorios y el tejido socio-productivo local en América Central'. *Revista Centroamericana de Ciencias Sociales*, 1 (1): 57–86.

Anheier, Helmut and Nuno Themudo, 2002. 'Organisational Forms of Global Civil Society: Implications of Going Global'. In *Global Civil Society 2002*, eds M. Glasius, H. Anheier and M. Kaldor, 191–216. Oxford: Oxford University Press.

Arana, Ana, 2005. 'How the Street Gangs Took Central America'. *Foreign Affairs*, 84 (3): 98–110.

Arias Peñate, Salvador, 1992. 'El contexto regional y mundial de la estrategia alternativa de desarrollo del Istmo Centroamericano'. In *Democracia sin pobreza: alternativa de desarrollo para el istmo centroamericano*, eds E. Stein and S. Arias Peñate, 21–82. San José: Departamento Ecuménico de Investigaciones.

Arquilla, John and David Ronfeldt, eds, 2001. *Networks and Netwars: The Future of Terror, Crime, and Militancy*. Santa Monica, CA: RAND.

ASOCODE [Asociación de Organizaciones Campesinas de Centroamérica para la Cooperación y el Desarrollo], 1999. Documento para la discusión sobre el 'Proceso de reorganización y reorientación de ASOCODE', Managua, enero.

ASOCODE, 2001. Memoria: Encuentro regional de dirigentes campesinos centroamericanos, Tegucigalpa, 04 y 05 de abril.

Barham, Bradford, Mary Clark, Elizabeth Katz and Rachel Schurman, 1992. 'Nontraditional Agricultural Exports in Latin America'. *Latin American Research Review*, 27 (2): 43–82.

Barquero, Marvin, 2007. 'Países productores de café descartan caída en precios'. *La Nación* (San José), 17 November. http://www.nacion.com/ln_ee/2007/noviembre/17/economia1318924.html. Accessed 1 December 2007.

Bartra, Armando, 2005. 'Reinventando una identidad colectiva. Foros sociales y encuentros campesinos en Mesoamérica', http://osal.clacso.org/dev/imprimir.php3?id_article=95. Accessed 20 December 2007.

Bartra, Armando, 2006. 'III Encuentro del Moicam: Por el poder popular y la recuperación de las soberanías'. http://www.redmesoamericana.net/?q=node/380. Accessed 20 December 2007.

Bartra, Armando, n.d. 'La crisis mesoamericana'. http://www.redmesoamericana.net/?q=node/297. Accessed 20 December 2007.

Bebbington, Anthony, 2007. 'Social Movements and the Politicization of Chronic Poverty'. *Development and Change*, 38 (5): 793–818.

Bevington, Douglas and Chris Dixon, 2005. 'Movement-relevant Theory: Rethinking Social Movement Scholarship and Activism'. *Social Movement Studies*, 4 (3): 185–208.

Biekart, Kees, 1999. *The Politics of Civil Society Building: European Private Aid Agencies and Democratic Transitions in Central America*. Amsterdam: International Books and the Transnational Institute.

Biekart, Kees and Martin Jelsma, eds, 1994. *Peasants beyond Protest in Central America: Challenges for ASOCODE, Strategies towards Europe*. Amsterdam: Transnational Institute.

Borras, Jr, Saturnino M., 2004. *La Vía Campesina: An Evolving Transnational Social Movement*. Amsterdam: Transnational Institute, TNI Briefing Series No. 6.

Brockett, Charles D., 2005. *Political Movements and Violence in Central America*. Cambridge: Cambridge University Press.

Carazo Vargas, Eva, 2004. 'Implicaciones del TLC desde la perspectiva de la agricultura familiar campesina'. In *¿Debe Costa Rica aprobarlo? TLC con Estados Unidos: contribuciones para el debate*, eds María Florez-Estrada and Gerardo Hernández, 249–70. San José: Instituto de Investigaciones Sociales, Universidad de Costa Rica.

CCS-Chiapas (Coordinación de Comunicación Social, Gobierno del Estado de Chiapas), 2002. 'Comunicado de Prensa – Con la participación de la CNPA, UNORCA, CIOAC y CLOC en Tapachula, 5 de mayo'. http://www.ccschiapas.gob.mx/pagina_anterior/boletines/2002/mayo/bol1115.htm. Accessed 1 June 2002.

CEPAL (Comisión Económica para América Latina y el Caribe), 2006. *Anuario estadístico de América Latina y el Caribe, 2006*. Santiago: CEPAL.

Comaroff, John L. and Jean Comaroff, 1999. 'Introduction'. In *Civil Society and the Political Imagination in Africa: Critical Perspectives*, eds John L. Comaroff and Jean Comaroff, 1–33. Chicago, IL: University of Chicago Press.

De Bremond, Ariane, 2007. 'The Politics of Peace and Resettlement through El Salvador's Land Transfer Programme: Caught between the State and the Market'. *Third World Quarterly*, 28 (8): 1537–56.

Desmarais, Annette Aurélie, 2007. *La Vía Campesina: Globalization and the Power of Peasants*. Halifax and London: Fernwood Publishing and Pluto Press.

Díaz Porras, Rafael and Wim Pelupessy, 2004. 'Agricultores, consumidores y la mediación institucional en las cadenas agro-alimentarias globales en Centroamerérica'. *Revista Centroamericana de Ciencias Sociales*, 1 (1): 25–56.

ECM (Encuentro Campesino Mesoamericano), 2003. 'Acuerdos del III Encuentro Campesino Mesoamericano, Tegucigalpa, 19–21 de julio'. Observatorio Social del Agro, http://redmesoamericana.net/?q=node/129. Accessed 21 December 2007.

Edelman, Marc, 1998. 'Transnational Peasant Politics in Central America'. *Latin American Research Review*, 33 (3): 49–86.

Edelman, Marc, 1999. *Peasants against Globalization: Rural Social Movements in Costa Rica*. Stanford, CA: Stanford University Press.

Edelman, Marc, 2001. 'Social Movements: Changing Paradigms and Forms of Politics'. *Annual Review of Anthropology*, 30: 285–317.

Edelman, Marc, 2003. 'Transnational Peasant and Farmer Movements and Networks'. In *Global Civil Society 2003*, eds M. Kaldor, H. Anheier and M. Glasius, 185–220. London: Oxford University Press.

Enríquez, Elio, 2006. 'Boletín CECCAM'. Posted on http://www.laneta.apc.org, 21 November, 15:09:37. Accessed 2 January 2008.

FAOSTAT, 2008. On-line database, United Nations Food and Agriculture Organization, Statistics Division. http://faostat.fao.org. Accessed 4 January 2008.

Fiallos, Alvaro, 2002. 'Nicaragua: principales problemas y limitaciones del sector agrícola y rural'. Santiago, Chile: Oficina Regional para América Latina y el Caribe, Organización de las Naciones Unidas para la Agricultura y la Alimentación.

FIAN (Food First Information and Action Network), 2006. *Voluntary Guidelines on the Right to Adequate Food: From Negotiation to Implementation*. Heidelberg: FIAN.

Fischer, Edward F. and Peter Benson, 2006. *Broccoli and Desire: Global Connections and Maya Struggles in Postwar Guatemala*. Stanford, CT: Stanford University Press.

Fox, Jonathan, 2000. 'Assessing Binational Civil Society Coalitions: Lessons from the Mexico-US Experience'. Working Paper No. 26, Chicano/Latino Research Center, University of California at Santa Cruz.

Gacitúa, Estanislao and Shelton Davis, 2000. 'Introducción: Pobreza y exclusion social en América Latin y el Caribe'. In *Exclusión social y reducción de la pobreza en América Latina y el Caribe*, eds Estanislao Gacitúa and Shelton H. Davis, 13–23. San José: FLACSO and Banco Mundial.

Gaete, Marcelo, Dennis Monero and Romano Sancho, 2005. *Análisis de movimiento campesino en el contexto costarricense actual. Lecciones para VECO Costa Rica y la Red COPRALDE*. San José: VECO-COPRALDE.

Galián, Carlos, 2004. 'TLC CA-EU: El tiro de gracia a la agricultura centroamericana'. *Tunapa*, 23: 4–5.

Gauster, Susana and S. Ryan Isakson, 2007. 'Eliminating Market Distortions, Perpetuating Rural Inequality: An Evaluation of Market-Assisted Land Reform in Guatemala'. *Third World Quarterly*, 28 (8): 1519–36.

Hecht, Susanna B. and Sassan S. Saatchi, 2007. 'Globalization and Forest Resurgence: Changes in Forest Cover in El Salvador'. *Bioscience*, 57 (8): 663–72.

Hirschman, Albert O., 1970. *Exit, Voice, and Loyalty: Responses to Decline in Firms, Organizations, and States*. Cambridge, MA: Harvard University Press.

Holt-Giménez, Eric, 2006. *Campesino a Campesino: Voices from Latin America's Farmer to Farmer Movement for Sustainable Agriculture*. Oakland, CA: Food First Books.

ICO (International Coffee Organization), 2008. 'ICO Indicator Prices. Monthly and Annual Averages 2005–2007'. http://www.ico.org/prices/p2.htm. Accessed 4 January 2008.

Iniciativa CID (Iniciativa Mesoamericana de Comercio, Integración y Desarrollo), 2002. 'Campaña regional en torno al Tratado de Libre Comercio entre los Estados Unidos y Centro América'. Document provided by the Federación Nacional de Cooperativas, Nicaragua.

Isaula, Roger, 1996. *Honduras: El ajuste en el sector agrícola y la seguridad alimentaria*. Tegucigalpa: Centro de Documentación de Honduras.

Jonakin, Jon, 1997. 'Agrarian Policy'. In *Nicaragua Without Illusions: Regime Transition and Structural Adjustment in the 1990s*, ed. Thomas W. Walker, 97–113. Wilmington, DE: SR Books.

Kay, Cristóbal, 2001. 'Los paradigmas del desarrollo rural en América Latina'. In *El mundo rural en la era de la globalización: incertidumbres y potencialidades: X Coloquio de Geografía Rural de España de la Asociación de Geógrafos Españoles*, ed. Francisco García Pascual, 337–429. Lleida: Universidad de Lleida, Servicio de Publicaciones.

Kay, Cristóbal, 2005. 'Perspectives on Rural Poverty and Development Strategies in Latin America'. Working Paper Series No. 419. The Hague: Institute of Social Studies.

Keck, Margaret E. and Kathryn Sikkink, 1998. *Activists beyond Borders: Advocacy Networks in International Politics*. Ithaca, NY: Cornell University Press.

Leanza, Gustavo, 2004. 'Análisis crítico de las estructuras y relaciones de poder vinculadas al sector agrario'. Guatemala: Programa de Naciones Unidas para el Desarrollo and Programa Dinamarca Pro Derechos Humanos para Centroamérica (manuscript).

Levenson, Deborah, 1989. 'Las Maras: violencia juvenil de masas'. *Polémica* 7: 2–12.

Lofredo, Gino, 1991. 'Hágase rico en los 90. ¿Usted todavía no tiene su ONG?' *Chasqui*, 39: 15–18.

Manwaring, Max G., 2005. *Street Gangs: The New Urban Insurgency*. Carlisle, PA: Strategic Studies Institute, US Army War College.

Marsden, Terry, 1995. 'Beyond Agriculture? Regulating the New Rural Spaces'. *Journal of Rural Studies*, 11 (3): 285–96.

Morales, Abelardo and Carlos Castro, 2002. *Redes transfronterizas: sociedad, empleo y migración entre Nicaragua y Costa Rica*. San José: FLACSO.

Morales Gamboa, Abelardo and Martha Isabel Cranshaw, 1997. *Regionalismo emergente: redes de la sociedad civil e integración centroamericana*. San José: FLACSO & Ibis-Dinamarca.

Moreno, Raúl, 2003. *The Free Trade Agreement between the United States and Central America: Economic and Social Impacts*. Managua: American Friends Service Committee.

Nelson, Paul, 2004. 'New Agendas and New Patterns of International NGO Political Action'. In *Creating a Better World: Interpreting Global Civil Society*, ed. Rupert Taylor, 116–32. Bloomfield, CT: Kumarian Press.

O'Brien, Robert, Anne Marie Goetz, Jan Aarte Scholte and Marc Williams, 2000. *Contesting Global Governance: Multilateral Economic Institutions and Global Social Movements*. Cambridge: Cambridge University Press.

Posas, Mario, 1985. 'El movimiento campesino hondureño: un panorama general (siglo XX)'. In *Historia política de los campesinos centroamericanos*, Vol. 2, ed. Pablo González Casanova, 28–76. Mexico City: Siglo XXI.

Proyecto Estado de la Región-PNUD (Programa de las Naciones Unidas para el Desarrollo), 2003. *Segundo informe sobre desarrollo humano en Centroamérica y Panamá*. San José: Proyecto Estado de la Nación.

Riechmann, Jorge and Francisco Fernández Buey, 1994. *Redes que dan libertad: Introducción a los nuevos movimientos sociales*. Barcelona: Ediciones Paidós.

Riles, Annelise, 2001. *The Network Inside Out*. Ann Arbor, MI: University of Michigan Press.

Robinson, William I., 2003. *Transnational Conflicts: Central America, Social Change, and Globalization*. London: Verso.

Ruben, Raúl and Francisco Fúnez, 1993. *La compra-venta de tierras de la Reforma Agraria*. Tegucigalpa: Editorial Guaymuras.

Ruben, Ruerd and Edoardo Masset, 2003. 'Land Markets, Risk and Distress Sales in Nicaragua: The Impact of Income Shocks on Rural Differentiation'. *Journal of Agrarian Change*, 3 (4): 481–99.

Rubio, Blanca, 2002. 'La exclusión de los campesinos y las nuevas corrientes teóricas de interpretación'. *Nueva Sociedad*, 182: 21–33.

Ruhl, J. Mark, 1984. 'Agrarian Structure and Political Stability in Honduras'. *Journal of Interamerican Studies and World Affairs*, 26 (1): 33–68.

Sandoval García, Carlos, 2002. *Otros amenazantes: los nicaragüenses y la formación de identidades nacionales en Costa Rica*. San José: Editorial de la Universidad de Costa Rica.

Santa Cruz, Wendy, 2006. *Una aproximación a la conflictividad agraria y acciones del movimiento campesino*. Guatemala: FLACSO.

Santos, Boaventura de Souza, 2006. 'The World Social Forum: Where Do We Stand and Where are We Going?' In *Global Civil Society 2005/6*, eds Marlies Glasius, Mary Kaldor and Helmut Anheier, 73–8. London: Sage.

Segovia, Alexander, 2004. 'Centroamérica después del café: el fin del modelo agroexportador tradicional y el surgimiento de un nuevo modelo'. *Revista Centroamericana de Ciencias Sociales*, 1 (2): 5–38.

Stirrat, R.L., 1996. 'The New Orthodoxy and Old Truths: Participation, Empowerment and Other Buzz Words'. In *Assessing Participation: A Debate from South Asia*, eds S. Bastian and N. Bastian, 67–92. New Delhi: Konark Publishers.

Tangermann, Klaus-Dieter and Ivana Ríos Valdés, eds, 1994. *Alternativas campesinas: modernización en el agro y movimiento campesino en Centroamérica*. Managua: CRIES and Latino Editores.

Tarrow, Sidney, 2005. *The New Transnational Activism*. Cambridge: Cambridge University Press.

Tilly, Charles, 1984. 'Social Movements and National Politics'. In *Statemaking and Social Movements: Essays in History and Theory*, eds Charles Bright and Susan Harding, 297–317. Ann Arbor, MI: University of Michigan Press.

Tiney, Juan, 2006. 'Antecedentes y perspectivas del Movimiento Indígena y Campesino Mesoamericano'. http://www.redmesoamericana.net/?q=node/406. Accessed 20 December 2007.

USAID and MFEWS [US Agency for International Development and Mesoamerica Famine Early Warning System], 2007. 'High Maize Prices Restrict Food Access – Guatemala Food Security Situation – January 2007'. http://www.fews.net/centers/innerSections.aspx?f=gt&pageID=monthliesDoc&m=1002304. Accessed 1 January 2008.

van Rooy, Alison, 2004. *The Global Legitimacy Game: Civil Society, Globalization and Protest*. London: Palgrave.

Vía Campesina, 2002. *Peasant Rights – Droits Paysans – Derechos Campesinos*. Tegucigalpa: Vía Campesina.

Vía Campesina, 2005. *Annual Report Peasant Rights Violation – Informe Anual sobre las violaciones de los derechos campesinos*. Jakarta: Vía Campesina.

Viales Hurtado, Ronny, 2002. 'Ruralidad y pobreza en Centroamérica en la década de 1990, el contexto de la globalización y de las políticas agrarias "neoliberales"'. In *Culturas populares y políticas públicas en México y Centroamérica (siglos XIX y XX)*, eds Francisco Enríquez Solano and Iván Molina Jiménez, 157–86. Alajuela, Costa Rica: Museo Histórico Cultural Juan Santamaría.

Whitaker, Chico, 2006. 'The World Social Forum: Where Do We Stand and Where Are We Going?'. In *Global Civil Society 2005/6*, eds Marlies Glasius, Mary Kaldor and Helmut Anheier, 66–72. London: Sage.

4 La Vía Campesina and its Global Campaign for Agrarian Reform

SATURNINO M. BORRAS JR

ORIGINS OF VÍA CAMPESINA

Today, the 'Global Campaign for Agrarian Reform' (GCAR) by Vía Campesina and its allies has gained importance in the global land policy-making scene. Vía Campesina's land agendas and demands (Vía Campesina 2000a, 2000b; Vía Campesina n.d.; Rosset 2006) constitute a serious counter-argument to the neoliberal doctrine (see, e.g., Broad 2002; Mayo 2005), a veritable alternative 'voice' from below (see, e.g., Appadurai 2006), representing marginalized rural peoples in the world. Why and how was this voice constituted, how has it evolved and what are its future prospects in the struggle against neoliberal globalization?

It is often assumed that the emergence of contemporary global justice campaigns such as the GCAR were the inevitable result of neoliberalism's onslaught. Yet there are numerous neoliberal policies that did not spark contentious campaigns by social movements, as has been shown in the literature, suggesting that other factors were at play. In particular, we contend that the socioeconomic and political transformations brought about by neoliberal globalization and the changes in the international political opportunity structure, among other factors, played a significant role in determining the timing and framing of the launch of the GCAR.[1] Calling this process 'internationalization', Tarrow (2005, 8) suggests that it is marked by three factors: '(i) an increasing horizontal density of relations across states, governmental officials, and non-state actors, (ii) increasing vertical links among the subnational, national and international levels, (iii) an enhanced formal and informal structure that invites transnational activism and facilitates the formation of networks on non-state, state and international actors'. These factors were crucial in the case of Vía Campesina. But as discrete processes they each took time to evolve. Only after some time had passed did Vía Campesina take the momentous step to launch a campaign on land issues that would be

---

[1] Tarrow (1994, 54) has defined political opportunities as 'the consistent (but not necessarily formal, permanent, or national) signals to social or political actors which either encourage or discourage them to use their internal resources to form a social movement'. He has also identified four important political opportunities: access to power, shifting alignments, availability of influential elites and cleavages within and among elites. Refer also to his later explanation about the need to bring in the notion of 'threats' (Tarrow 2005, 240).

marked by 'sustained organizing efforts', 'durable network' and 'collective identity' (see Tilly 2004, 3–4; Tarrow 2005, 6–7).

During the past two decades, nation-states in developing countries have been transformed by a triple 'squeeze': globalization, (partial) decentralization and the privatization of some of its functions (Fox 2001). Central states remain important, but are transformed (Keohane and Nye 2000; Scholte 2002; Sassen 2006; Gwynne and Kay 2004). The scope, pace and direction of this transformation, including its agrarian restructuring component (see Bernstein 2006; Friedmann 2004), have been contested by different actors (McMichael 2006, this collection; Patel 2006). The changing international–national–local linkages that structure the terms under which people accept or resist the corporate-controlled global politics and economy present both threats and opportunities for the world's rural population. The co-existence of threats and opportunities has prompted many rural social movements to both localize further (in response to state decentralization) and to 'internationalize' (in response to globalization). The seemingly contradictory pressures (of globalization and decentralization) that are having such an impact on the nation-state are thus also transforming rural social movements. As a result, one sees the emergence of more horizontal, 'polycentric' rural social movements that at the same time struggle to construct coherent coordinative structures for greater vertical integration – the emergence of contemporary 'transnational agrarian movements' (TAMs).[2]

Against this backdrop, Vía Campesina has evolved as an international movement of poor peasants and small farmers from the global South and North. Initiated by Central, South and North American peasant and farmers' movements and European farmer's groups, Vía Campesina was formally launched in 1993. Existing transnational networks of activists located in peasant movements and non-governmental funding agencies in the North facilitated the earlier contacts between key national peasant movements, most of which had emerged already in the 1980s. Vía Campesina currently represents more than 150 (sub)national rural social movement organizations from 56 countries in Latin America and the Caribbean, North America, (Western) Europe, Asia and Africa.[3] Since its birth, Vía Campesina's main agenda has been to defeat the forces of neoliberalism and to develop an alternative revolving around the concept of 'food sovereignty' (see Rosset 2006; IPC for Food Sovereignty 2006).[4]

Vía Campesina's positions and forms of action on key issues have differed fundamentally from its mainstream counterpart and rival, the International Federation of Agricultural Producers (IFAP). Founded in 1946 by associations of small to big

---

[2] In this chapter, we treat Vía Campesina loosely as a 'transnational movement', 'transnational network' and 'transnational coalition', following the useful explanation on these categories by Khagram et al. (2002, 9). Vía Campesina exhibits the features of all these categories depending on particular campaigns or circumstances.
[3] It focuses on seven issues: (i) agrarian reform, (ii) biodiversity and genetic resources, (iii) food sovereignty and trade, (iv) women, (v) human rights, (vi) migrations and rural workers and (vii) sustainable peasant's agriculture.
[4] This notion of food sovereignty also resonates with Bello's concept of 'deglobalization' (Bello 2002).

farmers mainly from developed countries, IFAP became the sector organization for agriculture that has claimed and made official representation to (inter)governmental agencies. Neoliberal policies generally have not adversely affected many of its constituents, at least not financially. While not a homogeneous network economically, IFAP's politics tend to be dominated by its economically and financially powerful and politically conservative members (Desmarais 2007). Since the 1990s, IFAP has also recruited or allowed entry of some organizations of poor peasants from developing countries, no doubt partly in reaction to the emergence of Vía Campesina. Vía Campesina has a highly heterogeneous membership not only in class, gender and ethnic terms; the ideological persuasions of its members vary as well. But in spite of apparent differences in terms of worldviews, political agendas and methods of work, there are important unifying commonalities too. Chief among these is that most of Vía Campesina's mass base more or less represents sectors in the global North and South that are already economically and politically marginalized. It is this profile that differentiates Vía Campesina from IFAP.

As an 'actor' on the world stage, Vía Campesina has gained recognition as the main voice of organized sectors of marginalized rural peoples, thus eroding IFAP's previous hegemony. At the same time, like any entity that seeks to aggregate, organize and represent a plurality of identities and interests, Vía Campesina constitutes an evolving 'arena of action', one where a movement's basic identity and strategy may be contested and (re)negotiated over time. This dual character helps to make Vía Campesina an important institution of and for national–local peasant movements, but a complex entity for external observers and actors to comprehend and deal with (Borras 2004). This discussion on the dual character of transnational movement is similar to the notions of 'network-as-actor' and 'network-as-structure' by Keck and Sikkink (1998, 7; see related discussions in Guidry et al. 2000, 3; Batliwala and Brown 2006). The GCAR is best seen from this perspective of the movement's dual character.

The remainder of this chapter is organized as follows: the next section examines GCAR, focusing on why and how it has emerged to become an important campaign. GCAR's impact is then analyzed. The fourth section examines pending contentious issues within Vía Campesina and possible future trajectories of GCAR, followed by concluding remarks.

GLOBAL CAMPAIGN FOR AGRARIAN REFORM (GCAR)

Six years after its founding, Vía Campesina launched GCAR in 1999–2000, at a time when land reform was coming under attack by neoliberals. Land policy was being resurrected on official agendas of international institutions and many nation-states for a variety of often conflicting reasons (see, e.g., Akram Lodhi et al. 2007; Rosset et al. 2006; Bernstein 2002; Byres 2004a, 2004b; Griffin et al. 2002; Ghimire 2001; Borras et al. 2007a). Departing from the classic land reform debate, the renewed interest in land policy has been dominated by a pro-market orientation. The Vía Campesina campaign is a direct reaction to the neoliberal model, the 'market-led agrarian reform' (MLAR).

For mainstream economists, the problem with past land policies was the central role of the state in (re)allocating land resources, leading (in their view) to distortions of the land market, resulting in 'insecure' property rights and investments in the rural economy. They often point to problems in public/state lands (e.g., lacking clear private property rights) as 'proof' of the undesirable effects of state intervention in the land market. In their view, what is needed instead are clear, formal private property rights in the remaining public lands in most developing countries and transition economies (see De Soto 2000;[5] see also World Bank 2003; but see Nyamu-Musembi 2007 and Cousins 2007 for critical insights in the African context). Similarly, from the neoliberal perspective, the 'failure' of state-led land reforms in private lands is attributable to the methods of land acquisition (e.g., expropriation and coercion) that were resisted by landlords.

Clearly, landlords have subverted the policy, evading coverage by subdividing their farms or retaining the best parts of the land. They have also launched blistering legal offensives that have slowed, if not prevented, much land reform implementation. Here, there can be no room for disagreement; the historical record prevents that. The point of departure is landlord resistance to land reform – should it be evaded or confronted? (See related arguments about the nature of expropriation offered by Chonchol (1970).) Neoliberal economists see landlord resistance as something to be avoided at all costs and, arguably, it is from this core belief that the neoliberal land reform model for private lands has been constructed. This model thus posits 'free market forces' as the most desirable mechanism for (re)allocating land resources, envisioning a process that is necessarily privatized and decentralized. Frequently referred to as MLAR, the model inverts what it claims to be key features of the conventional 'state-led' model: from expropriationary to voluntary; from statist-centralized to privatized-decentralized implementation and so on (Borras 2003; Borras et al. 2007a).

The rise of this neoliberal land policy model juggernaut in the 1990s certainly did not go unnoticed, in spite of vain efforts by proponents to camouflage it as 'anti-poverty community-based' or 'negotiated' land reform, or to repackage it as 'legal empowerment of the poor'. Yet it is important to note that the response to this policy among key state and societal actors in the land reform issue arena has been decidedly mixed, and less oppositional than one might have expected. Among those who do oppose the model, however, Vía Campesina is the undisputed leader and the GCAR was devised largely as the main vehicle for this opposition globally.

In undertaking this campaign, Vía Campesina has had to refine its initial take on the land issue, while developing and consolidating a 'human rights-based approach' to land. Indeed, the global framework of Vía Campesina's position on

---

[5] Hernando de Soto, a Peruvian economist, has argued that the potential capital of the poor is land, but that most of this is 'dead capital', having no individual private land titles that can be used as collateral in financial transactions. And so, investors and banks feel insecure to transact with the poor. The solution is to generate private individual land titles for the remaining public lands in developing countries. He has inspired many economists worldwide, is currently chairing the 'Commission for the Legal Empowerment of the Poor' (CLEP) with secretariat support from the UNDP, which advocates the implementation of de Soto's idea.

land has been evolving over the years, with the 2006 joint declaration with the IPC[6] for Food Sovereignty being the most comprehensive and systematic version. During the first few years of GCAR, the main call was a demand to drop MLAR. Eventually, the network's position evolved to include a demand for the adoption of their 'human rights' framework and alternative vision. This suggests that the campaign has aspired to define and articulate its own interpretation of the meaning and purpose of land and land reform, as a step toward constructing an alternative vision.

Vía Campesina aspires to neither 'sink' (i.e. too localized) nor 'float' (i.e. too globalized) in this effort, but rather to 'verticalize' collective action (in the manner described by Fox 2001; Edwards and Gaventa 2001) by connecting local, national and international groups. Looking more closely, the emergence of GCAR involved five interlinked processes: (i) a swift externalization of national–local issues; (ii) the forging of transnational allies; (iii) the forging of a common frame and target, (iv) the opening up of faster, cheaper cross-border communication and transportation and (v) increasing autonomy and capacity to combine forms of collective action.

Swift Externalization of National–Local Issues

MLAR was carried out in countries that are important to Vía Campesina, directly affecting organizations that are influential within the network. It is largely for this reason that the 'externalization' (i.e. 'the vertical projection of domestic claims onto international institutions or foreign actors' – Tarrow 2005, 32) of national–local land issues has been so swift within Vía Campesina. In varying degrees, MLAR has been carried out in Brazil, Colombia, in Central American countries, the Philippines, South Africa and Namibia. Negotiations were attempted in other countries such as Nepal. In all these countries, only Namibia does not figure in the radar of Vía Campesina, at least not yet. Central and South America, especially Brazil, are bastions of influence of Vía Campesina, with Brazil's MST[7] as one of the most influential groups within the global movement. When MLAR was introduced in Brazil in 1997 through the Projeto Cedula da Terra (PCT), it quickly ran into MST base areas on the ground. At the national level, PCT gained prominence, partly because of the favourable endorsement by landlords (Navarro 1998; Borras 2003). At this point MLAR promoters were in a triumphant mood, claiming successes in different countries. But rural social movements and their allies in Brazil were convinced that PCT would not deliver gains for redistributive reform, and would only undermine the existing efforts by the state and by the landless movement (Sauer 2003). Their opposition was cemented in the National Forum for Agrarian Reform, a national forum of all the major (competing) agrarian

---

[6] International Planning Committee. It originated from the civil society group formed for the 1996 World Food Summit in Rome.
[7] Movimento dos Trabalhadores Rurais sem Terra (Movement of the Landless) – see Wright and Wolford (2003).

movements in Brazil, including MST, CONTAG[8] and FETRAF.[9] The National Forum demanded an investigation of the PCT through the World Bank Inspection Panel, but the request was denied, twice, on technical grounds (see Vianna 2003). And despite problematic outcomes, MLAR continued and was even expanded (Pereira 2007; Medeiros 2007), making the threat more real in the eyes of MST and other Vía Campesina members in Brazil, and helping to push a rapid externalization of the Brazilian issue onto the international scene.

Meanwhile, MLAR was incorporated into peace accords in several countries in Central America in the mid-1990s. Guatemala, Honduras and El Salvador, for instance, witnessed the introduction of versions of MLAR, including one close to the textbook model (Guatemala; Gauster and Isakson 2007) and also a state–market hybrid (El Salvador; De Bremond 2007). Central America, however, was the birthplace of Vía Campesina,[10] at least informally. It was where, at a conference in Managua in 1992, the first concrete idea of establishing the global movement was discussed by not only Central American peasant leaders, but also others from outside the region. The Central American peasant coalition, ASOCODE,[11] was already virtually defunct by the time MLAR began gaining ground in these countries (Edelman 1998, 2003; also in this collection), but the region still remained host to some of the relatively active, relatively organized and articulate land-oriented groups in Vía Campesina.[12] It is no coincidence that, from 1996 to 2004, the global secretariat of Vía Campesina was hosted by the Honduran organization, COCOCH[13], and led by COCOCH's director Rafael Alegria. Alegria served as Vía Campesina's general coordinator during this period. In addition, the GCAR's international secretariat was run by seasoned cadres from the Nicaraguan organization, ATC.[14] As a result, land issues in Central America were rapidly externalized onto the global stage and picked up seriously by allies.

Meanwhile, the Philippines was the 'gateway' for Vía Campesina into Asia in 1993, when outsiders' contacts in the region were limited to KMP[15] in the Philippines and KRRS[16] in India. The Philippine movement facilitated further contacts in most parts of the region, and KMP was an influential member (at least until around 2004). And so when MLAR was first introduced in the Philippines in 1996, and then during the 1999 negotiation for a pilot project, the Philippine MLAR got into the radar of Vía Campesina and the GCAR, although the anti-MLAR activities were spearheaded by non-members of Vía Campesina (Franco 1999). Meanwhile, MLAR became the defining framework for the

---

[8] Confederação Nacional dos Trabalhadores na Agricultura.
[9] Federação dos Trabalhadores na Agricultura Familiar.
[10] The official birthplace is Mons, Belgium.
[11] Association of Central American Peasant Organizations for Cooperation and Development.
[12] For an earlier (optimistic) background on ASOCODE, see Biekart and Jelsma (1994).
[13] Honduran Coordinating Council of Peasant Organizations.
[14] Asociación de Trabajadores del Campo.
[15] Kilusang Magbubukid ng Pilipinas.
[16] Karnataka State Farmers' Association.

post-apartheid compromise agrarian reform in South Africa, although it is another kind of a state–market hybrid (Lahiff 2007; Ntsebeza and Hall 2006; Walker 2003). In the absence of a national peasant movement (at least until 2000), work around land reform in South Africa would be taken up by pockets of activists and NGOs, the most prominent of which was the now defunct National Land Committee or NLC (Mngxitama 2005). When the Landless Peoples' Movement (LPM) was born in 2000 with the help of NLC, it quickly became a Vía Campesina member (Greenberg 2004) at the same time that the South African land issue was quickly taken up by the GCAR.

To some extent, externalization occurred out of necessity: where significant resistance to the model was mounted by Vía Campesina members, initial efforts at influencing national governments did not yield the movements' desired outcomes, forcing peasant movements to externalize their campaigns, which enabled them to then come back to their national governments with greater power. This pattern validates the 'boomerang effect' advanced by Keck and Sikkink (1998, 12–13).[17] Meanwhile, many national contestations around MLAR that were externalized onto the Vía Campesina agenda and campaign have a common feature: most of them involved important Vía Campesina members. Where Vía Campesina does not have a presence, or where Vía Campesina allies do not have a network, MLAR issues, however problematic, tend not to be picked up by transnational land reform activists. For example, although MLAR was carried out in Namibia (van Donge et al. 2007), or a variant of MLAR in Egypt (Bush 2007), the cases were not taken up in the GCAR because there were no Vía Campesina members in these countries.

Availability of Transnational Allies

Alliances with groups with relevant political and logistical resources are necessary. For peasant movements in the South, well-connected NGOs, funding agencies and sympathetic academics, under certain terms and conditions, hold the greatest potential as allies. But such alliances do not spring up automatically with the appearance of an issue on the horizon, no matter how urgent it may be for the movements concerned. Instead, pre-existing (in)formal networks between individuals and groups usually play a critical role in laying the groundwork for more expansive cross-national and inter-sectoral/inter-network coalition-building. And once achieved, it does not remain static.

In the case of Vía Campesina, two alliances around GCAR are important. The first involves FIAN (FoodFirst Information and Action Network), a human rights activist network composed of individuals and groups located in both the

---

[17] 'When channels between the state and its domestic actors are blocked, the boomerang pattern of influence characteristic of transnational networks may occur: domestic NGOs bypass their state and directly search out international allies to try to bring pressure on their states from outside' (Keck and Sikkink 1998, 12–13). But Marc Edelman explains that the boomerang effect works better for some kinds of demands than others (see Edelman in this collection).

global North and South. Founded in the 1980s, FIAN focuses on putting flesh to the international UN convention on economic, social and cultural rights and particularly the 'right to food'. For FIAN, the 'right to land' is a necessary prerequisite to the right to food. Before forging an alliance with Vía Campesina in 1999–2000, FIAN was able to launch its own intermittent campaigns. But while these earlier initiatives were certainly important to the participating national groups and to FIAN's global advocacy, they were far less than what was needed to put FIAN's message in the corridors of global power. FIAN needed the organized force and global spread of Vía Campesina. For its part, Vía Campesina members confronted on the ground by MLAR lacked a 'master frame' that could link their campaign into the 'rights talk' that was fast gaining ground worldwide during this period (De Feyter 2005) and was the most logical counter-argument against MLAR. But not all human rights advocates have an understanding of agrarian issues; not all agrarian reform advocates have an understanding of human rights law and methodology. Each network was thus recognized as complementary to the other's work and a global alliance was forged. The alliance has managed to remain mutually beneficial and reinforcing since then, despite occasional tensions. As the campaign gained momentum, its activists quickly realized that a simple 'expose and oppose' and 'agit-prop' (agitation-propaganda) approach would be insufficient to defeat the MLAR threat. Success would require solid arguments backed up by evidence and more solid propositions regarding an alternative. This latter concern prompted a process of campaign reframing that would eventually result in the 'agrarian reform-based food sovereignty' call of today. In this framing, Vía Campesina found another strategic ally in LRAN (Land Research and Action Network), a global network of individuals and research think tanks working on the issues of land, food politics, agroecology and trade, originally hosted by non-governmental research organization Food First Policy Institute in Oakland, California, although it later became autonomous. This broadening of the campaign framework made GCAR more accessible and attractive to other (trans)national activist networks working around broader issues of food and the environment. Finally, it is worth noting that by the late 1990s, most non-governmental funding agencies had opened up new 'global programmes', alongside their more established country programmes, in response to the burgeoning field of transnational activism, and were seeking new partners to fund. The new global programmes became a key resource for TAMs like Vía Campesina as well as for the GCAR.

Common Meaning, Common Target

The very 'meaning' of land has been evolving within GCAR over time, at the same time that the World Bank's aggressive promotion of MLAR in the 1990s was contributing to the construction of a common meaning of 'land'. Vía Campesina activists agreed that 'land is critical to peasants' livelihoods, but that effective control of these resources is monopolized by the landed classes, and so the need to redistribute this to landless peasants; and MLAR will not be able to do this, in fact, it may even undermine such an effort' (see, e.g., Vía Campesina 2000a). This was the earliest shared understanding within Vía Campesina. Since then, within the network and without necessarily departing from its original philosophical moorings, understanding of the meaning and purpose of land (and consequently the nature of the global campaign itself) has continued to evolve as a result of ongoing efforts to link the issue of land with the broader issues of food sovereignty, the environment and other development issues (especially after LRAN joined). At present, Vía Campesina's involvement with the much broader and looser "International Planning Committee" (IPC) for Food Sovereignty (which includes pastoralists, fisherfolk and other sectors that are not particularly strong in Vía Campesina) appears to once again be re-orienting the former's take on land even further.[18]

Finally, a critically important factor that facilitated the making of common cause and made possible the initial emergence of the GCAR amidst such diversity was the existence of a clear, common target (or culprit). Certainly the World Bank, as MLAR's inventor-promoter-funder, provided a concrete, high-profile target whose 'villainy' was also relatively easy to explain to the different subjective forces and broader publics that the campaign hoped to sway.

Faster and Cheaper Cross-Border Communication and Transportation

Breaking the monopoly on information, communication and mobility by national governments and international financial institutions is a critical and favourable change in the 'political opportunity structure' for (TAM) activists (Keck and Sikkink 1998; Tarrow 2005; Bob 2005). It likewise partly accounts for the emergence of GCAR. When Vía Campesina was formed in 1993, electronic mail was just beginning to be introduced in the NGO world, and back then there were very few peasant movements that were able to use the new technology.[19] Instead, the fax – expensive and cumbersome (especially in handling bulk electronic documents) – remained the dominant mode of communication between Vía Campesina members until roughly the end of the 1990s. The advent of free web-based email and free access to documents on more and more websites opened up new opportunities for (trans)national peasant groups to communicate quickly and to find and share crucial information. In more recent years, Skype and text messaging have become yet another important, relatively affordable means by which TAM activists can easily connect and communicate with each other across

---

[18] Vía Campesina involvement in the much broader (in terms of representation and ideological persuasions) IPC for Food Sovereignty seems to be transforming both Vía Campesina and the IPC members through not-always-so-smooth interactions with other rural sectors, types of associations (IPC includes NGOs) and political-ideological differences (IPC includes less radical groups). This is based on the author's own observation of some of the key events participated in by the IPC, as well as based on a comprehensive semi-structured interview of Antonio Onorati, the global focal person of the IPC (June 2007, Berlin).
[19] It can be recalled that HTML code was only beginning to achieve widespread use in the mid-1990s. Prior to that there were text-based Internet pages and no web-based email systems (I thank Marc Edelman for reminding me about this).

an enlarged space–time continuum. And then there is cheaper air transportation, which has expanded the mobility of GCAR activists, enabling them to meet each other (and their 'enemies') face-to-face in global gatherings, to witness each other's national–local conditions, and even to literally stand together in solidarity: at a picket line or mass demonstration here, or human rights fact-finding mission there. The vastly increased opportunities for direct encounters and communications at all levels has had (and will continue to have) a profound effect on movement dynamics (at all levels), and certainly deserves more focused attention than we can give here. One important effect however has been that the monopoly by governments, big NGOs and development agencies on information related to global land policy-making has been eroded. This was clearly evident in the Philippine case, when in 1999 the World Bank began negotiating with the Philippine government to introduce MLAR, highlighting supposed MLAR successes in Latin America and South Africa. Sceptical Philippine activists, using email, quickly contacted colleagues in the United States and Latin America to ask for alternative views. Within days, the activists were armed with documents that showed the exact opposite picture of what the World Bank was claiming (Franco 1999; Borras et al. 2007a). A similar process would unfold over and over again in other countries too, with activists receiving relevant data and alternative analyses from movements elsewhere or accessing information on the web, enabling them to strengthen their advocacy positions at crucial moments.

Autonomy and Capacity to Combine Forms of Collective Actions

Collective actions carried out at international political spaces often bring TAMs face-to-face with international institutions, with some of which they have a previous history of interaction, while others they do not. In either case, the threat of 'co-optation' hounds TAM activists. It is useful here to distinguish between two concepts: 'independence' and 'autonomy'. Independence is often seen as a choice in absolute terms – groups either allow themselves to be co-opted by these international institutions, or they do not, and are thus insulated from any form of external interference or influence. By contrast, autonomy is 'inherently a matter of degree' and refers to the amount of external influence in the agrarian movements' internal decision-making. In this view, an organization may have relationships with other entities, but what matters is the *terms* of those relationships (based on Fox 1993, 28).

For Vía Campesina, the struggle for autonomy is fought on two fronts: with (inter)governmental international institutions and with NGOs. As explained in a recent organizational document, 'We do not have a choice as to whether we interact with others who are engaged in our arena – but we have a choice on how we work to effect the changes we desire' (Vía Campesina 2000b). It elaborates: 'Vía Campesina must have autonomy to determine the space it will occupy with the objective of securing a large enough space to effectively influence the event' (2000b; for background discussions see also Tadem 1996; Batliwala 2002). Meanwhile, when and how to use direct action and mobilization as a form collective action, and in the service of what broader political strategy, is a question that seems to be addressed in a rather open-ended and tentative manner within Vía Campesina, and internationalizing collective actions around land issues is not easy for the network as well (Vía Campesina 2004, 48–9).

The search for the appropriate tactics and forms of actions is linked to their inevitable interaction with global (inter)governmental institutions. The choice of what types of tactics and actions to take depends in part on what types of global institutions they interact with. The nature of a particular institution does matter for the calculation of Vía Campesina. In general, they tend to favour the UN system that adheres to a 'one country–one vote' representation mechanism, which helps to explain its critical but collaborative relationships with some groups within the Food and Agriculture Organization of the UN (FAO) and the International Fund for Agricultural Development (IFAD). But consistent with their basic framework, Vía Campesina more or less automatically takes a confrontational, 'expose and oppose' stance against international financial institutions, e.g., World Bank, viewing these institutions as the cause of, not the solution to, the problems of peasants and farmers.

Vía Campesina has been quite skilful in combining diverse forms of actions. It has launched confrontational actions against TNCs and their domestic partners, using militant forms of actions such as land occupation, torching of GM crop field sites, and marches in major cities. At the same time it has collaborated with pockets of allies in a few agencies on selected issues, and engaged in negotiation, collaboration and even joint initiatives. It is from this perspective that the strategies employed in GCAR can be better understood. It has taken an 'expose and oppose' position against MLAR and the World Bank, using coordinated and simultaneous militant forms of local–national–international actions, including 'agit-prop', public shaming and local land occupations. Meanwhile, it has undertaken collaborative work with some allies within IFAD and FAO in developing common documents, conferences and projects. These two broadly different types of approaches, each with their own type of media projection as well, are seen as mutually reinforcing. Negotiations with other agencies would be weak without the real threat that Vía Campesina can actually resort to militant forms of actions against them; conversely, purely 'expose and oppose' actions without intermittent negotiations would project the movement as unreasonable. A careful balancing act is required in the use of these forms of actions and good media work (for related discussions, see Hertel 2006; Bob 2005), within and beyond GCAR, and this is something Vía Campesina has been able to do relatively effectively.

In short, GCAR emerged largely because five interlinked factors associated with a change in the international political opportunity structure were present: swift externalization of national issues, emergence of allies with political and logistical resources, forging of a common meaning in the campaign and the emergence of a common concrete and easy target of the campaign, the emergence of faster and cheaper cross-border communication and transportation, and the attainment of greater degree of autonomy and capacity to combine forms of collective actions. Without the presence of all of these factors it is doubtful that the GCAR could have been launched, validating Smith and Johnston (2002, 8), who argue that while 'increased global integration generates *potential* sources of unity for political movements', other complementary factors are necessary for this potential to be realized.

INITIAL IMPACT

A few years ago, Baranyi et al. cautioned land reform observers: 'One should not underestimate the impact that the Global Campaign for Land Reform headed by Vía Campesina might eventually have on international policy debates in this regard' (2004, 47). How do we proceed to get a reasonable view of the impact of Vía Campesina's global campaign? What Keck and Sikkink offer is relevant: 'Networks influence politics at different levels because the actors in these networks are simultaneously helping to define an issue area, convince policymakers and publics that the problems thus defined are soluble, prescribe solutions, and monitor their implementation' (1998, 201). This means that 'We can think of networks being effective in various stages: (1) by framing debates and getting issues on the agenda, (2) by encouraging discursive commitments from state and other policy actors, (3) by causing procedural change at the international and domestic level, (4) by affecting policy; and (5) by influencing behaviour changes in target actors' (1998, 201). Looking more closely at Vía Campesina's campaign based on these five dimensions is better done while also keeping in mind the precaution offered by Tarrow: 'advocates of transnational activist networks have highlighted many successful instances of successful intervention on behalf of actors too weak to advance their own claims. In an internationalized world, we are likely to see more of such intervention, so it is important to look at it without illusions. Transnational intervention fails more often than it succeeds' (2005, 200).

Framing Debates and Getting Issues on the Agenda

The GCAR's impact has been significant with regard to (re)framing debates and getting issues on the agenda. The GCAR resorted to simplified framing of the campaign: unidimensional economic perspective versus multidimensional functions of land, land as commodity versus land as a common community resource, voluntary land sales versus expropriation-based land reform, claims of MLAR's success versus counter-claims of failures, and so on. Vía Campesina raised these issues in the context of the global debates, at the same time that its network members actually mobilized on the community level, making its advocacy well-grounded, empirically informed, and thus powerful. Today, the issues raised by Vía Campesina have become key themes in the global debates on land policies. One example of this impact could be seen in the International Conference on Agrarian Reform and Rural Development (ICARRD) – its content, and its process before, during and after the March 2006 event (see http://www.icarrd.org). It was a bold move for the FAO, or a section within FAO, to mainstream discussion of land reform at a time when most international agencies did not want to even use the same phrase in their discourse; even bolder was its decision to let Vía Campesina have an important role in it. Further evidence can be seen in the Farmers' Forum process at IFAD, where a relatively progressive 'land reform' framework has been mainstreamed in the official discourse.

Encouraging Discursive Commitments from State and Other Policy Actors

A low degree of impact can be seen in this area. Aside from the official commitments from some groups within FAO and IFAD, the campaign was not able to solicit favourable commitments even at the level of promises from other agencies. But this is partly explained by the fact that GCAR has not really engaged with many institutions. Vía Campesina's campaign has not engaged, collaboratively or confrontationally, with bilateral agencies. This has important implications because bilateral agencies have more funds, and the significance of this can be seen from three interlinked dimensions. First, these agencies have their own land policies and directly carry them out in local and national settings. There is a plurality of international agencies engaged in land policies, not just the World Bank, FAO and IFAD (Palmer 2007). Second, these bilateral agencies often provide funds to multilateral agencies, and so are influential actors in the latter. Third, many of the bilateral agencies also have co-financing schemes with non-governmental funding agencies that provide money to (trans)national agrarian movements. Having defaulted from any significant engagement with these agencies can thus, arguably, partly account for the low degree of achievement of the campaign in terms of soliciting commitments from these actors. Finally, at the local–national level, the impact is even more marginal in terms of getting official commitments from state actors, e.g., the Lula administration has even expanded MLAR in Brazil, while the South African government has stuck it out with a hybrid MLAR.

Causing Procedural Change at the International and Domestic Level

A low degree of impact can be detected in this area. Some change could have been achieved if the Brazilian movements' demand for a World Bank Inspection Panel on MLAR had gone further than just the filing. Greater degrees of transparency, participation and accountability in the agencies' policy-framing processes have also been demanded, particularly from the World Bank and the European Union (FIAN-Vía Campesina 2004), but to no avail. Nonetheless, the campaign was able to push for some procedural changes related to IFAD and FAO processes, especially in terms of expanding 'invited' political spaces which could be occupied by Vía Campesina members and allies. For example, IFAD's interface mechanism with civil society used to be dominated by IFAP and NGOs through non-conflictive, generally de-politicized 'partnership' mechanisms.[20] With the global campaign, Vía Campesina's entry into this space has ended the monopoly of these politically

---

[20] For a more general discussion on this issue, refer to Harriss (2002).

conservative groups – and politicized the process of interaction.[21] In Brazil, we can also call it a 'procedural change', i.e. the latest MLAR version (only those lands that are not subject to expropriation would be qualified for MLAR). But the effect has been negative for Vía Campesina in this country: the procedural change became the reason for MLAR expansion and for CONTAG to endorse and participate in the MLAR (see, e.g., Vianna 2003), breaking from the previous unity based on MLAR opposition within the National Forum for Agrarian Reform. One effect of this was the demobilization of the forum leading to its current state of affairs, described by one observer as in a 'momentum of rapid fragmentation' (Sauer 2007).

Affecting Policy

It is in the area of policy reform where the impact of Vía Campesina's campaign has been the most marginal so far. The campaign has not yet been able to force a policy shift among the agencies promoting MLAR and other neoliberal land policies, and it certainly has not been able to effect the adoption of their alternative propositions in substance, either at the international or national levels. The World Bank is on the defensive, but it has not (yet) dropped its MLAR. For its part, the World Bank's hopes are pinned on producing a successful case in Brazil, and it continues to argue that current problems with MLAR in some countries are merely operational and administrative. Meanwhile, it remains a challenge for Vía Campesina to relate with FAO and IFAD because of underlying tensions within these institutions between broadly anti- and pro-redistributive reform forces, which can produce erratic, contradictory positioning over time, including on land issues. Though important, the ICARRD is just one of several global venues that matter for global land policy political dynamics, and a relatively weaker one compared to those controlled by the bigger funders/players: the World Bank, EU and bilateral agencies. But the ICARRD was a major political achievement for Vía Campesina. Whether the momentum can be sustained and translated into policy reforms remains to be seen. Unfortunately, the apparent weakening of allies within FAO post-ICARRD (due to funding cuts and internal reorganization) does not bode well. At IFAD, Vía Campesina's allies are located mainly in the Policy Division, a relatively weak division politically, mainly because they do not control the funds and do not directly interface with country partners. For its part, the more powerful operations division of IFAD, where the main fund is directly handled, still lacks a coherent position on land reform, and has maintained broadly pro-market tendencies.[22]

---

[21] In this context, there are interesting dynamics between movements involving Vía Campesina, but these cannot be treated fully here. For a useful general discussion on this theme related to NGOs, see Jordan and van Tuijl (2000).

[22] For example, Kay (2006, 491) found out that in Latin America and the Caribbean over the years IFAD has generally followed the market-oriented land policies promoted by the World Bank. Hence, while Vía Campesina is able to gain ground at the global level with its alliance with IFAD's key policy experts, and so on, it may lose some ground at the local and national settings if the land policies that are carried out by IFAD country programmes are contradictory to the Vía Campesina vision.

Influencing Behaviour Changes in Target Actors

While the campaign has contributed to behaviour changes on target institutions, these have not necessarily favoured the GCAR. For example, the World Bank took the substantive and procedural issues raised by Vía Campesina relatively seriously. But in the countries where MLAR is underway, the changes have not necessarily been positive. In the Philippines, for example, the MLAR agenda was simply repackaged and resold to a new national government (see Borras et al. 2007b). In Brazil, the World Bank fine-tuned the framework and implementation guidelines, won over the Lula administration, and recruited CONTAG to its project. When issues of transparency and accountability are raised, the World Bank is a master at recruiting other friendly civil society groups to participate in the MLAR process by way of two global electronic consultations and the regional consultations which it can later point to as proof of having promoted 'participation' by 'civil society'. A problematic process protested by GCAR due to its being 'non-transparent' and 'manipulative' (FIAN-Vía Campesina 2004). Predictably, the positive behaviour changes that have occurred have been limited to some sections within the FAO and IFAD.

Finally, there is one especially urgent area where, unfortunately, the GCAR does not seem to have been able to effect positive behaviour changes so far, and that is in the area of rural violence and human rights violations against peasant land rights claimants. Campaigns to stop rural violence have been carried out worldwide as part of GCAR, but so far have too little effect in terms of ending the violence (for background discussions see, e.g., Vía Campesina 2006; Franco 2007; De Carvalho Filho and Mendonça 2007).

CONTENTIOUS ISSUES AND FUTURE TRAJECTORIES

Keck and Sikkink remind us that 'Transnational advocacy networks must also be understood as political spaces, in which differently situated actors negotiate – formally or informally – the social, cultural, and political meanings of their joint enterprise'. They continue, 'Power is exercised within networks, and power often follows from resources . . . Stronger actors in the network do often drown out the weaker ones, but because of the nature of the network form of organization, many actors . . . are transformed through their participation in the network' (1998, 3; see also Della Porta et al. 2006, 20). Such a perspective is complemented by the explanation put forth by Tarrow that 'Transnational activists are often divided between the global framing of transnational movement campaigns and the local needs of those whose claims they want to represent' (2005, 76). He argues that 'Global framing can dignify and generalize claims that might otherwise remain narrow and parochial. It signals to overworked and isolated activists that there are people beyond the horizon who share their grievances and support their causes.' He cautions, however, that 'by turning attention to distant targets, it holds the danger of detaching activism from the real-life needs of the people they want to represent' (ibid.; see also Bob 2005, 195). Tarrow concludes that 'transnational activism will be episodic and contradictory, and it will have its most visible impact on domestic politics' (2005, 219).

These are reminders that global–national/local links, representation and accountability are not unproblematic, despite what some TAM activists would claim. Of course, the everyday politics of movement-building, if anything, are about finding strategic unities amidst diverse experiences. But although understandable, the tendency to emphasize '*unity* in diversity' (while downplaying diversity) at times risks ignoring latent tensions that warrant attention.

There are three dimensions in particular that Vía Campesina activists might consider looking into further: class differences, ideological differences and the network's growing but still limited representation of the plural interests and identities of the rural poor. To a large extent, the hemispheric South–North divide is being sufficiently addressed by Vía Campesina in its discourse about transcending potential and actual differences through cross-border solidarity (see, e.g., Bové 2001, 96; Stedile 2002, 99–100; Desmarais 2007), while even gender differences are being addressed, albeit slowly, by agrarian movements both at the national (in the case of Brazil, see Deere 2003) and international levels (for background discussions see Vía Campesina-FIAN 2003; Monsalve 2006; Razavi 2003). Here in particular, at the international level, Vía Campesina indeed should be seen as a good example for its establishment of parity representation between men and women in its most powerful decision-making body, the International Coordinating Committee (ICC).

Class Differences

Vía Campesina is heterogeneous in terms of its base. A rough estimate of the class profile of Vía Campesina reveals the following: (i) landless peasants, tenant-farmers, sharecroppers and rural workers mainly in Latin America and Asia; (ii) small and part-time farmers located in (Western) Europe, North America, Japan and South Korea; (iii) family farms in the global South, including those in Africa as well as those created through successful partial land reforms, such as those in Brazil and Mexico; (iv) middle to rich farmers mainly, but not solely, in India; and (v) semi-proletariat located in urban and peri-urban communities in a few countries such as Brazil and South Africa.[23] The most numerous, most vibrant and politically influential groups within Vía Campesina are the Latin American block, the (Western) European group and a few Asian movements. This influence is partly reflected by, or has resulted in, a global leadership power distribution that tends to reinforce the American-European influence. Half of the membership of the ICC, an 18-person body (as of 2007), comes from Latin America and the Caribbean.[24]

---

[23] For this group, refer for example to the explanation made by Stedile (2002).
[24] The past couple of decades have witnessed the resurgence of rural social movements in Latin America (see, e.g., Veltmeyer 1997), including indigenous people's movements (Yashar 2005) in a scale and degree of political radicalization seen only in a few places in Asia and Africa during the same period. This largely accounts for the natural dominance of the Latin American contingent in the global leadership body of Vía Campesina.

The organization's African membership is growing, but still relatively small and highly heterogeneous in itself, ranging from the mainly peri-urban landless people in South Africa to small-scale farmers in Mozambique. The most consolidated organization in the region with an organizational and political orientation closest to its American-European and Asian counterparts is UNAC.[25] There are two vibrant members in West Africa, namely CNCR[26] in Senegal and CNOP[27] in Mali (plus CPM[28] in Madagascar). However, unlike nearly all other Vía Campesina members, CNCR (and CPM) simultaneously maintains its membership with the Vía Campesina rival, IFAP.

The movements from Latin American and some Asian countries are the most vocal groups within Vía Campesina in the GCAR. In Latin America, among the most recognized voices are those of the MST in Brazil and COCOCH in Honduras. In Asia, movements from the Philippines and Indonesia (especially when the global secretariat of Vía Campesina was moved to Indonesia in 2004; see Peluso et al. this collection), and recently some groups from South Asia, while important in their own right are not (yet) as cohesive or powerful as the solid Latin American block, perhaps for a combination of reasons, including significant linguistic diversity and ideological differences. Nonetheless, together, the Latin American and Asian landless peasant and rural workers' movements (plus perhaps the LPM before it contracted), were the main force behind the push for Vía Campesina to carry out land reform as a strategic global campaign.

The combined force of these groups was so influential that it prevailed even when another powerful group within Vía Campesina – i.e. KRRS – initially objected to land reform being a major campaign. India's KRRS, whose main mass base is middle and rich farmers, was decisively overruled on this matter. It is relevant to elaborate on KRRS. This group has been engaged since the 1980s on anti-TNC and later on anti-GM crops campaigns. Many of these campaigns have been dramatic in form and so got the media spotlight (Scoones this collection; Herring 2007). This campaign connects well with Northern advocacy against GM crops. As such, KRRS has become a critical actor in Vía Campesina's global anti-GM crops and anti-TNC campaigns. It has become extremely influential within the global movement and in turn earned the role of being the 'gatekeeper' in South Asia. But KRRS consciously evades issues that could bring sharper class issues. M. D. Nanjundaswamy, the leader of KRRS (who died in early 2004) explained earlier that: 'we cannot divide ourselves into landlords and landless farmers, and agitate separately, for the agitation will have no strength nor will it carry any weight' (Assadi 1994, 215). It is not surprising therefore that 'the KRRS opposes legislative ceilings on rural land while simultaneously advocating limits to the ownership of urban industrial property' (Assadi 1994, 213). Moreover, writing more than ten years ago, Assadi explained that 'both the [Maharashtra-based

---

[25] União Nacional de Camponeses or National Peasants' Union.
[26] Conseil National de Concertation et de Coopération des Ruraux.
[27] Coordination National de Organisations Paysannes (CNOP).
[28] Coalition Paysanne de Madagaskar.

Shetkari Sanghatana and the Karnataka-based KRRS] have not only not condemned atrocities against tribals and the segregation of Dalits but in some instances the perpetrators of such actions have themselves been their own members' (1994, 215).

What the KRRS case reveals is that serious class-based differences exist within and between movements that are (un-)affiliated with Vía Campesina. It would be equally relevant to use a class analytic lens to examine the various Bharatiya Kisan Union state organizations affiliated (or not) with Vía Campesina, and for this it would be useful to consult the various studies in an earlier volume edited by Brass (1994). These class-based differences have profound implications for the way campaign demands are framed and representation is constructed within a movement. In the case of KRRS, a significant proportion of the organized section of the rural-based exploited social classes not only in India but in South Asia more generally were excluded from the Vía Campesina process, either because KRRS blocked their entry into Vía Campesina or they refused to participate in the process where the 'gatekeeper' was KRRS.[29] Some of these organizations were able to gain entry into Vía Campesina much later. When Nanjundaswamy died in early 2004, 16 organizations from South Asia joined or were allowed entry into Vía Campesina a few months after. To date, a significant number of organizations of the landless rural poor in India have remained outside Vía Campesina, partly due to the continuing influence of KRRS and partly due to the political and ideological complications that emerged and developed in the late 1990s.[30]

Yet the above situation is not the first and is not unique in the history of Vía Campesina. The first serious fall-out in Vía Campesina was Nicaragua's UNAG.[31] UNAG was one of the key founders of Vía Campesina, a convenor of one of the original pillars of Vía Campesina, i.e. the Central American ASOCODE, and was host to the global solidarity conference in 1992 in Managua from which the most concrete idea of building Vía Campesina took shape. UNAG has also been a member of IFAP, reflecting a closer affinity to fellow middle to rich farmers' network and to issues more concerned about government support services, production and trade issues, and credit facility via bilateral and multilateral donor agencies. This is in contrast, for example, to the concerns of another Nicaraguan founding organization, the farmworkers' association, ATC[32] with landless people's issues and demands such as wages and land. When the conflict erupted between the then emerging leaders of Vía Campesina and the facilitating Dutch NGO (PFS) over the nature and orientation of Vía Campesina (this NGO was advocating, among others, that Vía Campesina members should instead

---

[29] Information about this is based on numerous conversations of the author with key movement leaders within Vía Campesina and from various groups in India over the years.
[30] This problem is captured in the issue raised by a close Vía Campesina ally, who said that 'In India, a higher caste of farmers joined Vía Campesina, and now the lower castes are kept out of Vía Campesina. How to fix this?' (anonymous close ally interviewed in Rosset with Martinez 2005, 37).
[31] Unión Nacional de Agricultores y Ganaderos.
[32] Asociacion de Trabajadores del Campo.

just join IFAP), UNAG sided with the Dutch NGO, chose to leave Vía Campesina, and remained in IFAP. While the incident appears to have been the usual 'turf-related' intra-movement political conflict, a closer look reveals a deeper class-based fault-line.

In short, taking a closer look at Vía Campesina, we see class-based differences within and between national movements.[33] Class-based differentiation of groups within Vía Campesina partly validates the official claim by movement leaders that their problems and oppressors are the same, but at the same time it demonstrates that this assertion is only partly correct: rich farmers could be the oppressors of farmworkers; land reform is an issue to be resisted by rich farmers, high price for food products is a good policy for food surplus-producing farmers, bad news for food-deficit rural households, credit facilities and trade issues may not be a critical issue for landless subsistence rural workers who do not have significant farm surplus to sell anyway, wages are not favoured issues by middle and rich farmers but a fundamental issue to rural workers, and so on. Indeed, they are all 'people of the land', yet they have competing class-based interests. Acknowledging such differences, rather than ignoring or dismissing their significance, is an important step toward finding ways to ensure truly inclusive and effective representation in decision-making and demand-making.

Ideological and Political Differences

Another source of tension is political ideology, which tends to be less talked about by TAM activists. Vía Campesina is a broad coalition of groups with diverse ideological orientations, including (i) varying strands of radical agrarian populists, (ii) various types of Marxists, (iii) radical groups with anarchist tradition, (iv) radical environmentalists and (v) feminist activists. Many groups and individuals fall somewhere in between these broad categories, while others have overlapping orientations, e.g., populist-feminist, and so on. Still others do not have any clear ideological provenance at all, or do not have well-developed ideological positions. The degree of ideological differences varies from one case to another. This diversity in ideological orientation is found not only *between* movements – compare, for example, Bangladesh's orthodox Marxist group BKF[34] with the unorthodox radical group SOC[35] of Andalucía, Spain. Diversity is found also *within* movements, especially the larger ones, such as MST of Brazil and CNCR of Senegal. However, the global leadership is currently dominated by a coalition of all these significant currents, with a radical agrarian populist tendency being the dominant current. It is important to note, however, that the overwhelming majority of the national movements within Vía Campesina are part of the wave

---

[33] Some movements, especially large ones, are also class-differentiated. Although there is no specific reference to CNCR, the work by Oya (2007) on class-differentiated rural accumulation processes among farmers in Senegal indirectly suggests that CNCR has a highly differentiated base, with the leadership influenced by the more affluent ones (also based on a personal discussion with Oya).
[34] Bangladesh Krishok Federation.
[35] Sindicato Obrero del Campo.

of social movements that have broken free from paternalistic political party sponsorship and control.

Further illustration from the Philippines is relevant. Here, three movements are connected to Vía Campesina, but in varying ways, raising the question of how this may influence the terms of externalization of land issues into the global campaign of Vía Campesina. All of these groups have a mass base, or at least formal claims of a mass base, among poor peasants. The first is KMP, a Maoist-inspired legal peasant organization whose ideological position on land reform follows a more orthodox Marxist position, campaigning for the nationalization of land, advocating for state farms, although allowing for a transitional individual ownership (see, e.g., Putzel 1992, 1995; Lara and Morales 1990). KMP's call for 'genuine agrarian reform' means land confiscation without compensation to large landlords and free land distribution to peasants (KMP 1986). KMP rejects the state land reform law (Comprehensive Agrarian Reform Program or CARP) as 'pro-landlord and anti-peasant', a 'fake land reform'. KMP was one of the founding organizations of Vía Campesina and represented Asia in the ICC during the latter's formative years. KMP's campaign is to thrash CARP; it employs a mainly agit-prop method. The second is DKMP,[36] a group that broke away from KMP in 1993 due to ideological differences. The DKMP took a more radical agrarian populist position in terms of land reform, advocating the cause of small family farms. However, largely because of personality differences among its leaders, DKMP failed to rally up and consolidate its forces. By the second half of the 1990s, DKMP had shrunk to a handful of peasant leaders and pockets of rice farmers in Central Luzon. With a few land reform cases and modest support from a few NGOs, DKMP has been able to maintain a relatively weak presence. Partly because it has weakened over time, DKMP continues to navigate within the parameters of the state land reform law, but uses less mass movement and more contacts within NGOs and government offices to facilitate favourable decisions for its few land claims (Borras 2007).

Both KMP and DKMP remain Vía Campesina members, although in recent years, and partly due to ideological reasons, KMP has fallen from grace within Vía Campesina (this will be discussed later). As a result, one finds an ironic situation where one member organization with a relatively significant mass base (KMP) has been marginalized within Vía Campesina, while another member organization without any significant mass base (DKMP) has been mainstreamed within the global movement. The third group is UNORKA.[37] A very large chunk of the peasant movement that broke away from the Maoist-inspired movement in the early 1990s did not find it conducive to rally under the banner of DKMP. Instead, they eventually regrouped under a new umbrella organization, UNORKA. Formalized only in 2000, UNORKA quickly became the largest group directly engaged in land reform in the Philippines, and it remains so today with its roots in nearly 800 agrarian disputes across the country (Borras 2007). Its mass base is

---

[36] Democratic KMP.
[37] National Coordination of Autonomous Local Rural People's Organizations.

mainly among the landless peasants and rural workers, and like the MST in Brazil, UNORKA is using the state land reform law as an institutional context for their campaigns, navigating within the parameters of the law by stretching its limits as well as by employing militant but pragmatic mass mobilization strategy (Franco forthcoming; for Brazil see Meszaros 2000). UNORKA is eclectic in terms of ideological position on land: while taking a generally agrarian populist stance, it also has a significant base among rural workers, so its advocacy is not oriented exclusively towards small family-farm creation (Borras and Franco 2005; De la Rosa 2005). UNORKA wants to join Vía Campesina, but KMP objects and because of an organizational rule that essentially allows existing members to reject any applicant from its own country, to date UNORKA's formal entry into the network remains blocked. Recently, however, despite objections from KMP, Vía Campesina has begun inviting UNORKA to some gatherings as an observer. In the global campaign, FIAN works closely with UNORKA, while LRAN (through the Focus on the Global South) works with UNORKA and DKMP.

What the Philippine case shows is that even when there are no significant class-based differences between movements, ideological differences make for important cleavages between them (recall Landsberger and Hewitt 1970). Ironically, there is less commonality between KMP (or DKMP) and MST regarding land reform strategy, and more commonality between MST and UNORKA; and yet KMP and DKMP are in, while UNORKA is out of Vía Campesina.

The Philippines is not the only case to highlight serious ideological cleavages within and between movements. Coming back to South Asia, we can see how the KRRS issue was used by other left-wing peasant groups in India to partly justify the formation of a separate, competing movement in the region, the Asian Peasant Coalition (APC). Its current secretariat is hosted by KMP. The strength of the APC network lies in its class line in terms of organizing poor peasants and rural workers; as a result, their main base is to be found among the most destitute strata of the peasantry, and thus this network has the potential to sharpen the class analysis and related demands of Vía Campesina, as well as to expand Vía Campesina's representation in the region. The APC network could well have strengthened greatly the land reform campaign in Asia – if not, that is, for an ideological and political stance that tends to be extremely exclusivist and sectarian. From there, the relationship between APC and Vía Campesina has taken a downward spiral.[38]

Other differences are not exactly very ideological in nature, but more political. The tension between Vía Campesina members in Mexico is a good example, where UNORCA[39] seems to have emerged to become the 'gatekeeper' in the country, relegating other important movements (e.g., ANEC[40] and CNPA[41]) to

---

[38] Data and information on this are based on the author's series of informal discussions with Vía Campesina leaders and other movement activists in Asia over the years.
[39] Union Nacional de Organizaciones Regionales Campesinas Autonomas.
[40] Asociacion Nacional de Empresas Comercializadoras de Productores del Campo.
[41] Coordinadora Nacional Plan de Ayala.

the margins despite the latter's objections. This fault-line is partly rooted in differences in political strategies, e.g., in relating with state programmes. A related example is the difference between Brazil's MST and Senegal's CNCR in terms of relating with the state and international development institutions: MST takes a far more autonomous stance from and conflictual relationship with these institutions, including taking an anti-World Bank position, while CNCR includes several government-sponsored organizations and opts to combine negotiation and intermittent confrontation with these institutions, including collaborative engagement with the World Bank (McKeon et al. 2004). It is a similar contrast we get looking at Vía Campesina-global and CNCR on these issues. Underpinning such differences, of course, are particularities embedded in the social and political histories of the different countries from which Vía Campesina members hail. The emerging tension in Southern Africa partly reflects such differences as well, though of a different kind. The point is that ideological and political differences are significant, and cannot be taken for granted.

Political Representation

In a recent interview with the *Socialist Register*, MST's Stedile talked about national movements and cross-national alliances. In part, he said: 'We projected a shadow much bigger than what we really were, and we became famous for that. In fact, the MST as an organized force of the workers in Brazil is very small' (2007, 195). Furthermore, he said that 'African movements have a very low level of organization and are extremely poor, and many are still located at the tribal and local level. Few countries have a national movement' (2007, 214).

Such candour about MST's still limited representation of the landless in Brazil, as well as about the comparative strength of movements in the different regions, especially Africa, helps to situate Vía Campesina and the GCAR more realistically (see also the contribution by Wolford et al. in this collection). Despite Vía Campesina's dramatic and impressive rise as a major international agrarian movement, the extent of its representation still remains fairly limited when seen from a global perspective. Even in the national bailiwicks of leading Vía Campesina members, its organized base remains limited, as Stedile's comments about the MST in Brazil suggest. It is unlikely to be any better in other countries where Vía Campesina is present, notwithstanding the claims of its member organizations, which can sometimes be overblown. Meanwhile, the south of Asia is a vast sub-region, and to date the main stronghold of Vía Campesina is Southeast Asia (plus South Korea), which, although significant, does not constitute an organized majority there. For its part, South Asia is host to numerous militant movements of the rural working classes engaged in class-based struggles, but many of these movements are not formally integrated within Vía Campesina, while the leading Vía Campesina organization (KRRS) is a middle-rich farmers' movement that evades discussion of social classes, class exploitation and class struggles, and stands against land reform. No significant national agrarian movements today in either Central Asia or the Middle East come close to Vía Campesina's political orientation or have formal links to it. Meanwhile, very large countries with large rural working classes such as China and the former USSR are also out of the orbit of Vía Campesina, so that some types of peasant resistance are likely to be missed in Vía Campesina's discourse (see, e.g., O'Brien and Li 2006; Kerkvliet 2005; see also Le Mons Walker as well as Malseed in this collection). Finally, in the vast African region, there are only a very few (five) members of Vía Campesina, and these movements' representation of the rural working classes in their countries is even more limited than that of MST in Brazil. In short, though more significant than any other transnational agrarian movement, Vía Campesina directly represents only a small fraction of the global rural working classes (at least for now).

Bringing in the diversity of land issues not only in countries where Vía Campesina members are present, but also in the many countries where they are still absent (including the vast ex-socialist countries 'in transition' – see, e.g., Spoor forthcoming), will certainly complicate the current 'global issue-framing' and demand-making processes and dynamics. Meanwhile, the global land reform debate itself has not remained static, but is also evolving, complicating Vía Campesina's position even further, since one of the major policy battles around contemporary land policy issues is being, and will be, fought around the issue of 'formalization' of land rights in, and 'privatization' of, remaining public lands (see, e.g., De Soto 2000; Cousins 2007). It is not the classic land reform issue, but it affects perhaps even more segments of the rural population.[42] Most of the affected settings for such a campaign are precisely the regions where Vía Campesina's presence is very thin if not totally absent, such as Africa. The threat from formalization/privatization initiatives and its relations to the GCAR campaign-framing is captured in what the Mozambican leader Diamantino Nhampossa from UNAC said: 'We already had a thorough agrarian reform. In order for [GCAR] to help us, it must focus on the challenge we are facing – "counter-agrarian reform" under neoliberalism. If the [GCAR] keeps focusing on just being "against latifundio" [private large estates], then it is less relevant to us.' Nhampossa added, 'The World Bank is promoting a new wave of land privatization [in Mozambique], and that needs to be denounced. We think the [GCAR] needs to broaden its mandate, it needs to also be a campaign "in defense of land" . . . against privatization of land.'[43]

While GCAR has started to resort to 'global issue-framing' around this particular issue, in effect, there are no significant local/national campaigns to be

---

[42] For example, up to 90 per cent of Indonesia's agricultural lands are officially considered 'state forest land' (see, e.g., Peluso 1992).
[43] Interview in Rosset with Martinez (2005, 21–22). Vía Campesina has actually formally launched GCAR in Africa in January 2007 during the World Social Forum in Nairobi, essentially calling for land restitution and land redistribution, targeting the white commercial farms in Africa. But in countries where such an issue is still relatively 'hot' (e.g., South Africa, Zimbabwe, Namibia), Vía Campesina is nearly absent. An insider within Vía Campesina interviewed by the author admitted that the Nairobi declaration is more of an 'agit-prop' political statement than a launch of a real campaign. For excellent background discussions about contemporary land issues in Africa, see Peters (2004) and Berry (2002).

'externalized'. This explains why the GCAR has made significant inroads in the conventional land reform issues, but to date has not (yet) gained any significant ground around the campaigns on anti-formalization and anti-privatization of public lands. But if 'externalization' is indeed key to building coherent and durable transnational networks (as Tarrow argues), then Vía Campesina ought to expand its presence into these areas first. In a sense, Vía Campesina now faces a dilemma: should it put its time and resources into expanding its presence beyond its current limits and into areas targeted by the neoliberal 'formalization' agenda, or should it continue to oppose this agenda without a significant local base to back it up? In the end, whether and how GCAR would be able to reposition itself and make a significant impact on current and future broader land policy issues will most likely depend on making the two political processes, i.e. combining global issue-framing from above and initiatives to launch local/national campaigns from below towards externalization, mutually reinforcing – a 'sandwich strategy'. It is a challenge that may require Vía Campesina to rethink and recast some of its organizational rules, alliance-building, ideas about forms of resistance and collective actions, and perspectives on land, among others. For instance, it may not be politically prudent and creative to insist on purely 'national peasant movements' as an organizational requirement in linking up with groups in many settings where such movements are unlikely to emerge anytime soon. Some transitional measures are likely to be helpful, such as forging alliances and understanding with groups including local progressive NGOs. Waiting for national peasant movements – in the image and likeness of Vía Campesina's 'ideal' national members – to emerge in these settings is likely to take a very long time, if it will ever come at all.[44] What Nico Verhagen, Vía Campesina's senior staff, said in the context of Africa, if implemented, can perhaps open up some initial paths: 'Right now many organizations in Africa do not have very clear political positions, but that can change. For example, in the case of ROPPA,[45] the more they interact with the Vía Campesina the more they are radicalizing . . . Our strategy in Africa should be to open up spaces for dialogue with Vía Campesina, and invite everyone in.'[46]

CONCLUDING REMARKS

The GCAR launched and led by Vía Campesina and coordinated with its allies has gained significant ground transnationally in terms of putting the issue of land reform and its opposition to MLAR onto the official agendas of development agencies and civil society. It has reshaped the terms of the current policy and political debates. It has been able to construct an alternative land policy vision. However, it has not gained any significant ground in terms of actual favourable

---

[44] In this context, the literature of everyday forms of peasant resistance can offer some relevant insights, see, e.g., Scott (1985), Kerkvliet (2005) and O'Brien and Li (2006). Refer also to Chavez and Franco (2007) for a relevant discussion about civil society in Africa.
[45] Réseau des Organisations Paysannes et de Producteurs de l'Afrique de l'Ouest.
[46] Interview in Rosset with Martinez (2005, 30).

policy reforms, internationally and in national settings, nor has it resulted in significant procedural changes or caused favourable behavioural changes among key actors in development institutions. Part of the reason why GCAR has been mounted and sustained was primarily due to successful processes of 'externalization' of local/national issues and campaigns. More specifically, the campaign which was mainly an anti-MLAR campaign, got mainstreamed quickly within Vía Campesina because MLAR was carried out in countries where influential members of Vía Campesina are situated and affected by MLAR.

But the GCAR has also revealed latent cleavages and fault-lines within Vía Campesina that are class-based and ideological. Confronting, not backing away from, such issues may contribute towards further ideological, political and organizational consolidation within Vía Campesina. What Paul Nicholson, a farmer from the Basque (Spain) and a key Vía Campesina leader, explained in 2004 about the movement's principles in alliance-building provides a fundamental starting point:[47] 'The alliance between farmers, men and women peasants, with rural workers, is a fundamental alliance in the rural world.' Taking seriously what Nicholson said is to recognize class issues, among others, within the movement.

Moreover, the GCAR has also exposed important weaknesses of Vía Campesina in terms of its current spread worldwide, and so its actual capacity to represent diverse interests of various groups in different settings. An understanding of these issues will help clarify why there has been no similar significant impact made by Vía Campesina in opposition to the land rights formalization and land privatization policies targeted towards public lands which are being carried out to a large extent in national–local settings where Vía Campesina's presence is thin if not totally absent. For Vía Campesina to be able to reposition its leadership in land issues worldwide, i.e. to deepen and widen the scope of its campaign, the option is not an either/or choice between 'global issue framing from above and then diffuse this nationally/locally' or 'local/national campaigns from below then externalize this onto international level'. Rather, perhaps the most promising option would be to adopt a 'sandwich strategy' to simultaneously push for these two processes from above and from below.

ACRONYMS USED

ANEC (Asociación Nacional de Empresas Comercializadoras de Productores del Campo)
ASOCODE (Association of Central American Peasant Organizations for Cooperation and Development)
ATC (Asociación de Trabajadores del Campo)
BKF (Bangladesh Krishok Federation)
CNCR (Conseil National de Concertation et de Coopération des Ruraux)

---

[47] This was during a press briefing in São Paolo on 11 June 2004 on the occasion of the IV global assembly of Vía Campesina held in Brazil. http://www.viacampesina.org; Accessed 30 October 2007.

CNOP (Coordination National de Organisations Paysannes)
CNPA (Coordinadora Nacional Plan de Ayala)
COCOCH (Honduran Coordinating Council of Peasant Organizations)
CONTAG (Confederação Nacional dos Trabalhadores na Agricultura)
CPM (Coalition Paysanne de Madagaskar)
DKMP (Demokratikong Kilusang Magbubukid ng Pilipinas)
EU (European Union)
FAO (Food and Agriculture Organization of the United Nations)
FETRAF (Federação dos Trabalhadores na Agricultura Familiar)
FIAN (Foodfirst Information and Action Network)
GCAR (global campaign for agrarian reform)
ICARRD (International Conference on Agrarian Reform and Rural Development)
ICC (International Coordinating Committee – of Vía Campesina)
IFAD (International Fund for Agricultural Development)
IFAP (International Federation of Agricultural Producers)
IPC (International Planning Committee)
KMP (Kilusang Magbubukid ng Pilipinas)
KRRS (Karnataka State Farmers' Association)
LPM (Landless People's Movement)
LRAN (Land Research and Action Network)
MLAR (market-led agrarian reform)
MST (Movimento dos Trabalhadores Rurais sem Terra)
NLC (National Land Committee)
PCT (Projeto Cedula da Terra)
ROPPA (Réseau des Organisations Paysannes et de Producteurs de l'Afrique de l'Ouest)
SOC (Sindicato Obrero del Campo)
TAM (transnational agrarian movement)
UNAC (União Nacional de Camponeses or National Peasants' Union)
UNAG (Unión Nacional de Agricultores y Ganaderos)
UNORCA (Unión Nacional de Organizaciones Regionales Campesinas Autónomas)
UNORKA (National Coordination of Autonomous Local Rural People's Organizations)

REFERENCES

Akram-Lodhi, H., S. Borras and C. Kay, eds, 2007. *Land, Poverty and Livelihoods in an Era of Globalization*. London: Routledge.

Appadurai, A., 2006. 'Foreword'. In *Transnational Civil Society: An Introduction*, eds S. Batliwala and D.L. Brown, xi–xv. Bloomfield: Kumarian Press.

Assadi, M., 1994. '"Khadi Curtain," "Weak Capitalism," and "Operation Ryot": Some Ambiguities in Farmers' Discourse, Karnataka and Maharashtra, 1980–93'. In *New Farmers' Movement in India*, ed. T. Brass, 212–27. London: Frank Cass.

Baranyi, S., C.D. Deere and M. Morales, 2004. *Scoping Study on Land Policy Research in Latin America*. Ottawa: The North-South Institute; International Development Research Centre.

Batliwala, S., 2002. 'Grassroots Movements as Transnational Actors: Implications for Global Civil Society'. *Voluntas: International Journal of Voluntary and Nonprofit Organizations*, 13 (4): 393–409.

Batliwala, S. and D.L. Brown, eds, 2006. *Transnational Civil Society: An Introduction*. Bloomfield, CT: Kumarian Press.

Bello, W., 2002. *Deglobalization: Ideas for a New World Economy*. London: Zed.

Bernstein, H., 2002. 'Land Reform: Taking a Long(er) View'. *Journal of Agrarian Change*, 2 (4): 433–63.

Bernstein, H., 2006. 'Once Were/Still Are Peasants? Farming in a Globalising "South"'. *New Political Economy*, 11 (3): 399–406.

Berry, S., 2002. 'Debating the Land Question in Africa'. *Comparative Study in Society and History*, 44 (4): 638–68.

Biekart, K. and M. Jelsma, eds, 1994. *Peasants Beyond Protest in Central America*. Amsterdam: Transnational Institute.

Bob, C., 2005. *The Marketing of Rebellion: Insurgents, Media and International Activism*. Cambridge: Cambridge University Press.

Borras, S., 2003. 'Questioning Market-Led Agrarian Reform: Experiences from Brazil, Colombia and South Africa'. *Journal of Agrarian Change*, 3 (3): 367–94.

Borras, S., 2004. 'La Vía Campesina: An Evolving Transnational Social Movement', *TNI Briefing Series* No. 2004/6. Amsterdam: Transnational Institute.

Borras, S., 2007. *Pro-Poor Land Reform: A Critique*. Ottawa: University of Ottawa Press.

Borras, S. and J. Franco, 2005. 'Struggles for Land and Livelihood: Redistributive Reform in Agribusiness Plantations in the Philippines'. *Critical Asian Studies*, 37 (3): 331–61.

Borras, S., C. Kay and E. Lahiff, eds, 2007a. 'Market-Led Agrarian Reform: Contestations and Trajectories'. *Third World Quarterly*, special issue, 28 (8).

Borras, S., D. Carranza and J. Franco, 2007b. 'Anti-Poverty or Anti-Poor? The World Bank Experiment in Market-Led Agrarian Reform in the Philippines'. *Third World Quarterly*, 28 (8): 1557–76.

Bové, J., 2001. 'A Movement of Movements? A Farmers' International?'. *New Left Review*, (12): 89–101.

Brass, T., ed., 1994. *The New Farmers' Movements in India*. London: Frank Cass.

Broad, R., 2002. *Global Backlash: Citizen Initiatives for a Just World Economy*. Lanham, MD: Rowman and Littlefield.

Bush, R., 2007. 'Politics, Power and Poverty: Twenty Years of Agricultural Reform and Market Liberalization in Egypt'. *Third World Quarterly*, 28 (8): 1599–616.

Byres, T., 2004a. 'Introduction: Contextualizing and Interrogating the GKI case for Redistributive Land Reform'. *Journal of Agrarian Change*, 4 (1&2): 1–16.

Byres, T., 2004b. 'Neo-Classical Neo-Populism 25 Years On: Déjà vu and Déjà Passé: Towards a Critique'. *Journal of Agrarian Change*, 4 (1/2): 17–44.

Chavez, D. and J. Franco, 2007. *Grassroots Democratization in Sub-Saharan Africa*. Amsterdam: Transnational Institute; Utrecht: ICCO.

Chonchol, J., 1970. 'Eight Fundamental Conditions of Agrarian Reform in Latin America'. In *Agrarian Problems and Peasant Movements in Latin America*, ed. R. Stavenhagen. 'Garden City', New York: Doubleday.

Cousins, B., 2007. 'More Than Socially Embedded: The Distinctive Character of "Communal Tenure" Regimes in South Africa and its Implications for Land Policy'. *Journal of Agrarian Change*, 7 (3): 281–315.

De Bremond, A., 2007. 'The Politics of Peace and Resettlement through El Salvador's Land Transfer Programme: Caught between the State and the Market'. *Third World Quarterly*, 28 (8): 1537–56.

De Carvalho Filho, J.L. and M.L. Mendonça, 2007. 'Agrarian Policies and Rural Violence in Brazil'. *Peace Review*, 19 (1): 77–85.

Deere, C.D., 2003. 'Women's Land Rights and Rural Social Movements in the Brazilian Agrarian Reform'. *Journal of Agrarian Change*, 3 (1&2): 257–88.

De Feyter, K., 2005. *Human Rights: Social Justice in the Age of the Market*. London: Zed.

Della Porta, D., M. Andretta, L. Mosca and H. Reiter, 2006. *Globalization from Below: Transnational Activists and Protest Networks*. Minneapolis, MN: University of Minnesota Press.

De la Rosa, R., 2005. 'Agrarian Reform Movement in Commercial Plantations: The Experience of the Banana Sector in Davao del Norte'. In *On Just Grounds*, eds J. Franco and S. Borras, 45–144. Amsterdam: Transnational Institute.

Desmarais, A., 2007. *La Vía Campesina: Globalization and the Power of Peasants*. London: Pluto.

De Soto, H., 2000. *The Mystery of Capital: Why Capitalism Triumphs in the West and Fails Everywhere Else*. New York: Basic Books.

Edelman, M., 1998. 'Transnational Peasant Politics in Central America'. *Latin American Research Review*, 33 (3): 49–86.

Edelman, M., 2003. 'Transnational Peasant and Farmer Movements and Networks'. In *Global Civil Society 2003*, eds M. Kaldor, H. Anheier and M. Glasius, 185–220. Oxford: Oxford University Press.

Edwards, M. and J. Gaventa, 2001. *Global Citizen Action*. Boulder, CO: Lynne Rienner.

FIAN-Vía Campesina, 2004. 'Commentary on Land and Rural Development Policies of the World Bank'. http://www.viacampesina.org, Accessed 25 November 2007.

Fox, J., 1993. *The Politics of Food in Mexico: State Power and Social Mobilization*. Ithaca, NY: Cornell University Press.

Fox, J., 2001. 'Vertically Integrated Policy Monitoring: A Tool for Civil Society Policy Advocacy'. *Nonprofit and Voluntary Sector Quarterly*, 30 (3): 616–27.

Franco, J., 1999. 'Market-Assisted Land Reform in the Philippines: Round Two – Where Have All the Critics Gone?'. *Conjuncture*, 11 (2): 1–6. Manila: IPD.

Franco, J., 2007. 'Again, They are Killing Peasants in the Philippines: Lawlessness, Murder and Impunity'. *Critical Asian Studies*, 39 (3): 315–28.

Franco, J., forthcoming. 'Making Land Rights Accessible: Social Movement Innovation and Political-Legal Strategies in the Philippines'. *Journal of Development Studies*.

Friedmann, H., 2004. 'Feeding the Empire: The Pathologies of Globalized Agriculture'. *Socialist Register 2005: The Empire Reloaded*, eds L. Panitch and C. Leys, 124–43. London: Merlin Press.

Gauster, S. and R. Isakson, 2007. 'Eliminating Market Distortions, Perpetuating Rural Inequality: An Evaluation of Market-Assisted Land Reform in Guatemala'. *Third World Quarterly*, 28 (8): 1519–36.

Ghimire, K., ed., 2001. *Land Reform and Peasant Livelihoods*. London: ITDG.

Greenberg, S., 2004. 'The Landless People's Movement and the Failure of Post-Apartheid Land Reform'. Durban: KwaZulu-Natal, School of Development Studies.

Griffin, K., A.R. Khan and A. Ickowitz, 2002. 'Poverty and Distribution of Land'. *Journal of Agrarian Change*, 2 (3): 279–330.

Guidry, J., M. Kennedy and M. Zald, 2000. 'Globalizations and Social Movements'. In *Globalizations and Social Movements: Culture, Power, and the Transnational Public Sphere*, eds J. Guidry, M. Kennedy and M. Zald, 1–34. Ann Arbor, MI: University of Michigan Press.

Gwynne, R. and C. Kay, 2004. *Latin America Transformed: Globalization and Modernity*. London: Arnold.

Harriss, J., 2002. *Depoliticizing Development: The World Bank and Social Capital*. London: Anthem Press.

Herring, R., 2007. 'The Genomics Revolution and Development Studies: Science, Poverty and Politics'. *Journal of Development Studies*, 43 (1): 1–30.

Hertel, S., 2006. *Unexpected Power: Conflict and Change Among Transnational Activists*. Ithaca, NY: Cornell University Press.

IPC for Food Sovereignty, 2006. 'Agrarian Reform in the Context of Food Sovereignty, the Right to Food and Cultural Diversity: "Land, Territory and Dignity"'. http://www.icarrd.org, Accessed 15 October 2007.

Jordan, L. and P. van Tuijl, 2000. 'Political Responsibility in Transnational NGO Advocacy'. *World Development*, 28 (12): 2051–65.

Kay, C., 2006. 'Rural Poverty and Development Strategies in Latin America'. *Journal of Agrarian Change*, 6 (4): 455–508.

Keck, M. and K. Sikkink, 1998. *Activists Beyond Borders: Advocacy Networks in International Politics*. Ithaca, NY: Cornell University Press.

Keohane, R. and S. Nye Jr, 2000. 'Governing in a Globalizing World'. In *Visions of Governance for the 21st Century*, eds S. Nye Jr and J. Donahue, 1–41. Cambridge: Cambridge University Press.

Kerkvliet, B., 2005. *The Power of Everyday Politics: How Vietnamese Peasants Transformed National Policy*. Ithaca, NY: Cornell University Press.

Khagram, S., J. Riker and K. Sikkink, 2002. 'From Santiago to Seattle: Transnational Advocacy Groups Restructuring World Politics'. In *Restructuring World Politics: Transnational Social Movements, Networks, and Norms*, eds S. Khagram, J. Riker and K. Sikkink, 3–23. Minneapolis, MN: University of Minnesota University Press.

KMP, 1986. 'Program for Genuine Land Reform'. Quezon City: KMP (Pamphlet).

Lahiff, E., 2007. '"Willing Buyer, Willing Seller": South Africa's Failed Experiment in Market-Led Agrarian Reform'. *Third World Quarterly*, 28 (8): 1577–97.

Landsberger, H. and C. Hewitt, 1970. 'Ten Sources of Weakness and Cleavage in Latin American Peasant Movements'. In *Agrarian Problems and Peasant Movements in Latin America*, ed. R. Stavenhagen, 559–83. New York: Anchor Books.

Lara, F. and H. Morales, 1990. 'The Peasant Movement and the Challenge of Democratisation in the Philippines'. *Journal of Development Studies*, 26 (4): 143–62.

Mayo, M., 2005. *Global Citizens: Social Movements and the Challenge of Globalization*. London: Zed.

McKeon, N., M. Watts and W. Wolford, 2004. 'Peasant Associations in Theory and Practice'. Civil Society and Social Movements Programme Paper No. 8. Geneva: UNRISD.

McMichael, P., 2006. 'Reframing Development: Global Peasant Movements and the New Agrarian Question'. *Canadian Journal of Development Studies*, 27 (4): 471–86.

Medeiros, L., 2007. 'Social Movements and the Experience of Market-Led Agrarian Reform in Brazil'. *Third World Quarterly*, 28 (8): 1501–18.

Meszaros, G., 2000. 'Taking the Land into Their Hands: The Landless Workers' Movement and the Brazilian State'. *Journal of Law and Society*, 27 (4): 517–41.

Mngxitama, A., 2005. 'The National Land Committee, 1994–2004: A Critical Insider's Perspective'. Durban: University of KwaZulu-Natal Civil Society Centre Research Report No. 4, Vol. 2, pp. 35–82.

Monsalve, S., 2006. 'Gender and Land'. In *Promised Land: Competing Visions of Agrarian Reform*, eds P. Rosset, R. Patel and M. Courville, 192–207. Oakland, CA: Food First Books.

Navarro, Z., 1998. 'The "Cédula da Terra" Guiding Project – Comments on the Social and Political-Institutional Conditions of its Recent Development', http://www.dataetrra.org.br. Accessed 21 January 2001.

Ntsebeza, L. and R. Hall, eds, 2006. *The Land Question in South Africa: The Challenge of Transformation and Redistribution*. Cape Town: HSRC Press.

Nyamu-Musembi, C., 2007. 'De Soto and Land Relations in Rural Africa: Breathing Life into Dead Theories About Property Rights'. *Third World Quarterly*, 28 (8): 1457–78.

O'Brien, K. and L. Li, 2006. *Rightful Resistance in China*. Cambridge: Cambridge University Press.

Oya, C., 2007. 'Stories of Rural Accumulation in Africa: Trajectories and Transitions among Rural Capitalists in Senegal'. *Journal of Agrarian Change*, 7 (4): 453–94.

Palmer, R., 2007. *Literature Review of Governance and Secure Access to Land*. London: DFID.

Patel, R., 2006. 'International Agrarian Restructuring and the Practical Ethics of Peasant Movement Solidarity'. *Journal of Asian and African Studies*, 41 (12): 71–93.

Peluso, N., 1992. *Rich Forests, Poor People: Resource Control and Resistance in Java*. Berkeley, CA: University of California Press.

Pereira, J., 2007. 'The World Bank's "Market-Assisted" Land Reform as a Political Issue: Evidence from Brazil (1997–2006)'. *European Review of Latin American and Caribbean Studies*, (82): 21–49.

Peters, P., 2004. 'Inequality and Social Conflict Over Land in Africa'. *Journal of Agrarian Change*, 4 (3): 269–314.

Petras, J., 1997. 'Latin America: The Resurgence of the Left'. *New Left Review*, (223): 17–47.

Putzel, J., 1992. *A Captive Land: The Politics of Agrarian Reform in the Philippines*. New York: Monthly Review Press.

Putzel, J., 1995. 'Managing the "Main Force": The Communist Party and the Peasantry in the Philippines'. *Journal of Peasant Studies*, 22 (4): 645–71.

Razavi, S., 2003. 'Introduction: Agrarian Change, Gender and Land Rights'. *Journal of Agrarian Change*, 3 (1/2): 2–32.

Rosset, P., 2006. 'Agrarian Reform and Food Sovereignty'. A paper prepared for and presented at the Land, Poverty, Social Justice and Development Conference at the Institute of Social Studies (ISS) in the Hague, the Netherlands, 9–14 January 2006.

Rosset, P. with M.E. Martinez, 2005. *Participatory Evaluation of La Vía Campesina*. Oslo: Norwegian Development Fund.

Rosset, P., R. Patel and M. Courville, eds, 2006. *Promised Land: Competing Visions of Agrarian Reform*. Oakland, CA: Food First Books.

Sassen, S., 2006. *Territory, Authority, Rights: From Medieval to Global Assemblages*. Princeton, NJ: Princeton University Press.

Sauer, S., 2003. 'A Ticket to Land: The World Bank's Market-Based Land Reform in Brazil'. In *The Negative Impacts of World Bank Market-Based Land Reform*, eds F. Barros, S. Sauer and S. Schwartzman, 45–102. Brazil: Comissão Pastoral da Terra, Movimento dos Trabalhadores Rurais Sem Terra (MST).

Sauer, S., 2007. 'Rural Social Movements and Democratization in Brazil'. Paper presented at the Rural Democratization Workshop, Transnational Institute, Amsterdam, 5–6 October 2007.

Scholte, J.A., 2002. 'Civil Society and Democracy in Global Governance'. In *The Global Governance Reader*, ed. R. Wilkinson, 322–40. London: Routledge.

Scott, J., 1985. *Weapons of the Weak*. New Haven, CT: Yale University Press.
Smith, J. and H. Johnston, 2002. *Globalization and Resistance: Transnational Dimensions of Social Movements*. Lanham, MD: Rowman and Littlefield Publishers.
Spoor, M., ed., forthcoming. *Contested Land in the 'East': Land and Rural Markets in Transition Economies*. London: Routledge.
Stedile, J.P., 2002. 'Landless Battalions: The Sem Terra Movement of Brazil'. *New Left Review*, 15: 77–105.
Stedile, J.P., 2007. 'The Class Struggles in Brazil: The Perspective of the MST'. In 'Global Flashpoints: Reactions to Imperialism and Neoliberalism', *Socialist Register 2008*, eds L. Panitch and C. Leys. London: Merlin Press.
Tadem, E., 1996. 'Reflections on NGO-PO Relations'. Paper presented at the NGO Parallel Forum to the Vía Campesina International Assembly, April 1996, Tlaxcala, Mexico. Vía Campesina Tlaxcala unpublished documents.
Tarrow, S., 1994. *Power in Movement: Social Movements, Collective Action and Politics*. Cambridge: Cambridge University Press.
Tarrow, S., 2005. *The New Transnational Activism*. Cambridge: Cambridge University Press.
Tilly, C., 2004. *Social Movements, 1768–2004*. Boulder, CO: Paradigm Publishers.
Van Donge, J.K., with G. Eiseb and A. Mosimane, 2007. 'Land Reform in Namibia: Issues of Equity and Poverty'. In *Land, Poverty and Livelihoods in an Era of Globalization*, eds H. Akram Lodhi, S. Borras and C. Kay, 284–309. London: Routledge.
Veltmeyer, H., 1997. 'New Social Movements in Latin America: The Dynamics of Class and Identity'. *Journal of Peasant Studies*, 25 (1): 139–69.
Vía Campesina, 2000a. 'Struggle for Agrarian Reform and Social Changes in the Rural Areas', Bangalore, India. Jakarta: Vía Campesina.
Vía Campesina, 2000b. Draft 'Vía Campesina Position Paper: International Relations and Strategic Alliances', discussed during the III International Conference in Bangalore. Jakarta: Vía Campesina.
Vía Campesina, 2004. 'Debate on Our Political Positions and Lines of Actions: Issues proposed by the ICC-Vía Campesina for regional and national discussion in preparation for the IV Conference', in IV International Vía Campesina Conference, pp. 45–58. Jakarta: Vía Campesina.
Vía Campesina, 2006. 'Violations of Peasants' Human Rights: A Report of Cases and Patterns of Violence'. Vía Campesina document. Jakarta: Vía Campesina.
Vía Campesina-FIAN, 2003. 'International Seminar on Agrarian Reform and Gender: Declaration of Cochabamba'. Heidelberg: FIAN.
Vianna, A., 2003. 'The Inspection Panel Claims in Brazil'. In *Demanding Accountability: Civil Society Claims and the World Bank Inspection Panel*, eds D. Clark, J. Fox and K. Treakle, 145–66. Lanham, MD: Rowman and Littlefield.
Walker, C., 2003. 'Piety in the Sky? Gender Policy and Land Reform in South Africa'. *Journal of Agrarian Change*, 3 (1/2): 113–48.
World Bank, 2003. *Land Policies for Growth and Poverty Reduction*. Washington, DC: World Bank.
Wright, A. and W. Wolford, 2003. *To Inherit the Earth: The Landless Movement and the Struggle for a New Brazil*. Oakland, CA: Food First Books.
Yashar, D., 2005. *Contesting Citizenship in Latin America: The Rise of Indigenous Movements and the Post-Liberal Challenge*. Cambridge: Cambridge University Press.

5 'Late Mobilization': Transnational Peasant Networks and Grassroots Organizing in Brazil and South Africa

BRENDA BALETTI, TAMARA M. JOHNSON AND WENDY WOLFORD

INTRODUCTION

One of the most surprising developments of late twentieth-century globalization has been the increasing presence and visibility of landless peasant movements around the world. Even as small farming decreases in importance as a global economic sector and as people increasingly move out of the rural areas and into urban centres, landless workers and small farmers have come together to demand access to land and new economic development policies that prioritize sustainable local communities and food sovereignty. From Brazil to Mexico to South Africa and the Philippines, grassroots social movements and non-governmental organizations concerned with this broad range of issues have articulated their efforts through transnational meetings and organizations, such that there is now an easily identifiable 'Transnational Peasant Network' (hereafter the TPN) organizing the global struggle for rural peoples and their 'right to have rights' (Desmarais 2007; Edelman 1998).

This is a historic moment for transnational activism more broadly: global-scale movement networks have generated considerable excitement in scholarly and political arenas because of their potential for circumventing local or national-level politics in the name of universalistic goods such as human rights, access to land, environmental conservation, women's rights, etc. Such universals are deployed in part to counter the universal(izing) logics of orthodox globalization. Because we are only beginning to see and understand these networks, much of the early work on them focused on the empowering ability of place-based movements to 'scale up' (Smith 1993), largely taking the 'transnational' potential of social movements for granted. It is only recently that the diverse relationships and localized contexts in which these transnational networks are situated are being analyzed (see Edelman 2005a, 2005b; Featherstone 2003; Routledge et al. 2007). This is an important direction because transnational social networks are attempts to create something new – as David Featherstone says, they are 'generative' of a new form of politics – at the same time as they are a compilation of historical relationships and characteristics (Featherstone 2003). Although this may appear to be an obvious point, we think that it has considerable implications for the study of the TPN.

In this chapter, we analyze the political dynamics of the contemporary TPN through a comparison of two movements: the Brazilian *Movimento Dos Trabalhadores*

Rurais Sem Terra (MST) and the South African Landless People's Movement (LPM). Organized in 1984, the MST has the largest domestic following of any social movement in the transnational network and this has placed it at the forefront of the broader international struggle. For a brief period following its formation in 2001, the LPM also garnered considerable international attention for its potential to mobilize grassroots interests for agrarian reform in South Africa. We argue that both the localized and transnational contexts are important for an analysis of the two movements – neither the MST nor the LPM can be properly understood outside of the transnational context (specifically, the context of the TPN), and the way in which both movements have engaged with transnationalism can only be properly understood within the context of localized political, social and economic conditions in Brazil and South Africa respectively – but we go a step further to develop a theory of 'late mobilization' (after Alexander Gerschenkron's 1952 arguments about 'late development'). Our theory of late mobilization is an attempt to bring geography and history together in our study of transnational movements. Place and scale matter, but so does timing: within an inter-connected world system, what one country/region/group does at one time affects all of those that follow, often in very deliberate ways (also see McMichael 1990).

Within a transnational network, one of the most tangible benefits is the exchange of ideas, experiences and information (Keck and Sikkink 1998), but this exchange does not happen in a vacuum, rather it happens in the historically-situated, power-laden context of an uneven world system. In our comparison of the MST and the LPM, therefore, we argue that the latter is the 'late mobilizer', and as such it has experienced both advantages and disadvantages of engaging with the TPN. Organized in 1984, MST activists from Brazil helped to train, educate and inspire South African activists after the LPM formed in 2001, but key elements of the MST's success were inappropriate or unworkable in the South African context. On the one hand, the MST's experiences and advice helped the LPM to develop strategies and gain visibility; on the other hand, the LPM had difficulty translating its spectacular international success into an organic grassroots campaign. Ultimately, the transfer of movement knowledge from the MST to the LPM may have worked against the long-term success of the latter.

We argue that the MST owed its success in Brazil to three main elements: first, the MST developed through the leadership of young sons and daughters of small farmers – 'organic' leaders who maintained a strong ideological coherence as they struggled for a collective peasant future; second, the movement has persisted in using direct-action land occupations as the primary strategy for pressuring the government to expropriate land – these occupations provide the movement with a coherent public face as well as an opportunity for movement members to build solidarity; third, since its inception, the movement has insisted on maintaining its autonomy from the Brazilian state, ensuring that the leadership was not co-opted by the 'feckless' politics of Brazil's electoral system (Mainwaring 1995). The MST then successfully translated its national success into a new peasant internationalism. The movement broadened its message and widened its support base throughout the 1990s.

As the MST became an important transnational actor, movement activists travelled to places like South Africa to support and advise rural activists there. Unfortunately, the MST's main strategy for mobilization – the land occupation – backfired in the South African context. The nearby example of racial tension in Zimbabwe made occupying land politically explosive in South Africa. Occupations were also difficult for the LPM to organize because the movement was unable to draw on the MST's other two elements of success: a leadership developed through grassroots experience and autonomy from civil society organizations and the state.

These difficulties of translation are not a necessary effect of late mobilization, of course. And they are certainly not responsible for the success or failure of *land reform* in the two countries. Our argument specifically concerns social mobilization, and here we argue that transnational organizing cannot be seen as an unqualified good. Instead, researchers and movement activists need to try and understand under what conditions transnationalism offers advantages or disadvantages. Just as early development practitioners did not always examine the specificities of the context in which development was to take place (Ferguson 1994), we argue that the rush to embrace transnational organizing may lead to the neglect of geographical and historical specificities.

In the rest of the chapter, we describe the political and economic context of land distribution in Brazil and draw upon interviews conducted with MST activists in order to outline the three factors that contributed most to the movement's success. After outlining the MST's transformation and importance, we then describe the LPM from its origins and location within a bitter debate over land restitution and reform in the wake of the 1994 end to apartheid in South Africa. We analyze two interviews with critical movement leaders in order to develop and situate our analysis.

FROM SQUATTERS TO *SEM TERRA*: THE RISE OF THE MST IN THE 1990s

Brazil has one of the most unequal distributions of land ownership in the world. This inequality is a legacy of Portuguese colonization and plantation society, which introduced and embedded racialized social hierarchies into land ownership and labour relations (Schwartz 1985). Sugarcane production provided the economic engine of growth in the early colonial years, and land tenure patterns reflected the Portuguese crown's preference for large landholdings. Alongside the plantations, however, a small-holding peasantry was allowed to develop. In the early 1800s, small-holders fleeing the ravages of population growth and agricultural depression made their way to Brazil, cultivating and settling the country's southern frontiers. Although plantation owners tried to keep the new immigrants from acquiring their own land so that they would have labour for their crops (Viotti da Costa 2000), a peasant class developed in the south of Brazil as well as along the new settlement frontiers (in the north and centre-west). As late as the 1950s, the Brazilian population was predominantly rural (75 per cent), although 85 per cent of that population lived and worked on only 10 per cent of the land.

This inequality in land ownership generated ongoing rural discontent that erupted in the 1960s when popular mobilization for access to land spread throughout Brazil and Latin America. At this time, peasants and rural workers grew increasingly radical, organized by the rural trade unions, various factions of Brazilian Communist groups, and the Catholic Church (Forman 1975; Maybury-Lewis 1994; Pereira 1997; Santos and Costa 1998). Many governments in Latin America began to implement agrarian reforms at this time (de Janvry 1981; de Janvry et al. 2001; Grindle 1986), but in Brazil, timid attempts at reform threatened the powerful landed elite, and in 1964, a military coup installed an authoritarian dictatorship that would hold power for 21 years (Hall 1990; Reis 1990). Once in power, the military government developed a two-pronged strategy to address rural unrest: on the one hand, the rural poor (or, peasant agitators) were re-located to new frontiers of colonization, such as the Amazon rainforest region, and on the other hand, large-scale farms were targeted for state-led modernization in an effort to increase efficiency and reduce Brazil's reliance on small farmers.

By the late 1970s, the authoritarian state began to weaken. As the military gradually withdrew from power, landless peasants and rural workers began to form squatter settlements throughout the country, concentrated in the southern states of Rio Grande do Sul, Santa Catarina and Paraná (Fernandes 1999). With the restoration of democracy in Brazil in 1985, widespread demands for access to land returned with renewed force, becoming part of the 'national political culture' largely because of the formation of new social movements such as the MST (Gohn 1997; Novaes 1998). Activists with the MST mobilized rural workers to occupy unproductive plantations, using Article 186 of the Brazilian Constitution to argue for the right to 'unproductive' property that was not being used in a way that benefited the 'social good' (see Branford and Rocha 2002; Fernandes 1999; Wright and Wolford 2003). Over time, the MST developed a sophisticated 'repertoire of contention' (Tilly 1978) that included the symbolic appropriation of spaces such as landed property, government buildings and public thoroughfares.

In the late 1990s, the political will for agrarian reform in Brazil increased because of national and international outcry over two violent massacres. In 1995, ten landless squatters were killed in Corumbiara, Rondônia, and just one year later, 19 landless squatters were killed as they marched to the capital city of the state of Pará along the state highway. These two incidents – the second one captured on video-tape – were partly responsible for the increased visibility of rural poverty, violence and the MST (Ondetti 2001; Pereira 2003). In 1997, president Cardoso settled 80,000 families, almost twice as many as were settled in his first year in office (Cardoso 1997), and from 1994 to 1998 the annual budget for agrarian reform was more than quintupled (Seligmann 1998). Luis Inacio 'Lula' da Silva, who won executive office in 2002, promised to continue agrarian reform efforts, although the experience under this administration has been mixed (Deere and Medeiros 2007).

During the mid-1990s, as the MST gained national visibility, the movement also gained international visibility. Partly as a result of the international outcry over the massacres, the MST's organizational efforts and influence skyrocketed. Over the course of the 1990s, the MST went from being considered a small, marginalized movement of 'vagabonds' and squatters to one of the 'most powerful and well-organized social movement[s] in Brazil's history' (Petras 1997, 18). Today the movement has approximately two million members (although these numbers are contested, and some estimates are significantly lower) who are settled in government-funded land reform settlements or living in temporary 'encampments' awaiting final resolution of their claim to land. Over 17,000 people attended the MST's Fifth National Congress in 2007.

In internationalizing its struggle for land, the MST sees itself as representing all of those who have been marginalized by the global project of modern industrial capitalism. And through an extraordinary expansion of its base, the movement has become a significant actor in the social movement world system, engaging as part of the anti-neoliberal, anti-globalization movement. It has also been one of the most influential, active groups organizing Vía Campesina (Peasant Way), the international peasant umbrella movement.

THE METHODS TO THE RISE: LEADERSHIP, OCCUPATIONS AND AUTONOMY

There were many reasons for the MST's success in Brazil and internationally, but three stand out. First the MST developed very strong circuits of leadership 'professionalization', where new leaders move around the country, studying at MST formational centres and building new ties with local activists and movement members; second, the MST has continued to emphasize land occupations as its primary strategy, these occupations serve as sources of new leadership, movement solidarity and political presence; and third, the MST has insisted on maintaining its autonomy from both the state and established elements, such as the Catholic Church or non-governmental organizations, in ways that allowed it to develop apart from the personalistic and unreliable politics of official power.

Strategies for Success #1: Leadership

Many academics and observers have attributed the MST's rapid rise to its dedicated leadership (Petras 1997; Wright and Wolford 2003). Most MST leaders are recruited out of left-leaning institutions – such as organized religion, rural trade unions or the Worker's Party – or out of the land occupations themselves. Many of the earliest leaders were the sons and daughters of small farmers who were introduced to the movement through the Catholic and Lutheran Churches (Houtzager and Kurtz 2000).

Once the movement began organizing occupations, young men and women who participated learned of the movement's goals and ideology. They were some of the most committed members, and they were also – perhaps ironically – the ones least likely to win access to land when the government expropriated a property (Rosa 2007). The government was legally bound to prioritize families over single adults or youth, and so young men and women without partners or children were passed from occupation to occupation, radicalizing their political views as they went. Once recruited, MST leaders were expected to travel anywhere, anytime for the good of the movement and the struggle. This political mobility helped the movement to expand its membership from the south to the rest of the country by the mid-1990s.

As the MST grew, development of such young leaders remained a movement priority. This can be seen in the pages of the MST's own newspaper, published every month since the late 1980s. A monthly column in the *Jornal Sem Terra* (Landless Newspaper) highlights the grassroots organizing of its homegrown activists (*militantes*) in an interview with a different one each month from across the country. All of the activists tell similar stories of their initial contact with the movement through participation in occupations or demonstrations, falling in love with the movement, rising to leadership through formal training by the movement and then working as organizers. It is through the work of these leaders that direct action remains a primary movement strategy because, although the movement is strategically scaling up, these leaders continue to work locally in the struggle for land, mobilizing people for occupations, organizing in the settlements, and recruiting and training new leaders who then move to new places to carry out these same activities.

In June of 2007, we interviewed Julia (not her real name), an MST leader in the state of Paraíba who was an excellent example of the MST's dedicated leadership. We interviewed Julia in her office in João Pessoa, the capital of the state. The city sits in the humid tropical coast, surrounded by sugarcane and, increasingly, settlements. Julia was clearly exhausted. She had been up since 4:00 am arranging regional affairs for the transnational social movement, Vía Campesina. She had recently been appointed the MST representative for Vía Campesina in the state, and because it was just getting started there was a lot of work. Julia was also coordinating a visiting charitable group from Italy: nine young Italian men and women who had donated money to the MST had come to Paraíba to see the work their money had done. Reluctantly, Julia sat down for the interview, shaking out a cigarette and crossing her arms tightly in front of her.

Julia grew up far from João Pessoa in the southern state of Santa Catarina. Raised in a small village without electricity, she grew up working on her family's farm. Her family did not have much, she said, but she always worked hard and never missed a day of school. When she was old enough, she began to study religion with plans to become a nun, but exposure within the church to Liberation Theology inspired her to activism and she became a militant in the Pastoral Land Commission (*Comissão Pastoral da Terra*, hereafter CPT), a church movement that (among other things) works as a facilitator for the landless in their struggle for land. Two years later, however, Julia left the CPT simply because, as she said, 'the MST was born'.

Julia heard about the MST through her religious work, and in 1989 the movement sent her – along with 19 other activists from the south (including Jaime Amorim, the well-known state leader in Pernambuco who grew up across the street from Julia) – to the northeast to help to establish the first settlements in the region.

At the time, the MST was a young, small movement, and this was appealing to Julia because it was, in her words 'a movement *of* the people not *for* them' (although Julia, like many of the most visible movement leaders, had not herself acquired land through the movement). During the 1990s, Julia lived and worked in Bahia, Pernambuco, Maranhão and Paraíba, educating the movement's grassroots base and training its leaders. Even after 18 years of living in the northeast, far from her family, Julia is still willing to go wherever the movement needs her. It is through this model of moving leaders from place to place to educate and train other militants that the movement has attained – as she says, 'hegemony in all of Brazil'.

Recruiting and training new young leaders is a key strategy of the movement, and a primary responsibility for many leaders. Another northeastern leader that Brenda and Wendy interviewed explained the importance of passing leadership responsibilities on to the younger generation: 'it is a question of adaptation, if [we don't get new leaders], we will be getting old and the settlements will get old too, because they will start to do things that are based in the heads of the old people, you know. . . . The idea is to be preparing the [young people], for them to go ahead and assume the responsibilities'.

Strategy of Success #2: Land Occupations

The MST has developed a broad set of strategies to promote its agenda, but land occupations continue to be the movement's primary political weapon (Fernandes 2007). Occupations are well-planned events that generally follow a standard script. Activists hold meetings and organize among the 'base' for months before choosing an area for occupation. Usually the property chosen is one that is not fulfilling its social (Constitutional) function to be productive or one with uncertain title. Then activists prepare as the selected day arrives. They notify all of the participants, telling them to be ready at a moment's notice, to keep their bag by the door, ready for when the bus or van or truck rolls by. The occupation itself usually happens at night when there is less chance of detection. Once on the property, activists organize the construction of small black plastic tents where each family will be living until negotiations with the government (or with the landlord's private guns) begin. Occupations are often coordinated throughout a broader region – the state or even the country – in an attempt to scale up all of the localized demands for land. As one regional leader in the northeast said:

> We aren't the ones who decide how everything [in an occupation] goes. There are state guidelines too, you know. For example, when we occupied here [in Paraíba], we didn't only occupy the one property here. This is the capacity that we have: we occupied here and occupied other properties in other regions of the state simultaneously. They are planned actions, planned with other regions of the state so that we can have more power.

Occupations provide the space for movement production and reproduction: it is here that many (though certainly not all; see Wolford 2006) movement members realize what the movement is really about (Fernandes 2007). They may have joined the occupation because their friends or church organizers encouraged them to, or because they were desperate to do anything that might win them land, but once there, people are often caught up in the intensity of the political moment. Movement leaders hold meetings regularly during occupations in order to keep people busy and prepare for the possibility of eventually winning access to the land. During occupations young leaders are recruited. Both in our interviews and in our experience, leaders tell similar stories of first encountering the movement through an occupation that ended with their parents winning a plot of land and them beginning to work to organize the movement base.

Strategy of Success #3: Autonomy

When the MST held its first National Congress in 1985, there was discussion as to whether the movement should work from within the Catholic Church or strike out on its own. Prominent leaders later recalled church organizers themselves arguing for the movement's independence: the MST needed to develop its own struggle, one that was not tied to traditional alliances or old ways of 'doing politics' (Branford and Rocha 2002, 21). As the movement grew stronger, and as democracy in the country consolidated, there were more calls for the movement to work more closely with the government: two prominent rural sociologists argued that the MST was missing a historic opportunity to entwine rural workers' interests with the federal state (de Souza Martins 2003; Navarro 2002). To every suggestion of political affiliation, the MST has remained resolute. As the movement's most prominent leader, João Pedro Stedile, said, 'our analysis of the farmers' movements of Latin America and Brazil taught us that whenever a mass movement was subordinated to a party, it was weakened by the effects of inner-party splits and factional battles. . . . The movement had to be free from external political direction' (2002, 80).

The movement's project to build hegemony has involved building alliances with other organizations such as different religious organizations, rural and urban trade unions, and politicians. These alliances, however, generally function as goal-specific partnerships not integration or even affiliation. Support from other well-known organizations such as the Central Workers' Union (CUT), the Pastoral Land Commission (CPT), or well-known academics and politicians from the Workers Party demonstrates popular support for the movement, but the movement's autonomy has never yet compromised its autonomy.

The MST's commitment to autonomy can be seen in discursive claim of a lack of dependence on both civil society and the state. The MST claims autonomy from the state, but this does not mean that they do not negotiate with the state, or that the MST's occupations and settlements are not dependent on the state for resources. They receive food baskets in the occupations, land in settlements, credit for settlers, and political strength and support from the state, but they claim these things as rights rather than as aid or as favours. By negotiating with the state from an external rather than dependent position, the MST maintains its claim to autonomy.

Julia also talked about the movement's position with respect to the federal government. The movement's three biggest enemies, she said, 'are the executive, the legislature, and the judiciary'; in other words, the government itself. She also critiqued the inefficiency of the National Institute for Colonization and Agrarian Reform (INCRA), the main governmental agency charged with executing agrarian reform. Julia argued that INCRA was incompetent in terms of getting land expropriated and providing the settlers with agricultural assistance, largely because she believed that a majority of INCRA employees were against the project of distributive land reform.

Another activist in the northeast explained his distrust of the federal government in this way:

> Legally, INCRA could come [to the settlement] more, but we don't allow it. This is how it is: I could allow an INCRA employee to come here and say that this man who lives next door should stay in the settlement [note: this was in reference to an ongoing dispute on the settlement]. But we are the ones who should decide, because after this government employee comes here and says . . . that this other guy has rights, then the employee gets back in his car and goes away, and those of us who live here still have the problem. . . . And so the community has certain autonomy, and INCRA goes and paints and embroiders [or whatever it does] (Interview, July 2008).

LAND FOR THOSE WHO WORK IT TOO: THE SOUTH AFRICAN CASE

As in the Brazilian case, the control of land was the backbone of colonial policies in South Africa. In South Africa, however, racialized policies of dispossession carried over into the independent state and made it difficult for a significant or coherent land-owning peasantry to develop. As a result, the contemporary demand for land in South Africa is less organized around a return to farming per se than around a broader struggle for citizenship, political rights, labour reform and social welfare policies. This difference between the Brazilian and South African experiences made it difficult for LPM activists to copy the MST's strategies for success.

Although Europeans had been settling in southern Africa since the 1600s, the process of alienating natives from their land only began in earnest when diamonds were discovered in Kimberley in 1867 and gold was discovered in what is now Johannesburg in 1886. The discovery of mineral wealth in South Africa piqued British imperial interest in the region, and as a result, land in the surrounding region became more valuable. Three years after the formation of the Union of South Africa, the Natives' Land Act of 1913 was passed in order to control Black activity on and possession of land. This legislation restricted the area for lawful African occupation: unless special permits were acquired, native Africans were forced to live on reserves (Bantustans) that comprised a mere 7 per cent of the land area. The Land Act, followed by new legislation in 1936, also eliminated sharecropping and rent tenancy, replacing them with labour tenancy (Sihlongonyane 2005; Thwala 2006). Many scholars point to the Land Acts as decisive moments in the proletarianization of the South African peasantry. Proletarianization was critical for the development of capitalist agriculture, as dispossessed peasants often became wage workers on White farms or mine workers (Ntsebeza 2007). Dispossession also reduced competition among White farmers and created a class of wage labourers to work on White-owned farms and in mines. Although the Land Acts may not have successfully dispossessed all Africans or stopped them from acquiring land, it did have the effect of further differentiating the African peasantry into a class of landowners and evicted peasants who moved into the freehold lands, many of whom became migrant labourers (Hart 2002). Unequal access to land was therefore an essential component of the political economy of South Africa (Thwala 2006).

Black migration to urban areas was also highly regulated and, for the most part, discouraged. The Bantu Consolidation Act of 1946, which identified Africans who were allowed to remain in urban areas for over 72 hours, created passbooks to record and monitor the movement of Black South Africans into the cities. This segregation was solidified in 1948 when the National Party gained power and created the Homeland System. The Homeland System was comprised of ten territories that separated the Black population by ethnicity in the hopes that the territories could serve as potential African nations. A new Native Laws Amendment Act of 1957 tightly controlled the recruitment and movement of Black workers from reserves in rural areas to urban employers in 'white South Africa'.

Cousins (2006) argues that 'de-peasantization' in South Africa was intentional and an integral part of segregationist and Apartheid policies throughout the twentieth century. Whether or not it was intentional, subsistence farming among rural people became less common as a primary source of livelihood and people in the reserves were increasingly forced to rely on non-agricultural work. Between the 1960s and 1980s, the situation worsened as approximately 1.75 million rural people were displaced, removed primarily from White-owned farms (65 per cent) and African freehold land (35 per cent), effectively eliminating even labour tenancy and squatting (Hart 2002, 89). Although 70 per cent of people in the rural homelands had access to land, more than 50 per cent of those people had access to less than one hectare, which discouraged subsistence farming, and encouraged livelihood strategies in rural labour and mining (Ntsebeza 2007).[1]

The rural population in South Africa today continues to be economically differentiated with localized groups exploiting different bundles of livelihood strategies including: agriculture, migration, agricultural labour, remittances, formal employment, pensions and micro-enterprises (Cousins 2006). Because of this, several researchers have argued that the agrarian question in South Africa should be 'delinked' from agriculture and connected to a broader range of demands

---

[1] In the 1980s, the average population density for the homelands was 151 people per square kilometre, as compared to 19 people per square kilometre in the rest of South Africa. Where White South African areas had housing surpluses, Black areas had housing deficits (Thwala 2006).

(Hart 1996, 269; Cousins 2006). People do not fit easily into the category of worker or peasant – or even a hybrid of the two – but, as Henry Bernstein has pointed out for the Global South more generally, are forced to make their living 'across different sites of the social division of labour: urban and rural, agricultural and non-agricultural, wage employment and self-employment' (Bernstein 2004, 205).

Today, 13 years after the fall of Apartheid, the land question in South Africa is still a critical one, and the reality of land ownership and control is still divided along racial lines. Land reform legislation has been seen as one method of balancing out the inequalities of Apartheid and including previously marginalized communities in the new 'Rainbow Nation' of South Africa. After the first democratic elections in 1994, the Mandela administration proposed a land reform programme centring on three elements: restitution (land returned to claimants who could prove that they were dispossessed of their land after the Native Lands Act of 1913), redistribution (land the government purchases at market value and redistributes to landless claimants), and tenure security (assurance that people living on White farms will not be arbitrarily evicted). The goal was to redistribute 30 per cent of agricultural land in the first five years following a 'willing-seller, willing-buyer' system; however, land reform in South Africa has been painfully slow, with only 3 per cent of the land having been redistributed by 2001 (Adams and Howell 2001).

During Apartheid, the struggle for land was incorporated into (and secondary to) the broader struggle against Apartheid, such that the 'land demand was not articulated in and of itself, but rather as a symbol of the lack of political democracy and the racism of the apartheid regime' (Greenberg 2004b, 15; Greenstein 2003). While the majority of the South African population is landless, most of them identify politically in other ways because of the way in which resistance was constructed under Apartheid. The fact that the struggle for land took a backseat to the broader liberation struggle has made organizing a landless struggle in the post-Apartheid context difficult. To a certain extent, building a landless movement would require re-contextualizing longstanding demands – for political participation, access to resources etc. – around the land issue (Greenberg 2004b).

In a 2001 article, Lahiff and Cousins suggested that the government's relative inattention to rural land reform is due, in part, to the absence of a grassroots landless voice to place pressure on the government in order to force this issue. That same year, however, the Landless Peoples' Movement (LPM) formed to challenge the South African government's land reform policies and approach, and to force land distribution into the national and international spotlight. In 2004, the LPM recorded a membership (loosely defined) of approximately 100,000 people, 90 per cent of whom were located in rural areas (Alexander 2004). These membership numbers are difficult to verify and may be inaccurate, but the claim to such numerical strength was an important strategy for the LPM to gain visibility. By the time we began talking to activists and scholars about the LPM in September of 2006, there was a general consensus that the LPM was in decline, with some going so far as to declare the movement dead.

In what follows, we analyze the rise and decline of the LPM as read through the experiences of two figures involved in the movement from its inception: Mangaliso Khubeka, a founder and national organizer of the LPM, and Ricado Jacobs, staff member at the Surplus People's Project in Cape Town, a non-governmental organization that formed in the 1980s to help communities organize against the ongoing state-led relocations of Black South Africans, and member of the advisory committee for the formation of the LPM. Mangaliso's interview illustrates the LPM's approach to the three central concepts that we argued above were key to the success of the MST: political autonomy, land occupations and organic leadership. Ricado's interview illustrates how these factors are shaped by the South African context. While we recognize that the opinions of these figures are not universally shared by all actors with the movement, we believe these interviews not only highlight the mentalities of key figures within the LPM at the current conjuncture, but their words also reflect the trajectory the movement has taken in its short history.

THE RISE OF THE LANDLESS PEOPLE'S MOVEMENT

We interviewed Mangaliso in November of 2006 at the Surplus People's Project office in the Western Cape. He was travelling with other LPM members based in KwaZulu-Natal as well as with several MST visitors from Brazil. Their tasks were to network with LPM chapters in the Western Cape while trying to recruit other landless communities into the movement. Mangaliso was a founding member of the LPM: he became involved with landless mobilization while resisting eviction from his home on a farm outside of Newcastle in the province of KwaZulu-Natal. He and several others formed a small local committee that became a province-wide organization called the Tenure Security Coordinating Committee (TSCC) after meeting with the Association for Rural Advancement (AFRA). Mangaliso and other members of the TSCC felt as though their complaints had gone unheeded by local and provincial government officials. There was a need, they felt, for the formation of an organization 'to be the mouthpiece of the people'.

In August of 2001, South Africa hosted the United Nations World Conference Against Racism and Discrimination, an event that focused international eyes on both global and South African social movements. Special interest groups met in Durban for weeks leading up to the event. This conference 'provided space and opportunity for resource mobilization for the first joint national action of new independent community movements', and it was then that the Landless People's Movement was formed out of disparate local landless groups (Greenberg 2004b, 18). During this period the LPM, along with a number of established organizations joined under the banner of the Durban Social Forum to promote a 'Landlessness=Racism' campaign and foreground the plight of the landless (Ntsebeza 2007).

In 2005, Mangaliso described the LPM as being comprised of 'the poorest of the poor, people not having their own land' (Interview), many of whom, like Mangaliso, were evicted Black farm workers. In another article (n.d.) Mangaliso articulated the desires of LPM members: 'they asked me what it is I want. I want to plough. . . . We don't have land. We see no sustainable development; people are hungry We have no land or jobs' (Martorell 2007). The World Summit on Sustainable Development held in Johannesburg in 2002 provided another opportunity for NGOs to bring together landless members from all over the country to form the second Landless People's Assembly. Five thousand delegates from the LPM joined 25,000 members from the Social Movements Indaba (an umbrella organization for South African social movements created to foster solidarity among grassroots campaigns working on similar issues) to march against neoliberal policy trends in the South African government. Mangaliso's comments reflect sentiments that range from general disillusionment with the South African government and the slow pace of land reform: 'there is nothing good about land reform in South Africa', to outright indignation, 'black people, they don't have rights in South Africa. They don't have rights. It's only on paper that they are saying we have rights'.

On 11 June 2007, we interviewed Ricado Jacobs, a staff member at the Surplus People's Project (SPP). Currently, the SPP advocates for rural mobilization and agrarian reform. Ricado has been with the SPP for many years and was a member of the National Land Committee (NLC), the organization instrumental in the establishment of the Landless People's Movement. A self-proclaimed activist, Ricado states that even after the dissolution of the NLC many former members were still active within the Landless People's Movement and served as an advisory committee, organizing political education for LPM constituencies. When asked about the LPM's current situation, he responded that because of the slow pace of land reform, the political situation was still ripe for a landless movement and that since 1994 the demands of the poor had not been met, particularly in rural areas. He attributed this to neoliberal agricultural restructuring and the South African government's subsequent preference for a Market-Led Agrarian Reform (MLAR) over the grassroots-led alternative (see Borras 2003 for background discussion).

THE LPM IN INTERNATIONAL NETWORKS

Shortly after the formation of the LPM in 2001, the movement began forging ties with international peasant movements like Vía Campesina and successful landless movements in other countries such as the MST in Brazil. Unlike the MST, the LPM was an international movement from its inception because of the support and advice it received from international peasant and landless organizations. As the national organizer of the LPM, Mangaliso travelled to Brazil five times between 2001 and 2006. He also travelled to Mexico and the United States, proclaiming that 'everywhere, there are problems about land. So we are connected to the world as LPM' (Interview). The MST is particularly significant to the South African landless movement because of the similarities between the two countries in agrarian structure, such as a dual agricultural economy, concentrated land ownership with racial overtones, and vast inequality. Such similarities may have worked to obscure some of the differences between the two places. For instance, Brazil's movement was founded by peasants who articulated their struggle for land in terms of the right to land and labour. In South Africa, the right to land was identified more with a right to equal rights. While the two movements had similar goals – land redistribution – the historical foundations for unequal distribution and the philosophical justification for redistribution were profoundly different.

According to Mangaliso, members of the MST visited South Africa every year between 2003 and 2006. They came to meet with LPM members, share their experiences with land reform in the Brazilian context, and offer insights on how to strengthen landless mobilization. In 2005, two MST activists spent three months in South Africa 'to teach the activists of the LPM how mobilizations were performed in Brazil' (Rosa 2007, 4). According to Ricado, the MST's primary objective while visiting South Africa is to 'support the LPM and share experiences . . . Part of the traveling around is to assess and look at the state of the movement, and have debriefing sessions so that the [LPM] can improve'. When funds are available, the LPM sends its members abroad to meet with the MST to learn 'how they do things in Brazil' (how to work with the soil, for example). By becoming members of Vía Campesina, Mangaliso felt that the LPM was connected to a global movement of 'people who are on the side of agriculture' and who could 'speak with one voice' against the injustices of global capitalism. Many South African scholars were equally optimistic: 'it is in organizations like these [the Landless People's Movement] that our nation has come alive and it is here that the real fight to defend and deepen our democracy is being fought' (Desai and Pithouse 2003).

Ricado also recognized the benefits of forming ties with peasant movements across the globe, but he argued that when the LPM was founded, rural social movements like the MST were so strong that there was an attempt to 'build primarily a Latin American-ist myth' in a South African context. This, he argues, was a fundamental weakness within the movement: while the Brazilians built their movement around Brazilian history and heroes, the philosophy and practices of the LPM were not rooted in the history of *South African* peasant mobilization. Ricado also suggested that South African NGOs had been heavily influenced by what he called the 'celebrities of the anti-globalization movement', and, in fostering the LPM, tried to apply Western theories to a South African context.

In what follows we argue that the LPM attempted to mimic the MST's success in mobilizing the rural poor by carrying out land occupations. The strategy was difficult to reproduce, however, both because of the particular political-economic context in South Africa and also because the LPM did not have the other two elements that were so crucial to the MST's success: organic leadership with a grassroots connection and political or financial autonomy.

Unsuccessful Strategies: Land Occupations

Shortly after its formation, the LPM began to develop increasingly radical tactics in its attempts to challenge the South African government. According to Amanda

Alexander, a visiting scholar at the Centre for Civil Society at the University of KwaZulu-Natal, LPM members first approached the government bureaucratically, by requesting to meet with government officials, sending letters and marching to government offices. However, after continued frustration with governmental inaction, movement leaders began to threaten to occupy land 'as one method of redistributing land through the self-activity of the landless, [identifying] unproductive, unused or underused land and land belonging to abusive farmers as the focus for initial redistribution' (Greenberg 2004b, 2). Actually, in the Landless People's Charter of 2001, the LPM calls for the redistribution of 'all unproductive, unused, vacant or indebted private land'. If their demands are not met, the movement's charter states that the LPM will 'launch a campaign to occupy vacant land and state land, and return to our land'.

In South Africa, however, land occupations are politically explosive because of the disastrous economic and social results of land occupations in neighbouring Zimbabwe. Since 2000, the land reform policies of Zimbabwean president, Robert Mugabe, have had a profound impact on the debates surrounding South Africa's approach to land reform. In the run-up to the presidential elections of February 2000, war veterans of Zimbabwe's struggle for independence, fed-up with the slow pace of land reform there, decided to take matters into their own hands by invading farms owned by White farmers. When White farmers appealed to the government, their requests were ignored. Mugabe, who was at risk of losing his presidency, condoned the invasions, eventually glorifying them as a blow to former colonial oppressors.

These actions led to the displacement of White commercial farmers and Black farm workers that resulted in a food crisis and famine in the region. In five years, Zimbabwe, once the region's breadbasket, is now commonly referred to as the region's basket case. The international community turned to South Africa, as the region's economic and political leader, to speak out against Mugabe's actions on land reform; however, because of South Africa's political 'obligations' to Mugabe (stemming from assistance Mugabe's government provided during the struggle against Apartheid in the 1980s), Mbeki did not strongly criticize Mugabe's actions. South Africa's unwillingness to criticize Mugabe early on has affected its economic situation.

Many members of the LPM vocally supported Mugabe's land expropriation programme and tried to use the Zimbabwe situation to give weight to their own threats to occupy land (Greenberg 2004b, 30). The South African Communist Party supported the LPM, writing on their website that:

> The LPM seeks a Land Summit that broadens our thinking as a society about land reform to include the lessons of other countries and continents. In particular, the LPM would like to place on the agenda the highly-successful land occupations model employed by the MST and other organisations in Brazil, as well as the Constitutional 'social obligations clause' that makes this possible in many Latin American countries. Essentially a 'use it or lose it' principle of land ownership and access, a social obligations

clause could empower the landless to identify, occupy and use land that is currently unused or unproductive. The LPM also considers the land of abusive farmers, and the land of absentee owners to be socially unproductive, and calls for this land to be made available for redistribution and use by the landless. (AIDC 2004)

Mangaliso seconded this statement when he argued that 'we want our people who are staying on farms to start cultivating the land now. [We don't care] that the farmers are saying "stop cultivating on our land". Those who are not afraid of farmers must start cultivating the land' (Interview). The LPM tried to leverage the Zimbabwe situation by threatening to occupy land, but the strategy backfired on them. In response to the economic downturn associated with support for Zimbabwe, the South African government, worried that foreign investor confidence would falter at any signs of Zimbabwe-style land grabs, came down hard on early attempts at occupation implementing a zero-tolerance policy and cutting off a major potential strategy of the LPM. In a press statement released in March 2004, the ANC declared: 'South Africans will not tolerate hooliganism that is only aimed at misleading people and creating chaos and discord. Those with designs to deliberately flout the law and occupy land will be met with the full might of the law. . . . If the LPM has legitimate concerns regarding the land restitution process, these can and should be dealt with through the appropriate government departments' (cited in Alexander 2004).

Of course, land occupations in South Africa are nothing new. At the height of the anti-Apartheid struggle, land occupations were staged throughout the country as part of a tactic to render it ungovernable. Most often these occupations were symbolic and temporary (Mngxitama 2005, 17–18). Greenberg (2004a) describes land occupations in the Apartheid era as 'weapons of the weak'. However, in this current conjuncture, according to Ricado, land occupation strategies that have been effective in Brazil are not politically viable in South Africa.

> So now we want to adopt the [strategies of the] MST and you hear people threaten about land occupations. People are always making these silly statements in the media . . . But it's not going to happen, you know, because there is no political process that could accompany [land occupations]. Even though the Constitution of Brazil allows for the 'social function of land' . . . it was a result of peasant resistance that they made the [1946] constitutional amendment (Interview).

Ultimately, Ricado argued that there were different understandings of how one could (legitimately) claim land in South Africa and Brazil.

Unsuccessful Strategies: Local Leadership

As demonstrated in the MST section of the chapter, movement leadership in Brazil is carefully cultivated at the grassroots level in the encampments. In one recent examination of leadership within the MST, Marcelo Rosa argues that for MST activists, the 'conduct of life obtains meaning from the social experience of landlessness' (2007, 7). In contrast, Rosa argues that leaders of the LPM conduct their lives outside of the sphere of landlessness. They tend to be older activists who have had previous experiences working within political parties and/or NGOs. They often have outside commitments that include other occupations and family obligations (unlike the MST, the LPM cannot provide its leaders remuneration to sustain their families). Greenberg's assessment of the LPM in 2004 found that, on paper, the LPM is structured 'at the top' with a national council, provincial councils and branch structures on paper. Most council members have no secure access to land. At the grassroots level; however, the LPM is 'mainly unstructured' (Greenberg 2004b, 21). Greenberg speculates that the attempt on the part of activists and development workers to create 'a tightly defined structure' for the LPM as a social movement was premature (2004b, 21).

Ricado argued that the objectives and actions of the LPM 'should be rooted in the everyday struggles of the people', but this is complicated because the base of the movement is 'from 40s upwards' and the leadership of the movement is ageing. 'This is a tough process requiring time and energy. Therefore the movement should focus on re-politicizing South African youth.' Ricado deliberately compared the LPM and the MST: the MST, he noted, had 'effectively resolved' this issue of young, energetic leaders by grooming young people into the movement at a young age and giving them major responsibilities within the encampments. He argued that the LPM leadership had to span national, local and international networks, and incorporate people who could organize and articulate the aspirations of members. A key problem for the LPM, according to Ricado, is that leadership is seen as part of the middle-class intelligentsia because NGO staff members have been so influential in the direction the LPM has taken. According to Ricado, even the South African state criticized the landless movement for not having a leadership that was connected to the grassroots base: 'the State essentially . . . says . . . "on whose behalf are you talking? Where are these people that you are representing?"' (Interview).

Unsuccessful Strategies: Political Autonomy

Perhaps the most important difference between the MST and the LPM is the ability for each movement to maintain its autonomy from both the state and organized elements of civil society. In Brazil, the MST adamantly cultivated its independence, using resources from the Brazilian government and the Catholic Church but never ceding its independence in exchange for funding. In South Africa, however, the LPM was heavily dependent on civil society organizations from the beginning. When the movement's relationship with these organizations began to sour and funding became more difficult, the LPM had few resources of its own to draw upon.

In the early post-Apartheid era, civil society organizations like the NLC were key political organizations. The relationship between state and civil society –

mediated through these organizations – went from one of 'opposition and conflict' to 'partnership and cooperation' (Greenberg 2004a; Alexander 2004; Greenstein 2003). The new government funded NGOs to implement developments projects that were compatible with the government's vision for a new South Africa. This partnership often made it difficult for NGOs to openly criticize or oppose the government (Alexander 2004): many organizations were compelled to professionalize their staff members in order to actively pursue those projects and activities prioritized by donor organizations and the government (Greenberg 2004a, 25).

These processes of politicization and professionalization alienated civil society organizations from the constituencies they were responsible for serving and shifted priority away from grassroots organizations. A racialized division of labour also developed within the NGOs. For example, NLC fieldworkers were mostly Black activists from the local communities, while programme coordinators and researchers were mostly White. This racial segmentation exacerbated internal tensions (Mngxitama 2005).

Paradoxically, these processes of depoliticization, professionalization and alienation provided the political space for new social movements to develop (Barchiesi 2004, 4). President Mbeki's first term in office 'coincided with a resumption of social movement politics and activism which articulates a common rejection of the ANC's conservative turn in macroeconomic policy' (Barchiesi 2004, 23; Greenstein 2003). This upsurge of new social movements, however, built upon the legacy of strong civil society organizations. Some scholars argued that the social movements continued to be too closely linked to civil society institutions that held the purse strings and ultimately this encouraged 'gradualism over radical courses of action' (Alexander 2004, 49–50; Mngxitama 2004).

This changing relationship between the state, civil society and social movements lays the foundation from which to examine the national, regional and international circumstances and support that led to the rise of the LPM. As previously mentioned, 'the LPM was a creation of these NGOs from top to bottom, inspired by the MST, seeking to encourage a response from the families of Black rural labourers that were being expelled from rural areas' (Sihlongonyane 2005, cited in Rosa 2007, 4). The LPM relied on the NLC in particular for organizational and financial support. Andile Mngxitama of the NLC argues that implicit in these relationships was the idea that 'without donor funding there can be no possibility of struggle' (2005, 41). Likewise, middle-class intellectuals working with the movement through civil society connections shape 'the image of the movement' through their national and international resources and connections (Greenberg 2004b, 25).

Ricado argued that the NLC was a conflicted institutional host for the LPM. The NLC leadership disagreed over how best to approach the government's political and economic stance after 1994. Greenberg (2004b) describes this conflict as a split between what he calls the antagonistic faction within the NLC, those who saw redistribution through mass occupations as the best method to gain access to land and the faction that supported critical engagement with the government. Andile Mngxitama (2005), who served with Ricado in the NLC, describes this debate as a struggle between the welfarist-reformist tendencies within the committee, and the mass action tendency, especially around the issue of land occupations and the attitude toward new government. These internal tensions within the NLC were transmitted to the LPM, and Ricado attributed the break-up of the NLC to this clash of ideas and ideologies and disputes over the nature of dissent and relationship with the government.

As factions within the LPM proposed increasingly radical strategies (such as land occupation) that opposed the government rather than operating within the legal framework (established bureaucracy) for negotiation, NLC members and affiliates who preferred a less radical approach began to separate themselves from the movement. Because the LPM relied heavily on the NLC for funding and financial management, office space, and media and legal assistance (Alexander 2004), when the NLC dissolved the movement lost its primary source of organizational and financial support. In an interview, Ricado suggested that the NLC had been partly responsible in allowing financial dependency to develop: 'But I think as NGOs, we have to really look at ourselves and how we create certain dependencies and orientations among people and communities, because this is the difficulty that [the LPM] are sitting with to organize and mobilize.... And so we have to look at the role of the NGOs in that' (Interview). Ricado argued that the LPM needed to be able to define itself better as a movement by developing a 'program of action, a philosophy, and an outlook ... that will inspire rural peasants or the landless to stand up and start engaging in [open] struggles' by speaking to the everyday realities of landless South Africans.

CONCLUSION: LATE MOBILIZATION IN A TRANSNATIONAL WORLD

In the end, what do we gain from this comparison of the MST and the LPM? They are clearly two very different movements, despite their common participation in the transnational peasant network. And yet, it is only in hindsight that their differences stand out. When the LPM first formed in 2001, the movement bore many similarities to the MST: the LPM drew on a long history of agrarian repression, and the movement also articulated a popular anti-state sentiment that allowed its activists to win support from a broad range of people throughout the South African countryside and cities. Upon forming, the LPM became an immediate – and seemingly natural – partner with the MST. Brazilian activists sympathized with the long history of landlessness and dispossession in Brazil, and they supported the LPM's work by sending activists to South Africa and hosting the LPM's leaders in return. It seemed self-evident that the LPM would be an important part of the transnational peasant network along with the MST. Only five years later, however, the LPM had essentially disbanded and key scholars and activists no longer regarded it as a potential actor in the South African debate over land inequality and poverty. How can we understand this transition?

We do not attempt to provide a comprehensive explanation for the LPM's failure in this chapter. Rather, we explore the relationship between the LPM and the MST in an attempt to better understand the dynamics of the transnational peasant networks – and how these dynamics may have shaped the LPM's trajectory. A side-by-side comparison of the movements (Matrix 1) suggests that key differences between the two made it difficult for knowledge to be passed among them through network connections. This presents a potential problem for progressive politics because the transfer of knowledge is considered to be one of the main benefits of such networks.

Matrix 1: **A Comparison of the MST and the LPM**

	The MST	The LPM
International participation and representation	Became internationalized over the course of its development.	International visibility from its inception.
Demand for redistribution	Based on the right to property for those who work it (labour theory of property); land for small farmers to work.	Based on demand for racial equality; land for residence, citizenship, housing and rights.
Relationship to NGOs	NGOs provided institutional support but the MST always kept its distance from NGO authority and developed diverse sources of funding.	Founded by NGO leaders and maintained close ties, especially for funding.
Basis of leadership	Organic leadership from the grassroots; cultivated in occupations and through prior political activism.	Most visible leaders are NGO transplants, although there is grassroots participation.
Sources of funding	Settlements (internal funding), sale of MST products, NGOs – particularly international NGOs, state assistance.	Primarily NGO funding, both national and international.

In conclusion, there are three key arguments in this chapter. Our first and most basic argument is simply that transnational networks – no matter how progressive or how alternative – are not flat. They, like the world system more generally, are riddled with inequalities and power differences. At the same time, actors within transnational networks or movements are located within quite distinct historical trajectories, no matter the universal calls for 'peasant', 'landless' or 'worker' solidarity. Our second argument is that as a result of historical specificities, the study of transnational networks needs to be located at multiple scales, both the 'globalized' and the 'localized'. At both scales, we argue that moral economies of production and social reproduction are important, as is the very nature of the places where transnational networks meet up and touch down. Third, and finally, we argue that this sort of critical transnational analysis requires a fairly specific method. The study of histories in the present at multiple scales requires both critical ethnography (Hart 2003, 2004) that is sensitive to the intimate nature of social relationships and what Philip McMichael (1990) calls 'incorporated comparison', where cases are compared both for their differences/similarities and for their inter-connecting relationships. Such methods operationalize Gerschenkron's theory of late development, or in the case of peasant social movements, late mobilization. Understanding the precise mechanisms by which transnational networks come to favour 'early mobilizers' versus 'late mobilizers' requires more ethnographic attention to the networks themselves than we have done in our own research. Again, we would like to stress that the disadvantages of late mobilization are not pre-determined or automatic. The influence of the transnational networks on particular movements is always situated within the concrete histories and contingent workings of the world system.

REFERENCES

Adams, M. and J. Howell, 2001. 'Redistributive Land Reform in Southern Africa'. *Natural Resource Perspectives*, 64.

AIDC (Alternative Information and Development Centre), 2004. 'Landless People's Movement (Gauteng): Discussion Document on Forging a United Front with the SACP'. http://www.aidc.org.za/?q=book/view/363, Accessed 8 January 2008.

Alexander, A.S., 2004. '"Not the Democracy We Struggled For": The Landless People's Movement and the Politicization of Urban-Rural Division in South Africa'. Unpublished BA Honours Thesis. Boston, MA: Harvard University.

Barchiesi, F., 2004. 'Classes, Multitudes and the Politics of Community Movements in Post-Apartheid South Africa'. Centre for Civil Society, Research Report 20: 1–41.

Bernstein, H., 2004. '"Changing Before our Very Eyes": Agrarian Questions and the Politics of Land in Capitalism Today'. *Journal of Agrarian Change*, 4 (1/2): 190–225.

Borras, S. Jr, 2003. 'Questioning Market-Led Agrarian Reform: Experiences from Brazil, Colombia and South Africa'. *Journal of Agrarian Change*, 3 (3): 367–94.

Branford, S. and J. Rocha, 2002. *Cutting the Wire: The Story of the Landless Movement in Brazil*. London: Latin American Bureau.

Cardoso, F.H., 1997. *Reforma Agrária: Compromisso de Todos*. Brazil: Presidencia da Republica and Secretaria de Comunicação Social.

Cousins, B., 2006. 'Agrarian Reform and the "Two Economies": Transforming South Africa's Countryside'. In *The Land Question in South Africa*, eds L. Ntsebeza and R. Hall, 220–45. Capetown: HSRC Press.

De Janvry, A., 1981. *The Agrarian Question and Reformism in Latin America*. Baltimore, MD: Johns Hopkins University Press.

De Janvry, A., E. Sadoulet and W. Wolford, 2001. 'The Changing Role of the State in Latin American Land Reforms'. In *Access to Land, Rural Poverty, and Public Action*, eds A. de Janvry, G. Gordillo, J. Platteau and E. Sadoulet, 279–303. Oxford: Oxford University Press.

De Souza Martins, J., 2003. 'Representing the Peasantry? Struggles for/about Land in Brazil'. In *Latin American Peasants*, ed. T. Brass, 300–35. London: Frank Cass Publications.

Deere, C.D. and L.S. de Madeiros, 2007. 'Agrarian Reform and Poverty Reduction: Lessons from Brazil'. In *Land, Poverty and Livelihoods in an Era of Globalization: Perspectives from Developing and Transition Countries*, eds A.H. Akram-Lodhi, S. Borras Jr and C. Kay, 80–119. London: Routledge.

Desai, A. and R. Pithouse, 2003. '"But We Were Thousands". Dispossession: Resistance, Repossession and Repression in Mandela Park'. *Journal of Asian and African Studies*, 39 (4): 239–69.

Desmarais, A.A., 2007. *La Vía Campesina: Globalization and the Power of Peasants*. London: Pluto Press.

Edelman, M., 1998. 'Transnational Peasant Politics in Central America'. *Latin American Research Review*, 33 (3): 49–86.

Edelman, M., 2005a. 'When Networks Don't Work: The Rise and Fall and Rise of Civil Society Initiatives in Central America'. In *Social Movements: An Anthropological Reader*, ed. J. Nash, 29–45. Oxford: Blackwell.

Edelman, M., 2005b. 'Bringing the Moral Economy Back in . . . to the Study of Twenty-first Century Transnational Peasant Movements'. *American Anthropologist*, 107 (3): 331–45.

Featherstone, D., 2003. 'Spatialities of Transnational Resistance to Globalization: The Maps of Grievance of the Inter-Continental Caravan'. *Transactions of the Institute of British Geographers*, 28: 404–21.

Ferguson, J., 1994. *The Anti-Politics Machine: 'Development', Depoliticization, and Bureaucratic Power in Lesotho*. Minneapolis, MN: University of Minnesota Press.

Fernandes, B.M., 1999. *MST, Movimento Dos Trabalhadores Rurais Sem Terra: Formação e Territorialização*. São Paulo: Editora Hucitec.

Fernandes, B.M., 2007. 'The Occupation as a Form of Access to Land in Brazil: A Theoretical and Methodological Contribution'. In *Reclaiming the Land: the Resurgence of Rural Movements in Africa, Asia and Latin America*, eds S. Moyo and P. Yeros, 317–41. New York: Zed Books.

Forman, S., 1975. *The Brazilian Peasantry*. New York: Columbia University Press.

Gerschenkron, A., 1952. 'Economic Backwardness in Historical Perspective'. In *The Progress of Underdeveloped Areas*, ed. B. Hoselitz, 3–29. Chicago, IL: University of Chicago Press.

Gohn, M.d.G., 1997. *Os Sem Terra, ONGs e Cidadania*. São Paulo: Cortez.

Greenberg, S., 2004a. 'The Landless People's Movement and the Failure of Postapartheid Land Reform'. *Centre for Civil Society Research Report*, 26: 1–40.

Greenberg, S., 2004b. 'Post-Apartheid Development, Landlessness and the Reproduction of Exclusion in South Africa'. *Centre for Civil Society Research Report*, 17: 1–42.

Greenstein, R., 2003. 'Civil Society, Social Movements and Power in South Africa'. RAU Sociology Seminar Series, Fourth term, 12 September: 1–40.

Grindle, M., 1986. *State and Countryside: Development Policy and Agrarian Politics in Latin America*. Baltimore, MD: Johns Hopkins University Press.

Hall, A.L., 1990. 'Land Tenure and Land Reform in Brazil'. In *Agrarian Reform and Grassroots Development: Ten Case Studies*, eds R. Prosterman, M. Temple and T. Hanstad, 205–35. Boulder, CO: Lynne Rienner Publishers.

Hart, G., 1996. 'The Agrarian Question and Industrial Dispersal in South Africa: Agro-industrial Linkages through Asian Lenses'. In *The Agrarian Question in South Africa*, ed. H. Bernstein, 245–77. London: Frank Cass.

Hart, G., 2002. *Disabling Globalization: Places of Power in Post-Apartheid South Africa*. Berkeley, CA: University of California Press.

Hart, G., 2003. 'Denaturalizing Dispossession: Critical Ethnography in the Age of Resurgent Imperialism'. Paper prepared for Creative Destruction: Area Knowledge & the New Geographies of Empire conference, Center for Place, Culture & Politics CUNY Graduate Center New York, 15–17 April 2004.

Hart, G., 2004. 'Geography and Development: Critical Ethnographies'. *Progress in Human Geography*, 281 (1): 91–100.

Houtzager, P.P. and M. Kurtz, 2000. 'The Institutional Roots of Popular Mobilization: State Transformation and Rural Politics in Brazil and Chile'. *Comparative Studies in Society and History*, 42 (2): 394–424.

Keck, M. and K. Sikkink, 1998. *Activists Beyond Borders: Advocacy Networks in International Politics*. Ithaca, NY: Cornell University Press.

Lahiff, E. and B. Cousins, 2001. 'The Land Crisis in Zimbabwe Viewed from South of the Limpopo'. *Journal of Agrarian Change*, 1 (4): 652–66.

Mainwaring, S., 1995. 'Brazil: Weak Parties, Feckless Democracy'. In *Building Democratic Institutions: Party Systems in Latin America*, eds S. Mainwaring and T. Scully, 354–98. Stanford, CT: Stanford University Press.

Martorell, J., 2007. 'World Summit for Sustainable Development in Johannesburg: ANC Government Tries to Silence its Critics with Apartheid-Style Repression'. http://www.marxist.com/Africa/wssd_repression.html, Accessed 15 November 2007.

Maybury-Lewis, B., 1994. *The Politics of the Possible: The Brazilian Rural Workers' Trade Union Movement, 1964–1985*. Philadelphia, PA: Temple University Press.

McMichael, P., 1990. 'Incorporating Comparison within a World-Historical Perspective: An Alternative Comparative Method'. *American Sociological Review*, 55 (3): 385–97.

Mngxitama, A., 2005. 'The National Land Committee, 1994–2004: A Critical Insider's Perspective'. Centre for Civil Society, Research Report, 34 (2): 1–48.

Navarro, Z., 2002. 'Mobilização sem emancipação – as lutas sociais dos sem-terra no Brasil'. In *Produzir para Viver*, ed. B. de Sousa Santos, 189–232. Rio de Janeiro: Editora Civilização Brasileira.

Novaes, R.C.R., 1998. 'A Trajetória de uma bandeira de luta'. In *Política e Reforma Agrária*, eds R. Nonato and L.F. Carvalho Costa, 169–80. Rio de Janeiro: Ed. Mauad.

Ntsebeza, L., 2007. 'Land-Reform Politics in South Africa's Countryside'. *Peace Review: A Journal of Social Justice*, 19 (1): 33–41.

Ondetti, G., 2001. 'When Repression Backfires: The Rise of the Brazilian Landless Movement in the Mid-1990s'. Paper presented at Latin American Studies Association meetings, Washington DC, 6–8 September.

Pereira, A.W., 1997. *The End of the Peasantry: The Rural Labor Movement in Northeast Brazil, 1961–1988*. Pittsburgh, PA: University of Pittsburgh Press.

Pereira, A.W., 2003. 'Brazil's Agrarian Reform: Democratic Innovation or Oligarchic Exclusion Redux?'. *Latin American Politics and Society*, 45 (2): 41–65.

Petras, J., 1997. 'Latin America: The Resurgence of the Left'. *New Left Review*, 223: 17–47.

Reis, E.P., 1990. 'Brazil: 100 Years of the Agrarian Question'. *International Social Science Journal*, 50 (157): 419–32.

Rosa, M., 2007. 'Landless Experiences: Youth and Social Movements in Brazil and South Africa'. Unpublished Manuscript. Brazil: Federal University of Fluminese.

Routledge, P., A. Cumbers and C. Nativel, 2007. 'Grassrooting Network Imaginaries: Relationality, Power, and Mutual Solidarity in Global Justice Networks'. *Environment and Planning A*, 39 (11): 2575–92.

Santos, R. and L.F.C. Costa, 1998. 'Camponeses e Política no Pré-64'. In *Política e Reforma Agrária*, eds L.F.C. Costa and R. Santos, 13–41. Rio de Janeiro: Mauad.

Schwartz, S., 1985. *Sugar Plantations in the Formation of Brazilian Society: Bahia, 1550–1835*. New York: Cambridge University Press.

Seligmann, R., 1998. *PROCERA: Programa Especial de Credito Para Reforma Agrária*. Brasilia, DF: Instituto Nacional de Colonização e Reforma Agrária.

Sihlongonyane, M.F., 2005. 'Land Occupations in South Africa'. In *Reclaiming the Land: the Resurgence of Rural Movements in Africa, Asia, and Latin America*, eds S. Moyo and P. Yeros, 142–64. London: Zed Books.

Smith, N., 1993. 'Homeless/Global: Scaling Places'. In *Mapping the Futures. Local Cultures, Global Change*, eds J. Bird, B. Curtis, T. Putnam, G. Robertson and L. Tickner, 87–119. London: Routledge.

Stedile, J.P., 2002. 'Landless Battalions: The Sem Terra Movement of Brazil'. *New Left Review*, 15: 77–104.

Thwala, W.D., 2006. 'Land and Agrarian Reform in South Africa'. In *Promised Land: Competing Visions of Agrarian Reform*, eds P. Rosset, R. Patel and M. Courville, 57–72. Oakland, CA: Food First Publications.

Tilly, C., 1978. *From Mobilization to Revolution*. Reading, MA: Addison-Wesley.

Viotti Da Costa, E., 2000. *The Brazilian Empire: Myths and Histories*. Chapel Hill, NC: University of North Carolina Press.

Wolford, W., 2006. 'The Difference Ethnography Can Make: Understanding Social Mobilization and Development in the Brazilian Northeast'. *Qualitative Sociology*, 29 (3): 335–52.

Wright, A. and W. Wolford, 2003. *To Inherit the Earth: The Landless Movement and the Struggle for a New Brazil*. Oakland, CA: Food First Publications.

6 Mobilizing Against GM Crops in India, South Africa and Brazil

IAN SCOONES

INTRODUCTION

This study explores the local and transnational character of mobilization dynamics around GM (genetically modified) crops in India, South Africa and Brazil over the ten-year period to 2005. The story starts in the mid-1990s with the first concerted attempts to introduce GM crops. In India, transgenic material – Monsanto's Bt product[1] – was first imported in 1995 and, in 1989, the first biosafety regulations were approved. However, Bt cotton crops were not approved for commercial release until March 2002, although several years of illegal plantings of 'pirated' GM cotton had occurred previously.[2] In South Africa, the first Bt cotton trials were started by Monsanto in 1992 under the apartheid regime, with transgenic cotton released for commercial use in 1997. In 1998 yellow Bt maize was also commercialized, with Bt white maize following in 2000. The South Africa GMO Act was passed in 1997, but did not come into force until the end of 1999. In Brazil the 1995 Biosafety Law created the regulatory authority CTNBio (the National Biosafety Technical Commission), which in 1997 approved the first trial of Monsanto's 'Roundup Ready' herbicide-tolerant GM soya. This was only formally approved for sale in 2002, following long wrangles in the courts. In March 2005, a new Biosafety Law was ratified giving CTNBio the authority to approve plantings.

GM crops were introduced into very different agrarian settings in the three countries. In Brazil, GM soya was smuggled across the border from Argentina, and used extensively by large-scale commercial producers, without paying the premium price that Monsanto was charging their competitors (da Silveira and de Carvalho Borges 2005). In South Africa, similarly large commercial concerns were strong advocates of GM maize, particularly as a route to cost reduction, as subsidies to the largely white commercial farm sector were progressively reduced. GM crops only later became linked to smallholder farming, especially

---

[1] Bt cotton is a transgenic product based on the insertion of the Cry1a gene from the bacterium *Bacillus thuringensis*. It confers resistance to some insect pests, notably the bollworm *Helicoverpa armigera*.
[2] As one reviewer of this chapter correctly pointed out, the term 'pirated' GM cotton is a misnomer in the India context given that there is no patent protection in place.

with Bt cotton in the Makhatini Flats area of KwaZulu Natal (Thirtle et al. 2003). Only in India were GM crops primarily smallholder crops in the cotton farming areas, where they were adopted on a massive scale even before formal regulatory release. Here the incentive to reduce the costs of pesticide use on cotton was high, and Monsanto was keen to move beyond their traditional focus on agro-chemicals into the seed sector (Glover 2007). Thus, different groups of farmers were primarily concerned with planting GM crops – large-scale commercial producers of soya and maize in Brazil and South Africa, and the often relatively well-off smallholder growers of cotton, able to afford the inputs, in India and (to a lesser extent) in South Africa.

The arrival of GM crops has, however, not occurred quietly. Over the past decade a storm of protest has erupted in all three countries, fuelled by a variety of groups, intense media interest and a growing global debate. Debates have centred on commercialization decisions for GM crops, scares about the prospects of 'terminator' genes,[3] alongside wider concerns about patenting, corporate control and the consequences of globalized trade relations. In many respects the framing of opposition to GM crops was similar in all three countries, but there were important contrasts, derived from local considerations and contexts, as will be shown below. In all three countries this period saw an array of anti-GM groups emerge. Each had varying, and sometimes unclear affiliations and associations with farmers' organizations and movements. Sometimes there was clear alignment of political and material interests; sometimes these were less clear. A key question, explored below, is to ask, who were these new anti-GM groups speaking for and what interests were they representing? Thus, new alliances and networks were made at national, regional and global levels, creating a range of sites for mobilization; each with 'local' and 'global' dimensions, resulting, in turn, in a variety of intersecting forms of transnational and local mobilization. These local efforts have tapped into networks of global activism, connected by the Internet and extensive links between groups coming from agrarian, environmental and consumer campaigns, among others.

Anti-GM activists argued that, due to monopoly power, GM crops would result in the costs of inputs increasing and the diversity of seed choice declining, forcing poorer farmers out and allowing a form of uniform, corporate-capitalist agriculture to dominate. These risks would be compounded, they argued, by potential threats to biodiversity from the spread of GM genetic material, and consumers could be additionally at risk from potentially unsafe foods. Pro-GM advocates argued, by contrast, that GM seeds would reduce costs for farmers in a scale-neutral way, allowing rich and poor alike to benefit. By removing farmers from the burden of purchasing pesticides, for example, both health and economic benefits would result. No known health or environmental risks existed, they claimed, and, if governed by a streamlined regulatory system, all would

---

[3] 'Terminator' genes refer to Genetic Use Restriction Technologies (GURTs), where inserted genes prevent use in the following generations. Despite the furor over this technology, it has not as yet been released.

be well, and the benefits of a 'gene revolution' would be realized. These stylized arguments were played out in very different ways in different places, with different elements being emphasized, depending on the context. Across the case-study countries, diverse alliances were forged between those concerned with the protection of indigenous crop varieties in dryland farming areas, land activists arguing for radical land reform, nationalists interested in protecting the national economy and culture from outside influences and anti-globalization protestors, linked to a growing international network of activists.

Some commentators have dismissed anti-GM mobilizations as merely copycat responses by elite activists, using links with farmers' organizations as a way of raising funds (e.g., Paarlberg 2001). But are such efforts in reality more than this? Are these not new forms of legitimate political expression, ones replete with contemporary contradictions, but nevertheless an important contribution to democratic debate in a context where, because of the forces of neo-liberalism, alternatives have little space? This chapter argues that, in different ways, these anti-GM groupings are all examples of hybrid network forms of social activism, linking people, issues and politics in new ways around globally-defined issues, but always co-constructed in local contexts and through local political processes. By tracing connections from three national settings to the international arena, and by examining links between the three countries, the study assesses the degree to which anti-GM activism adds up to a transnational movement, generated by a process of 'globalization-from-below' (Appadurai 2000), and, in turn, also raises questions about processes of representation, and the relationships between anti-GM groups and a differentiated rural farming populace.

The chapter starts with an examination of mobilization dynamics, exploring comparatively both the transnational and local character of activism. This located approach focuses first on the 'messy, close up view of collective action' (Edelman 2001, 286), before drawing out broader insights on the structural, political and historical factors conditioning such practices. Across a series of different sites of engagement, the study asks: what issues are being raised and what demands are being made? How are these being framed and by whom? What connections are there between local and global sites of engagement? And, most critically, what do these practices of mobilization tell us about the transnational character of anti-GM mobilization, and its social and political bases.

For each country, the aim has been to focus on unfolding events over a ten-year period to 2005, and to set this analysis in a wider historical view of mobilization and politics within national, regional and global settings. The empirical material presented is largely based on semi-structured interviews with key players,[4] combined with archival analysis of documents – including official government records, activist organization publications and newspaper articles.

---

[4] A total of over 70 detailed interviews were carried out as part of this research in India (mainly February 2004), in South Africa (March 2004) and in Brazil (mainly April 2004). Work in India built on longer-term research between 2000 and 2005 on this theme (Scoones 2005a).

MOBILIZATION DYNAMICS: THREE COUNTRY CONTEXTS

India

In India, the debate about GM crops was brought to national media attention for the first time in 1998 around the so-called 'terminator' controversy.[5] Monsanto's importation of transgenic cotton raised fears that such a product would include – in the labelling of the global anti-GM activists, notably RAFI (Rural Advancement Foundation International, based in Canada)[6] and GRAIN (Genetic Resources Action International, based in Spain) – a terminator gene, which would prevent replanting and make farmers reliant year-on-year on the seed companies. This was emotive stuff, and highly effective in raising fears, despite Monsanto's strenuous denials. These fears touched many chords. For some, Bt cotton became symbolic of a much wider struggle against the dominance of multinational capital, particular forms of technological modernization and globalization more generally.

By mid-1998, a public relations battle was on. Monsanto launched a series of adverts in the press and, at the same time, NGO groups launched the 'Monsanto Quit India' campaign to heighten awareness (RFSTE 1998). Although it was well-known that field trials had been established, details of trial sites became public only in November 1998. The KRRS (*Karnataka Rajya Ryota Sangha*, the now split Karnataka farmers' movement) immediately announced the 'Cremate Monsanto' campaign. The late KRRS leader, Professor M. D. Nanjundaswamy, identified a series of slogans: 'Stop Genetic Engineering', 'No Patents on Life', 'Cremate Monsanto' and 'Bury the WTO'. He gave notice that all trial sites in the southern Indian state of Karnataka would be burned, with the media in attendance.[7] The US embassy, in turn, requested police protection for US companies in Bangalore, and the High Court of Karnataka ruled to protect sites and the property of the Mayhco seed firm (*Samykta Karnataka* 1998). By December, the KRRS threatened to launch a criminal case under the Union Seed Act in magistrates courts against Monsanto, as well as the state and central government, on the basis that trials were illegal (*Deccan Herald* 1998).

The media debate continued at a high pitch through 1999 and 2000, with plenty of opportunity for press commentary prompted by a number of workshops and consultations.[8] Monsanto was still in the spotlight following its attempts to import plasmids for transgenic crop research at their Bangalore research centre (*Economic Times* 1999; *The Hindu* 1999). There were by now some more concerted

---

[5] For example, see Omvedt (1998); Shiva (1998).
[6] Now known as the ETC Group, see http://www.etcgroup.org.
[7] According to press reports, the first burning took place on 28 November 1998 with the consent of the farmer. Activists from KRRS along with the (little known) 'Progressive Front', 'Action Front for the Untouchables', 'Karnataka Liberation Front' and the 'Organisation of the Landless', according to a press release from KRRS, Bangalore. The burning was also attended by a five-member team from the Geneva-based 'Global People's Action Group'.
[8] For example, the TERI 'stakeholder dialogues'; the National Science Summit in Bangalore; the MSSRF national consultation on GM plants, Chennai.

counter-moves by the pro-GM lobby, with interventions from non-resident Indian scientists (most notably C. S. Prakash, from Tuskegee University in the USA, who made several visits to India and managed to place a wide range of articles in the press), other farmer leaders (including Sharad Joshi of *Shektari Sanghatana* and Chengal Reddy of the Andhra Pradesh Farmers' Association)[9] and industry commentators, including a more measured contribution from Monsanto, which had commissioned a public opinion survey, unsurprisingly showing how farmers were in support of biotechnology. There remained very few local Indian scientists prepared to enter the fray, although the likes of the Nobel laureate and Green Revolution hero Norman Borlaug were not shy of offering their opinion in pieces placed in the Indian press.

Soon after Bt material was imported, objections were presented as a court petition by the Research Foundation for Science Technology and Ecology (RFSTE), headed by Vandana Shiva, disputing the form and content of the regulations (Shiva et al. 1999). Since then, the courts have seen continued action, with public interest litigation following thick-and-fast. The RFSTE petition involved extensive hearings at the Supreme Court and vast amounts of evidence. A Delhi High Court action by the Gene Campaign in 2001 claimed that the illegal sales of GM seeds in Gujarat were made with the knowledge of the government. A further petition by the Gene Campaign argued for the right to information disclosure on trial results under the Freedom of Information Act.

Other more direct forms of protest have also continued, with the KRRS active in crop-burning media events, and arguing for a five-year moratorium on GM seeds (*Economic Times* 2001), following the pattern of the established European 'freeze' campaign. Protests have occurred in a variety of places, including regular rallies and demonstrations at Monsanto's former India research headquarters at the Indian Institute of Science in Bangalore (AgBioIndia 2002). Events such as the citizens' juries in Karnataka in 2000 and in Andhra Pradesh in 2001 also provided foci for activists to denounce GM crops and the associated future for agriculture (ActionAid 2000; Pimbert and Wakeford 2002). Media interest remained high, with competing Internet-based services providing alternative views on the Indian scene.[10]

The formal release of Bt cotton in 2002 provoked more protests. Attempts at crop burning during 2002 had mixed results, with some farmers accepting compensation from KRRS protestors for the public destruction of their crop, while others firmly refused such advances and called in the police. In the last few years, protests have been more muted. In part this was because of the failing health of Nanjundaswamy, prior to his death in February 2004. With his grip on the KRRS faction that he controlled faltering, a less coordinated and energetic

---

[9] See *Financial Express* (1999, 2000, 2001).
[10] For example, see AgBioIndia (http://www.Agbioindia.org), supported by Delhi-based activist Devinder Sharma; more international lists (including bio_activists@iatp.org) with a more sceptical crops stance, and Monsanto India (http://www.monsantoindia.com) taking more pro-biotech positions.

campaign was evident. However, others remained active. Gene Campaign, for example, held a high-profile conference in Delhi, which argued for an overhaul of the regulatory system.[11] Meanwhile, Greenpeace, with its India office now located in Bangalore, was gearing up for consumer-based protests in shopping outlets, and eye-catching protests around regulatory discussions.

In 2003 and 2004 protests continued, but many activists had their eye on the three-year review of the Bt cotton results in 2005. Much was invested in providing alternative evidence based on surveys in the cotton areas, which would demonstrate the limits of the technology. Many campaign-focused NGOs also had begun to see the anti-GM campaign as inherently limiting, and were keen to provide the other side of the story, developing a narrative about possible alternatives. For example, the Karnataka Coalition Against GM Crops developed links between a range of groups, including those very much located within the alternative agriculture movement in India and beyond.[12]

South Africa

In 1997 the campaign and research group, Biowatch, was formed in South Africa by a small group led by a former university biological scientist, Rachel Wynberg. Biowatch had a relatively low-key start involving a series of workshops, debates and commissioned papers focusing on issues such as labelling and segregation.[13] By 2000 an allied organization, SAFeAGE (the South African Freeze Alliance on Genetic Engineering) had been launched, inspired by the European 'freeze campaign'. Led by a Cape Town based activist, Glenn Ashton, SAFeAGE raised the tempo with a media-oriented strategy of raising awareness of the GM issue. Regular newspaper articles, TV slots and appearances on talk radio meant that GM arrived as a public issue.[14] The emerging coalition of groups organized a number of high-profile events, pulling in global anti-GM luminaries like Vandana Shiva from India and Tewolde Berhan Egziabher from Ethiopia. To counter this, an industry-funded pro-GM organization, AfricaBio, was launched in the late 1990s (see Friedberg and Horowitz 2004), which attempted to occupy the scientific high ground, legitimizing its stance in terms of scientific expertise. Over this period, there were frequent confrontations between opposing sides on the GM debate, almost always with the same people on the platform.

The debate, however, was dominated almost exclusively by white, middle-class, well-educated activists with bases in Cape Town, Johannesburg or Durban. In the post-apartheid era, left-leaning white activists who had been part of the struggle in various ways before 1994 found themselves in an uneasy position. Largely excluded from the new government machinery, they found themselves in a variety of NGOs that had to refashion their existence to the context of the 'new'

---

[11] See genecampaign.org press release.
[12] Including Green Foundation, ICRA, AME and others.
[13] See http://www.biowatch.org.za.
[14] For an example, see *Business Day* (2003, 2004); *Mercury* (2003); *Mail and Guardian* (2000).

South Africa. Attempts have been made to reach out to a broader constituency through links to other organizations as part of building a firmer anti-GM coalition. Thus unions (e.g., the Food and Allied Workers Union), farmer groups (the Organic Agriculture Association of South Africa), consumer groups (the National Consumer Forum; the Safe Food Coalition), rights-based organizations (the Environmental Justice Networking Forum, EJNF), development and environment organizations (Environment Monitoring Group; the World Conservation Union), faith groups (including the South African Council of Churches; the Pietermaritzburg Agency for Christian Awareness; the Ecumenical Service for socio-economic transformation), conservation groups (Botanical and Wildlife Societies of South Africa) and green groups (Earthlife Africa; Earth Women; Trees for Africa) have all been involved in the campaigns. SAFeAGE argues that it is an alliance of over 200,000 people across South Africa.

Making links with international players has also been an important strategy. These have included the Malaysia-based Third World Network, RFSTE in India, and northern-based anti-GM activists associated with GRAIN, RAFI and others. The World Summit on Sustainable Development, held in Johannesburg in September 2002, was an important meeting point, with many parallel workshops devoted to GM issues, and South African participants from a range of organizations were very much involved. This international work has boosted confidence and legitimacy for work at home, which has become increasingly focused on strategic interventions in legal processes, combined with some profile-raising protest actions. In August 2002, Biowatch served court papers to the National Department of Agriculture, who were joined by Monsanto the following year as co-defendants in a case focusing on constitutional rights to access to information on GM trials and approvals. The court hearings were finally held during 2004 with a judgement in favour of Biowatch's case delivered in February 2005. A similarly protracted case involving the approval of Syngenta's Bt11 maize was filed by Biowatch in 2002. The appeal board case was only heard in 2004, with the appeal dismissed, although conditions on Syngenta were applied. In parallel, intensive work with the parliamentary portfolio committees of Agriculture and Land Affairs and Environment occurred, focusing on the bill amending the 1997 GMO Act. The new provisions were finally passed in 2004, prompting a full-scale response from Biowatch.[15]

Meanwhile, SAFeAGE – together with a few of the more activist environmental groups, such as Earthlife Africa, EJNF, Earth Women, Ekogaia – engaged in popular protests aimed at the retailing chains, highlighting issues of labelling in particular. Consumption boycotts – picking up on apartheid era protest tactics – were focused on outlets such as Pick 'n Pay and Woolworths. Trolley runs (where protestors fill supermarket trolleys with goods and request that they are checked for GM products) provided a focus for further media coverage.

---

[15] See details of all these actions at http://www.biowatch.org.za.

In 2005, following the court success by Biowatch, the pro-GM lobby group AfricaBio issued a vituperative statement entitled 'Thoughtless activists continue to misinform and mislead the public'.[16] They claimed that anti-GM activities were undermining human rights and were fuelled by newspaper editors who did not check their facts. The statement claimed that the anti-GM campaigns were part of:

> a well-orchestrated campaign financed to the tune of some $70 million a year by foundations, organic food interests, EU governments, and even UN agencies and programmes. It employs moratoriums and threats against agricultural imports from countries that grow biotech crops, complex and expensive requirements for labeling all GM ingredients and tracking them from seed to store shelf, even outright lies about the safety of biotechnology.

A decade ago, GM crops were barely a concern in South Africa. The government, together with industry and a small cabal of scientists, set the terms. Today, this has changed. A combination of high-profile court cases, on-going demonstrations, a growing media profile and long-term engagement with legislators, bureaucrats and scientists has meant that the GM debate has been opened up to greater scrutiny, even if impacts on decisions and policies have been limited.

Brazil

In Brazil much anti-GM activism has centred on the courts. In 1997, Greenpeace-Brazil filed an unsuccessful lawsuit against the importation of GMOs. In 1998, CTNBio, the government regulatory authority, approved five Roundup Ready soya varieties for controlled and monitored commercial release. The consumer organization, IDEC (*Instituto Brasileiro de Defesa do Consumidor*)[17] immediately responded by filing another lawsuit with a federal judge to prevent release. Eventually, an injunction against commercial release was issued in December 1998. Through 1999, protests increased with much media coverage, and attempted to get state-level governments to create obstacles to further expansion of GM cropping. In March 1999, Governor Dutra of Rio Grande do Sul state issued a decree requiring a special licensing system for GM companies in the state, and the Rio Grande government ordered Monsanto to provide detailed environmental impact assessments for its operations. Monsanto later successfully sought an injunction to prevent state government intervention in its research station activities. But, thanks to effective state-level agitations, the governor responded by declaring the state 'GM-free' later in the year. This decision was queried by the state legislature the following year, but nevertheless created the appropriate media profile, raised the tempo, and focused attention increasingly on state-level actions.

Meanwhile, at federal level a growing NGO and political party grouping – involving IDEC, IBAMA (the Brazilian Environment ministry's Institute for the Environment and Renewable Natural Resources) and the PT (Workers' Party) –

---

[16] See www.africabio.com/press/biowatch.pdf.
[17] See www.idec.org.br.

had sought legal intervention to reverse decisions being made by the Agriculture Ministry. The ministry authorized registration of GM soya varieties without a formal environmental impact assessment, which, the complainants argued, was required by law. In August 1999, a federal judge turned the temporary injunctions issued against planting of GM crops into an official court decision. Monsanto appealed against the decision, and continued to lobby hard.

IDEC and Greenpeace meanwhile began to emphasize consumer awareness and boycotts. This focused on the urban, middle-class consumers of the major conurbations, a significant group in Brazil. Drawing on IDEC's strong credentials as a consumer organization and Greenpeace's international experience with consumer boycotts, this approach began to capture media attention. In June 2000 they announced that 11 GM food products were on the shelves of Brazilian supermarkets. Through the latter part of 2000 and into 2001, Greenpeace led protests at supermarkets across the country (Reuters 2000b).

Challenging importation of GM products was another tactic deployed at this time. Greenpeace went to court to prevent the shipping from San Francisco of two large container boats of GM maize for poultry farming. This highlighted the growing trade in GM products, and the increasing dependence of Brazil on them. In July 2000 farmers, led by the Landless People's Movement (MST), attacked a ship in Recife containing GM maize from Argentina (Reuters 2000a).

All of these actions resulted in increasing frustration from government authorities. All cabinet ministers signed a note in favour of GM crops in July 2000, and President Cardoso signed a decree empowering CTNBio to authorize GM crops (Dow Jones 2000). But this only heightened the determination of the anti-GM activist groups, with blockades and injunctions against shipments intensified. With a loosely networked cluster emerging around the campaign for a GM-free Brazil (*Campanha Nacional por um Brasil livre de Transgênicos*), the range of groups involved expanded. The MST – particularly through its connections with the international peasant farmers' movement, Vía Campesina – became increasingly involved, and was able to mobilize farmers in numbers. The World Social Forum in Porto Alegre in January 2001 was an important focus for protest, and continued to be so in the follow-up events of 2002 and 2003, attracting many international activists from around the world. In 2001, over 1,000 MST workers invaded a Monsanto farm in Rio Grande do Sul, destroying five acres of GM soybeans. They were joined by the French farmer activist José Bové, who was arrested for participating. Anti-GM mobilization now hit the international press (*Financial Times* 2001) and attention increased with the ActionAid-facilitated citizens' jury in Fortaleza (Toni and von Braun 2001).

By this stage the campaign involved participation from an impressive diversity of organizations and constituencies – consumers (through IDEC), environmentalists (through Greenpeace and ESPLAR, *Centro de Pesquisa e Assessoria*), development actors (through ActionAid Brasil and AS-PTA, *Assessoria e Serviços a Projetos em Agricultura Alternativa*), education groups (FASE, *Federação dos Órgãos para Asistencia Social e Educacional* and INESC, *Instituto de Estudos Socio-Econômicos*), farmers' movements (through the MST and affiliates) and political organizations (through the PT). Each was able to focus on different areas of advocacy and action – some more direct and protest-oriented; some focused on practical demonstration of alternatives; some through the courts and the media. Thus the network that emerged around the simple slogan 'GM-free Brazil' created a 'discourse coalition' (cf. Hajer 1995), which was able to agitate across a range of different spaces.

During 2002 much hope was invested in the success of the PT in the elections. Many NGOs, activist groups and movement players hitched their aspirations – including what many had understood as a commitment to ban GM in Brazil – on the PT success. But when Lula was elected in October 2002, the jubilation was short-lived. Even before the new government began, the new Minister of Agriculture, Roberto Rodrigues – and a well-known supporter of agribusiness – spoke out in favour of GM crops (*The Guardian* 2002).

Alongside the courtroom wrangling and activist protest from 1998 onwards, GM crops were increasingly being planted in Southern Brazil, despite the bans, decrees and injunctions.[18] Aspirations for a GM-free Brazil were being substantially undermined. In 2003 the Lula government, recognizing the crop's existence in large quantities, faced a major dilemma – either continue to uphold the ban and destroy the crop, resulting in major demands for compensation, or allow its sale. A series of Presidential decrees allowed the sale of GM soya, despite it still having not been approved for planting (Reuters 2003a). De facto, Brazil was no longer GM-free.

At this time, a key battlefront became the revision of the Biosafety Law. Monsanto launched a vigorous PR campaign: cinema adverts proclaimed the benefits of GM crops, despite still being nominally illegal.[19] Monsanto also became increasingly frustrated at the loss of royalty revenues from illegally grown GM soya, and pursued this in the courts and in negotiations with major producers (*New York Times* 2003). A key sticking point was the authority of CTNBio, and the place of independent environmental assessments. Although early concessions were won on the draft by the anti-GM lobby, and long delays occurred, the government eventually recognized the full powers of CTNBio, and the bill was passed in March 2005.

Since then the tide has turned against the anti-GM position. With the legal battle now over, CTNBio moved quickly to approve Monsanto's Bollgard GM cotton. By March 2005 Monsanto had reached a royalty agreement with Brazilian soy producers, and a concerted attempt was on to eliminate the black market in seed sales, so ensuring Monsanto's market dominance, pending the release of local varieties from the state agricultural research organization, Embrapa. In the April 2005, the newsletter of the GM-free Brazil alliance vowed to fight on, but with the soya industry now formally committed to GM, the vision of a GM-free Brazil was now virtually dead and buried.[20]

---

[18] See Nature (2000); Dow Jones (2001).
[19] See Reuters (2003b); *Financial Times* (2004).
[20] Although the GM-Free Brazil newsletter of 20 December 2007 (see http://www.aspta.org.br) celebrated that GM crops had still not been formally approved by the end of the year.

GLOBAL CONNECTIONS, LOCAL ROOTS

So how do we make sense of these anti-GM campaigns in this period, in relation to their local, national and transnational contexts? Was this a flash-in-the-pan set of protests, driven by elite groups and financed by dubious (European) interests, as some claim (Paarlberg 2000)? Or has this period shown the emergence of a type of protest where relationships between politics (and so values and ethics) and knowledge (and not only mainstream scientific perspectives) are seen in a new light, built on transnational solidarities and shared objectives? With GM crops either hailing a bright future for smallholder producers, or certain doom, depending on your position, it is worth exploring how different strands of the GM debate have aligned with different agrarian interests and politics, in turn shedding some light on the class character of the GM issue.

Different sides of the debate have made great play of the position and interests of farmers – as victims (Shiva et al. 2000) or sometimes as heroes (Herring 2005, 2007). But who were these farmers, apparently at the centre of the debate, but too often silent in the discussions beyond some token representation by either side? As discussed earlier, GM crops arrived into very different agrarian contexts, but in all cases the main market was the relatively well-off, more commercially-oriented farmer: 'rich peasants' or 'capitalist farmers', to employ a classification used by some. In some settings, GM seeds were tried out by many different farmers, and in some parts of India there was little other cotton seed available (Stone 2007). But GM seeds were rarely easily afforded by poorer, more subsistence-oriented ('peasant') farmers (Shah 2005). Yet the pro-GM lobby assumes a positive view of the benefits of capitalist farming, even ultimately for poorer producers, given the assumed scale neutrality of the technology and the need to engage with global markets. The anti-GM lobby, by contrast, offers a much more sceptical perspective on the benefits of such an agrarian trajectory, arguing that the penetration of such forms of capitalist agriculture would result in increasing inequality, landlessness and impoverishment, and a further deepening of existing agrarian crises. For example, in the highly unequal dualistic agrarian economies of South Africa and Brazil, the propping up of large-scale commercial agriculture, in the face of on-going demands for land from the poor and landless, is seen as an affront to a commitment to democracy, freedom and development.

A key feature in all three countries has been the global connections that have linked anti-GM activists and debates. Campaigns at a national level have been reflected in global debates about anti-globalization, food sovereignty, farmers' rights and biodiversity, for example. In the loose networks making up global protesters, the GM issue had become a focus for a whole array of different issue-focused protests against the monopolization of knowledge and technology ownership through patents and the TRIPS agreement; for trade justice as part of the reform (or abolition) of the WTO; against the perceived depredations of multinationals (with Monsanto becoming a global target); or in relation to wider rights-focused campaigns around food, health and farming. For many, the battle over GM crops thus became a battle over a much wider agenda, encompassing the big issues of poverty, trade and human rights – with often a poorly-defined category of 'farmer' at the centre of each. This framing of the global debate intensely frustrated pro-GM industry lobbyists, government regulators and many scientists. They felt activists were conflating issues, smuggling in wider protests – and so politics, values and ethical standpoints – into what they saw as a relatively simple and narrow technical issue about GM crops. What was important, however, was the broadening of the frame by national campaigns to highlight these wider issues – of rights and social justice in the context of neo-liberal agrarian policies – beyond the technology itself (cf. Jepson 2002).

While each campaign took very different courses, deploying different strategies and tactics, and involving different actors and networks, all had links to this global domain. A number of international events over the last decade have been important in bringing activists together and consolidating links and networks across sites. The major anti-globalization protests at the WTO ministerial in Seattle in 1999 were a key moment, soon followed by the first World Social Forum (WSF) in Porto Alegre in 2001. Since then the WSF in Brazil in 2002, 2003 and 2005 has been an important meeting point, as was the Mumbai WSF in 2004 and the World Summit on Sustainable Development in Johannesburg in 2002, as well as global meetings of the Vía Campesina and other campaign groups and movements.

This globalization of protests had a number of consequences. It exposed people to a wider network, linking people through the Internet, email lists and meetings at fora and workshops. It allowed confirmation and support for positions that were often being fought in an isolated manner back at home. The growth of international anti-GM 'stars' provided a sense of occasion to an otherwise average protest event, garnering publicity and media coverage along the way. Thus the annual WSF events also became opportunities for staged protests – by José Bové (from France), Vandana Shiva (from India), Peter Rosset (from the USA), Percy Schmeiser (from Canada) and many others.

Perhaps no one typified this move to the global arena more than the late Professor M. D. Nanjundaswamy of the KRRS in India.[21] Through his engagement in the anti-GM campaign – and his links to global farmers' and anti-globalization movements, notably Vía Campesina – he moved from being a local politician and state-level farmer leader to an international figure. He increasingly moved in international activist circles, and was revered as a 'southern' farmer leader, a voice from the poor and marginalized.[22] In an interview he reflected: 'there is no difference between international, domestic and local issues these

---

[21] See the contributions in Brass (1995).
[22] Even before the GM campaigns, the KRRS had a history of high-profile, media-grabbing direct action, targeting inter alia the Miss World competition, Kentucky Fried Chicken and Cargill. The burning of the Cargill depot in Karnataka resulted in a good deal of condemnation in the press and among a wider group of activists because of the destructive manner of the protest.

days. Seeing international contexts is important for local activism. Seeing what happened in Seattle and Prague first hand was significant'.[23]

He was also almost continuously in the media spotlight, including the international press, with his well-staged burning of field trial sites, replicating – and indeed encouraging – the protest tactics of Greenpeace, Genetic Snowball and others in Europe. Such events were also often in the presence of international observers and activists, many now familiar names on the anti-GM global circuit. Over this period, the burgeoning Internet-based activist networks also propelled KRRS and Nanjundaswamy into the international arena, with his press releases copied to thousands of in-boxes throughout the world.

MOBILIZATIONS IN PRACTICE: SITES OF ENGAGEMENT

These global engagements were, however, very much influenced by processes of mobilization in particular national contexts. These took a variety of forms: some, as we have seen, were focused on formal, invited spaces where activists engaged in consultations with state agencies around regulatory reform; others were convened by activist networks. In India, for example, the set-piece 'tribunals' organized by RFSTE[24] occurred alongside Monsanto-led events. Both sides used similar tactics. For example, a major Delhi conference organized by the Gene Campaign in 2003 was disrupted by farmers bussed in by pro-GM groups (Suhai 2003). In both India and Brazil, experiments with citizens' juries were also organized by anti-GM campaign and development groups (see above). These events provided ideal media opportunities, with the results carried in both local and international media.

Yet much engagement by anti-GM activists was behind-the-scenes informal lobbying and networking. The well-educated, urban, middle-class profile of many activists meant they were also well-connected, and able to articulately put a case to senior ministers, civil servants and others. For example, in India activists effectively played off the central Department of Biotechnology, the Ministry of Environment and Forests and the Ministry of Health and, within agriculture, the Ministry of Agriculture of the Union government and some state government departments. With access to the right people, activists have thus influenced the process by careful briefing and exposure.

A key element of lobbying beyond the bureaucracy is to ensure that elected politicians are on board. If key elected figures back a position, they can call the shots, even in the face of reluctant civil servants. Tactics in Brazil have, as discussed, been very much hooked into electoral politics and the importance of state-level support for an anti-GM stance. The close involvement of movement actors in the PT campaign for the national presidency meant that activists had a strong sense (if not actually a firm commitment) that a Lula-led government would institute a ban on GM crops. That this did not happen was, as discussed

---

[23] Interview, M. D. Nanjundaswamy, Bangalore, 27 February 2001.
[24] See http://www.vshiva.net/archives/campaigns and Business Line (2000).

above, a major disappointment, but, nevertheless, resulting divisions within the Lula government have been exploited as a key bargaining route. During the passage of the Biosecurity law, MST activists camped outside the parliament in Brasilia for months, providing a continuous presence outside, while discussions were going on inside. In South Africa, the focus has been on influencing the parliamentary portfolio committees on Environment and Agriculture and Land Affairs. In South Africa, given its constitutional commitment to open forms of democracy, these spaces are relatively easy to access: any citizen can present arguments to the committee as evidence. But securing influence is another matter, which requires more astute lobbying. Again, political and personal differences are researched and exploited, personal connections are capitalized upon and briefings seed key arguments. For example, the Cape Town-based NGO, Environmental Monitoring Group, took the whole Agriculture and Land Affairs portfolio committee on a field trip to a 'sustainable agriculture' project, as part of an attempt to demonstrate that there are viable small-scale agricultural alternatives in the country.

If the political process is prone to exclusion, fickleness and often ultimate disappointment, then another route available in all three countries is an independent court system. Procedural compliance to statutory regulation and constitutional/rights-based challenges have both been an important part of anti-GM activism. Raising issues in the courts can be a critical route to heightening political – and media – awareness of an issue, and galvanizing politicians. The ability to pursue public interest litigation in India, for example, is seen by many as critical to a functioning democracy. That it is cheap and relatively easy to make a submission, and that the court is obliged to hear it (even if the plea is rejected) is seen as part of basic democratic rights. A day in court – or even an hour – is an opportunity to raise awareness, attract media attention, call witnesses, and to lay out arguments in public. In India many such cases have been submitted – demanding the release of information on trial data, full environmental assessments, the appropriate sequencing of regulatory testing and compensation for failed crops.[25] In Brazil the courts have been used very successfully, formally banning the planting of GM crops until 2005. This has allowed the consumer group, IDEC, together with Greenpeace, to argue that GM crops should not be planted unless a full environmental assessment is undertaken in accordance with existing environmental law. A process of procedural delay, even though ultimately unsuccessful, allowed the issue to remain in the public – and political – eye.

Generating knowledge and presenting evidence is seen very much as a key role for activist organizations, and not only in court cases. But the relationship between advocacy and research is often an uneasy one, as many activists

---

[25] For *Times of India* (1998), Gene Campaign also filed a public interest litigation petition with the Supreme Court in January 2004, arguing that the regulations be amended to comply with constitutional rights (see http://indiatogether.org/2004/jan/env-gmsyspil.htm;http://www.genecampaign.org). In addition to Supreme Court petitions, an increasing number of other PIL cases have been submitted to state-level High Courts.

acknowledge. A staff member from Biowatch in South Africa frankly admitted that: 'in the early years the data from our research was not good enough. We could not make the arguments'. In part, this was a matter of resources. As they went on to observe: 'Monsanto can produce data for any occasion. They can get university researchers from anywhere in the world to come and do work'.[26] In India, for example, following the formal release of Bt cotton in 2002, the competition between interpretations has been intense. Literally dozens of 'surveys' have been carried out, each proclaiming a different result (Scoones 2005a, 289–91).

But all such different findings are inevitably context-specific. Often the most effective storytelling is based on showing results on the ground. Projects focusing on small-scale sustainable agriculture have thus increasingly become demonstration sites and witnessing opportunities, and so part of the overall story. The narrative becomes not only one of impending disasters, but also one that encompasses a positive alternative. In all three countries, field-based agriculture practitioners are key elements in anti-GM networks. Taking people to project sites – politicians, regulators and scientists – is often very powerful.

All of these mobilization practices – whether informal lobbying, invited consultations, court cases or demonstration projects – of course interact. Different parts of the wider activist network in each country have focused on different strategies. But few subjects generate more debate than strategies and tactics for direct action protests. What might be destructive to a wider acceptance of ideas? What might undermine attempts at informal lobbying or court cases? Is non-legal direct action, including the destruction of property and crops, acceptable? Should protests be aimed at the media, raising a profile for other actions, or be an end in themselves? There are as many views on these issues as there are activists and groups, and no simple answers.

As noted earlier, in all three countries, direct actions and protests of different sorts have occurred at different moments over the past decade. In India the KRRS led the way with the destruction of Bt cotton field trial sites as early as 1998, preceding the wave of similar style protests in Europe. These were media events – staged, dramatic, worthy of copy and providing good photo opportunities. As in Brazil, they became part of the global activist tourist circuit. International activists were invited to witness the events, and report back to their networks. Their presence provided yet another media angle, and helped assure that any police actions were not too excessive. Nanjundaswamy's lust for publicity certainly paid off. KRRS protests between 1998 and 2001 were almost continuously in the media. Sometimes they targeted the Monsanto Research Centre in Bangalore, sometimes the Monsanto/Mayhco field trial sites, sometimes major international events (such as the Asian Seed Industry Congress) and sometimes the legislative assembly. Always, though, the press were informed in advance, and press releases from the tiny Bangalore KRRS office went out as

---

[26] Interview, Cape Town, March 2004.

e-mails across the world. These actions were seen by some as maverick and inappropriate, but no one could dispute that they raised the profile of the debate.

In Brazil it was the involvement of the MST that allowed a shift from conventional campaigning to more direct protest action. Linking the GM issue to the wider question of agrarian reform allowed land invasions to extend to invasions of Monsanto and other research stations, and include uprooting trial crops. As in India, these have been staged, symbolic events, which have been designed for the media spotlight. The WSF events in Porto Alegre have been a focus for such actions, with international anti-GM high-flyers joining the fray. Following tactics successful in Europe, Greenpeace initiated supermarket protests and leafleting, including trolley runs. Newspaper reports for each occasion meant that the GM issue was continuously in the papers, keeping up the pressure on the long, drawn-out legal deliberations.

As the discussion above has repeatedly highlighted, the media has a major influence on mobilization opportunities. Because of the way news is created and managed, it generates villains and heroes, iconizing some and demonizing others. Key articulate individuals – able to offer a good sound-bite or willing to write an op-ed piece to a tight deadline – are vital to journalists. A news story must be one with two sides, so with two groups – pro and anti – pitched in battle with each other, the David and Goliath narrative of local NGO and farmer activists battling against global multinational corporations can often be regurgitated in print.

That Monsanto too has had an effective national and global media sensibility adds to the ease with which news can be made. With any event, Monsanto in India, for instance, sends faxed briefings to a vast list of journalists. This is instant copy. With one phone call to the 'other side' a story is made. As one journalist who covered the GM issue for many years for a national newspaper commented, 'stories are easy. You get a fax from Monsanto, you ring up a campaign group for the other side and you write an article'.[27] Collectively the anti-GM movements also have a highly effective PR machine. Greenpeace, for example, countered the good-news story on China presented in *Science* (Huang et al. 2002) with the release of a Greenpeace-sponsored report from Nanjing University which showed how Bt cotton was not faring as well as the proponents suggested (Xue 2002). This was immediately picked up by GM activists in India and summaries of the findings were transferred through cyberspace to websites and anti-GM networks, and thence to in-boxes everywhere.[28]

UNDERSTANDING MOBILIZATION PROCESSES: ANTI-GM ACTIVISM IN CONTEXT

The discussion so far has offered a comparative description of the processes and practices of mobilization around GM crops across the three countries. In this

---

[27] Interview, Chennai, February 2002.
[28] For international lists and networks see http://ngin.tripod.com;http://www.gene.ch/genet;http://www.biotech_activists.iatp.org;http://www.gmwatch.org, among many others.

section the broader political-economic contexts for anti-GM mobilization are examined, exploring in turn political and economic contexts, particularly around transitions to democracy and the adoption of neoliberal economic reforms and the implications these have had for agrarian contexts; the wider array of debates to which anti-GM activism has been linked; and the way alliances and networks have been formed, including the social and political positioning of activists themselves. All these factors, together, fundamentally shape opportunities, tactics and strategies for mobilization processes.

Political and Economic Transitions

Today, India, South Africa and Brazil pride themselves on being strong, established democracies. The democratic transition in South Africa was in 1994 and in Brazil in 1986. Memories of the apartheid regime in South Africa and the dictatorships in Brazil are recent, and the practice and experience of democracy novel and much valued. India is the contrasting example, having established a parliamentary democracy at independence in 1947, which bar the Emergency period, has remained more-or-less robust since. However, despite this longevity, the threats to democracy are strongly felt. In all three countries it is difficult to escape the feeling – articulated during interviews with government officials, NGO activists and the liberal media alike – that the ability to protest, debate and organize is an important part of what the transition to democracy was about, whether in 1947, 1986 or 1994.

To varying degrees, all three countries have a federal structure (Heller 2001). India and Brazil are large countries, where sub-national states have populations and economies comparable to medium-sized countries. Sub-national politics have always been an important part of the picture, with relations between the states/provinces and the centre being an important dynamic. A uniting factor has been a constitutional umbrella setting out a broad set of rights and responsibilities of citizens, wherever they come from. In South Africa the constitution is held up as an icon of success of the negotiated settlement, and is celebrated as the most comprehensive and radical in the world (Habib and Padayachee 2000). Constitutional challenges to legislation or government action are a key feature of civil society responses in all three countries. An independent judiciary – or at least one where particular judges are seen to be sympathetic to arguments based on wider premises than those forwarded by government – is seen as central to the democratic state.

Democratic practice, however, is highly conditioned by changing economic factors. An understanding of institutions – of law, administration or regulation – cannot be separated from a broader assessment of political economy under conditions of neo-liberalism. The nature of the state – and with this the structure of the economy, and particularly agrarian conditions – has changed dramatically in all three countries. From 1991, India began liberalizing the economy, dismantling many of the state functions held so dear by the post-independent Nehruvian state. This ushered in a different form of politics, one based on the federal market economy, where what happens at the state level is as important as what happens in the centre. Dictates from the centre – whether in the form of regulations or statutes – have to be implemented to have meaning, and state capacities are increasingly limited (Rudolph and Rudolph 2001). For example, widespread illegal planting of GM soya (in Brazil) and GM cotton (in India) made a mockery of the central government regulations, resulting in embarrassing about-turns, temporary measures and time-limited decrees by the central state (Herring 2007).

With intense competitive pressures in the global economy, getting (usually foreign) investment is perhaps the main focus of policy for all governments, whether at national or sub-national levels. All else pales into insignificance, it seems, with deals, bargains and provisions struck with investors (including major GM companies) that are not necessarily subject to full democratic scrutiny. In the liberalized, globalized economies of the federal systems of India, South Africa and Brazil, the imperatives of the market often supersede those of ordinary citizens. And, at least until election time, the deepening crises of the agrarian economy go unnoticed. Such economic and political transitions are on-going, affecting the nature and possibility of democracy and protest, with major implications for what spaces are open or closed, and so the tactics and strategies of activism.

In all three countries, agrarian settings are in transition and with this agrarian politics and the alliances of interests. In India the old certainties of peasant politics no longer apply, with the electoral game being played out through a complex coalition of interests in rural and urban settings, with the rhetoric of the 'new economy' dominating everything. Being mixed up with the old politics of the rural areas, and the need to broker alliances with elite farming groups, GM crops have often been seen by political and business elites as a dangerous diversion, when the focus on biotechnology, as a driver of industrial growth, has been the dominant narrative (Scoones 2007). Nevertheless, the sustained crisis in the agricultural economy, and the persistent poverty of many rural areas, remains a concern, although how GM crops are supposed to address such issues, no one is sure. In Brazil and South Africa, by contrast, the persistent influence of commercial farming interests on the way policies are framed and agrarian politics are conducted is important. GM crops are seen by such groups as critical for their economic survival, in the face of pressures for redistributive land reform, subsidy removal and agrarian restructuring. To date, careful political manoeuvring and economic interests might have ensured the perpetuation of such large-scale commercial farming interests, but the anti-GM lobby, with its diverse coalition of interests and issues, is seen as a threat, challenging the assumptions of privilege and economic superiority of the large-scale commercial sector.

Given these divided political constituencies, the election cycle has been an important focus for mobilization activity in all three countries. In all cases, the current governments are coalitions, with a strong dominant party. In India, the 2004 elections saw the return of the Congress Party, together with a broadly left-oriented alliance. In South Africa, the African National Congress was returned with an enhanced majority in 2004 and continues to lead the tripartite alliance that took over following the fall of apartheid a decade ago. In Brazil, the PT-led alliance has maintained power from 2003. As emerging economies with global ambitions, all three countries, despite their more populist 'pro-poor' rhetoric, do not shift too far from a firmly pro-business and investment line. This results in some uneasy compromises between nationalist, socialist, or at least social democratic, rhetoric and actual policy and practice. In all three countries the incumbent governments have been strongly criticized for compromising too much with the forces of global capital in their rush to present themselves as viable international investment destinations. Thus in South Africa, the alliance between the ANC and the unions (COSATU) and the South African Communist Party is often fraught (Lodge 2003). The movements that backed Lula in his bid to become president in Brazil have become increasingly critical of the PT government. Pressure from the MST, for example, on the implementation of the promised agrarian reform policy heightened from 2004 to 2005, with mass invasions of farms. In India, many are perhaps more cynical about the political process. But the unexpected return of Congress to the centre in 2004 did raise the spectre of a rural backlash against an arrogant, urban-biased, elite-driven politics (Patel and Muller 2004). In all three countries, the politics of neoliberalism is thus a key factor in shaping mobilization strategies and state responses. GM crops are thus seen as a battleground for these debates, and the future of agrarian settings, with different interests pitted against each other in an intensely political debate. The attempts, particularly by pro-GM advocates, to narrow the discussion to ones of technology efficacy or biosafety, and so obscuring politics and interests, are not accepted by the anti-GM coalitions, who continuously emphasize the wider debate and the clear associations between a pro-GM position and particular interests – and so implicit visions of an agrarian future.

Linking Issues: Beyond the GM Debate

In discussing anti-GM mobilization, however, it must be remembered that in none of the three countries are GM issues anywhere near the top of the political agenda. They have raised their head at various points, attracting media attention and responses from politicians, but more as emblematic issues linked to other debates. For those in the thick of the GM debate this may be frustrating, and the effort is continuously to push the GM issue up the pecking order, highlighting its wider implications and making links to other more high-profile issues and concerns. Indeed, very often, the very same farmers mobilized by organized farmer movements – whether the KRRS in Karnataka or the MST in Brazil – are the same farmers planting GM crops illegally, or would try them out if they could.

Thus in Brazil the MST is able to mobilize farmers around the GM issue by linking it to the wider question of agrarian reform. The Vía Campesina movement, to which the MST is linked, talks, for example, of food rights and food sovereignty and the need for peasants to be independent of the clutches of global agribusiness.[29] For the marginalized rural poor in Brazil this chimes well with many of their concerns. Even when they often know little about GM crops, seeing Monsanto as the enemy, allied to a Brazilian state reluctant to engage in any meaningful rural reform, produces a convincing storyline to which people have signed up in numbers.

That this connection has failed to emerge in South Africa is perhaps a puzzle. A similarly disenfranchised rural populace, with comparable patterns of land inequality, has not resulted in a similarly vibrant movement around land reform, rural livelihoods and agrarian change. Here again, we must reflect on the core issues that do result in mobilization. In South Africa, where the political movement that became the ruling party emerged from the urban areas and the union movement, rural issues are not prioritized. Land reform, while important as part of government rhetoric, has not been the ANC's major policy priority. No equivalent of the MST has emerged; although the Landless People's Movement models itself on the MST, it has largely failed to generate mobilization on any scale (Lahiff 2003). The political elite sees large-scale commercial farming – although with transfers to black ownership – as the future for the agricultural sector, with GM crops very much part of the picture. Instead, priorities for activists have centred on issues of urban violence, responses to the HIV/AIDS pandemic and labour conditions – themes around which there has been significant civil society mobilization in the past decade.

In the campaigns against GM crops across all three countries, activists have linked GM crops to problems of indebtedness and increasing reliance on credit and loans from traders and seed companies, for example. They have also been linked to a dynamic of commercialization in the farm economy, with smaller farmers being side-lined in favour of large-scale units and contract farming, highlighting the political consequences of agrarian restructuring. The debate has also been linked to the erosion of local varieties and choice for farmers, and especially recycling of seed, and to the wider globalization debate, the WTO and the removal of quantitative restrictions on imports, and the fear of price collapses for commodities with the flooding of local markets. Yet none of these issues are of course *only* about GM crops; they apply just as well to most hybrid varieties, and to wider trends in the agricultural economy, wholly independent of the GM issue. And, in Brazil and India in particular, many farmers had long planted GM crops, despite a lack of regulatory approval. What is highlighted by these linkages and elisions, however, is a wider set of questions about choice, sovereignty and future livelihood options: all far more pertinent and challenging issues for political debate, and more concretely linked to people's tangible concerns, than the rather arcane scientific and technical debates about the impacts of GM crops and foods.

Thus across the three countries, the GM debate has been characterized by the strategic development of alliances and the linking of actors and organizations in

---

[29] See http://www.viacampesina.org;http://www.mstbrazil.org.

new, often fragile, coalitions. Most of these have focused broadly on the politics of agrarian change issues, which as we have seen, have had variable purchase on the political process. Galvanizing urban consumers has been even more challenging. In all three countries, consumer organizations have become involved in the GM debate, taking their lead from their European counterparts in developing awareness about food safety issues, mobilizing for food labelling, and consumer boycotts to hold supermarkets to account. But in India, South Africa and Brazil, most consumers do not have the interest or awareness of the discerning European shopper. There is a small but influential group of middle-class consumers who are prepared to pay a premium for non-GM food, and will argue for environmentally-friendly, locally-based production systems. However, for most urban – and indeed rural – consumers with low incomes, their concern is with prices of commodities, not their origin.

Again, concerns about GM foods are linked to other issues within activist discourses. In South Africa, consumers often raise questions about nutritional quality for AIDS sufferers. Baby food, increasingly purchased from standard retailers, is similarly a concern, where even poorer consumers are keen to ensure the highest quality and take no risks. Yet, as most consumer activists admit, raising awareness about food issues is an uphill struggle; although with the consolidation of retailing into a select number of supermarket chains, more purchase has been found for middle class, elite anti-GM activists by forcing labelling and in some cases bans.

Alliances and Networks: Gaining Legitimacy and Authority

Building alliances and networks to upgrade the priority of the GM issue has perhaps been most successfully developed in Brazil. Here, a hugely diverse range of organizations came together under the loose banner of the GM-free Brazil campaign (see above). Some of these organizations would not normally have been seen together; indeed each had a very different view of the most appropriate and legitimate strategy for opposing GM. But they have largely been able to avoid disagreeing on detailed strategy and tactics, which were left to each organization to decide, and focus on uniting under a simple banner. The coalition thus formed disagreed on many things, but focused only on areas of agreement. It was thus fairly fragile and had to be managed with care. The coordinators were well aware of this, and were able to manage the tensions effectively.

A similar pattern is evident in South Africa, although on a smaller scale. Within the broad grouping identifying with the anti-GM campaign, there are those who disagree strongly with direct action tactics, while there are those who feel civil disobedience and a more disruptive stance is probably the only way to go. They are able to work together because the network is not reliant on a single vision and strategy. Over time it has also evolved from a small group centred on Biowatch and SAFeAGE to a larger range of organizations, many of which do not have anti-GM activism as their sole focus. Most have a broadly environmental focus; others have a practical emphasis on sustainable or organic agriculture; others have a rights/justice orientation; and others are linked to the labour movement or are faith-based groups (see above).

In India, a broadly similar array of organizations are incorporated under the anti-GM umbrella, but there is no sense of a coordinated campaign. Various attempts at coming together on a common front have failed due to differences in views, but particularly personalities have been the cause. Each of the main groups presenting an anti-GM position over the past decade has strong individuals as leaders. Vandana Shiva is perhaps the most celebrated, heading the RFSTE. Devinder Sharma of the Forum for Biotechnology and Food Security is an equally effective campaigner and media commentator, but with a different style. Suman Sahai, convenor of the Gene Campaign, is different again too, with her emphasis on research and engagement with policy. And, finally, leader of the KRRS faction, the late M. D. Nanjundaswamy, was in a league of his own – maverick and astute, a supreme media-savvy publicist.

Yet, in particular places, out of the national spotlight and media game, there are other individuals and groups who operate on a more modest level, without the exposure and without (most of) the personality politics. In a survey of anti-GM activist organizations in the southern Indian city of Bangalore, I identified over 20 organizations with an explicitly stated anti-GM stance, including of course the KRRS (Scoones 2005a). These again clustered into those working practically in the field through demonstration projects on sustainable and organic agriculture, seed saving and biodiversity (Green Foundation, AME Foundation, Honey Bee Network; Organic Agriculture Network); those with a broader development focus (including the international NGO ActionAid and the local NGO MYRADA); those with an explicit rights focus (DISC – emphasizing food and worker rights; CREAT – emphasizing consumer rights and education); and those with an environment focus (Greenpeace-India; Environment Support Group). There were also other organizations, including political parties (notably CPI(M)) and academic networks (e.g. Association of Environmental Economists).

The legitimacy and authority of these anti-GM networks is, of course, a major issue. It is all well and good mobilizing a diverse group, creating a coalition around a simple narrative – a GM-free Brazil or a freeze in South Africa or a Monsanto Quit India slogan – but how easily dismissed is such a coalition? Does it have any chance of influencing those in power?

As discussed earlier, virtually all leading activists in all three countries are well-educated, urban-based and relatively well-off. As founders or key players in organizations, they are well connected to the funding world, often linked to a variety of international aid donors (HIVOS, ActionAid, Ford Foundation were among the most often mentioned). They all have had past experiences in activist/political arenas – in Brazil in struggles against the dictatorship in the 1970s and 1980s, and for Lula's election in 2002–03; in India in farmers', women's and environmental movements, and more broadly in anti-globalization efforts in recent times; and in South Africa in the struggle against apartheid, including union organization and consumer boycotts. Some have had forays into the formal electoral process themselves. What they all understand in great detail is the inner workings of the political process, and the functioning of the state bureaucracy. But does this elite character of the movement leadership matter? Does this, as some suggest, undermine the legitimacy of their claims? To answer these difficult questions requires looking in more depth at the origins and political positioning of these anti-GM groupings in context.

In South Africa, the recent history of the anti-apartheid struggle has shaped many anti-GM groups. The labour unions were of course central to the opposition to the apartheid regime, and the key union grouping COSATU is formally part of the tripartite alliance governing the country. That COSATU is notionally anti-GM has little impact directly, but that the anti-GM groupings can highlight the commitment of the unions to their cause carries much weight. The anti-GM networks are dominated by more conventional environmentalist groups. Some of these are fairly conservative and elitist, certainly historically, including the botanical and wildlife societies; others present themselves as radical, but in a 'deep green', rather Eurocentric way; others, such as the EJNF, work on environmental issues in the townships and have a more socially diverse membership base. Together, they offer an alternative storyline on South Africa's development, yet the degree to which they can influence events in the current South African context is limited, given the ruling party's strong commitment to a very different neoliberal development trajectory (Bond 2000).

In Brazil, a key stage in the development of the anti-GM network was the enlistment of the MST. Initially sceptical, the advantages soon became apparent to MST leaders. As an increasingly internationalized movement – with websites and support groups in the USA and Europe – this was a relatively easy step. Thus, with the arrival of activists by the plane load to Porto Alegre for the WSF festivities each year, MST was able to raise their international (and therefore local) profile and forge links with the hall-of-fame of international anti-GM activism. For the small group of NGOs, which to that point were the core network members, this was critical. They could no longer be criticized for just being unrepresentative, unaccountable, foreign-funded NGOs (which they were), but were linked to a mass movement with official and informal connections to the PT, allowing political clout well beyond what was possible before. Whether this argument for legitimacy and authority stands up to scrutiny is an open question, it is the symbolic importance of the alliance, which gives the stamp of approval in the eyes of many. By constructing an alliance across issues and organizations, the anti-GM movement was able to present an alternative voice, challenging the policies of the government on a variety of fronts, including the on-going attempts to formally introduce GM crops.

Something similar applies in India. Here too, the history of the social movements has been very significant – in struggles against rural oppression, around the time of the 'Emergency' and in day-to-day involvement in rights issues at local levels (Ray and Katzenstein 2005). Again, the symbolic importance of such movements is central. As discussed already, the KRRS in Karnataka has been at the forefront of anti-GM struggles in India. Yet the KRRS is essentially an alliance of relatively prosperous farmers from Karnataka interested in pushing their claims for farm subsidies (especially cheap electricity and water for irrigation) and price control (for both inputs and outputs), and the anti-GM position is not critical for many members (it has indeed been the focus of divisive debates and splits) (Assadi 2002). Engaging in wider campaigns about corporatization of India, anti-globalization, WTO and patents or GM crops have been add-on concerns, largely at the instigation of the charismatic and persuasive Nanjundaswamy. But knowing the importance of rural vote blocs, the state government must take them seriously. That large numbers of farmers turned out to support the demonstrations and actions is perhaps witness to the importance of the organization as an effective lobbyist on other issues, rather than a genuine commitment to getting rid of GM crops per se. But, as in Brazil, the symbolic association with a mass-based farmers' organization has been important for many other organizations linked to the anti-GM network. As small NGOs without mass membership, they must make a case for their speaking on behalf of farmers and the rural poor. Sometimes they do this modestly, speaking as intermediaries with experience of particular people and areas as a result of their ongoing field engagements through projects. Sometimes links are made to wider fields of mobilization, for instance around food rights, which has managed to mobilize people from diverse walks of life. Sometimes somewhat extravagant claims are made that, because of links to particular groups, the NGO does speak on behalf of large categories of people (the poor, women, tribes and so on).

Thus the politics of representation remain a complex issue in the anti-GM networks across the three countries. Some key questions arise from this assessment: does the fact that multiple contradictions and inconsistencies exist, and allegiances and alliances are sometimes only temporary and shallow, hiding deeper divisions and disagreements, undermine the legitimacy and significance of the anti-GM stance? By exposing such complexities in the politics of the anti-GM movement, does this automatically suggest that, in some way, the pro-GM position is, by default, correct? The concluding section will argue that a more sophisticated appreciation of agrarian movements and struggles over GM crops is needed, one that values the importance of injecting diverse arguments and contrasting positions into an open and democratic debate that, without such inputs, would inevitably be closed down and captured by particular, narrow interests.

CONCLUSION

The anti-GM mobilizations discussed in this chapter are, as we have seen, a much looser, more fragile network-based form of interaction (Castells 1997) than the way 'classic' social movements are often portrayed (McAdam et al. 2003). These networks are linked to global players and debates and have a distinctly transnational character, yet are always located in local struggles and political processes. They galvanize selective and strategic alliances among different and diverse groups around a variety of issues. Some would dismiss these as incoherent and poorly substantiated, but together they often add up to an alternative perspective on agrarian futures to the standard neoliberal line, even if sometimes poorly articulated and partially contradictory. Such positions are the result of complex, hybrid coalitions of interests and ideas, and, as discussed, do not represent a particular, defined set of (class) interests. With their global connections and elite, educated, urban leaderships, they can be seen as often very detached from rural realities and agrarian struggles. But the resonances and connections are definitely there, as we have seen, and the strength of their appeal, and the political force that they potentially have, lies in the way such connections – between local, national and global issues; rural and urban; producer and consumer; elite and poor – are constructed and mobilized.

Despite the critique often levelled at anti-GM groups in the developing world, such activism cannot be read simply as being copied from and directed by events elsewhere. While sharing and interchange have been important features of anti-GM mobilizations, the local and country-focused efforts have, as shown above, taken on very distinct characters. Through creating strategic alliances and linking the GM debate to other broader concerns, it has been possible to insert the GM issue into local, national and global political agendas in different ways. The GM issue therefore has become iconic: representative of a wider set of struggles, and the focus for mobilizations across multiple divides and locales.

But such efforts have, as discussed, distinct limitations. First, and admitted by most if not all activists, is the issue of representation. Who do such loosely networked activist groups speak on behalf of? While some engage in contortions demonstrating that they are 'of the people'; others are more sanguine. They argue that by generating a debate – creating a 'discursive space' – they are opening up a discussion, which would otherwise be closed down. This, they argue, is essential for democracy, and so justified on its own terms. The challenge is to encourage deliberation across diverse spaces, enlisting others in discussion. When the mainstream political and economic discourse under conditions of neoliberalism is so dominated by the concerns of engaging in the global market economy, the openings for wider democratic deliberation are often slim, making the role of (often unrepresentative) NGOs and campaign groups key in making real the claim of a vibrant democracy.

Second is the issue of impact. How successful have these activist strategies been? Does activism make a difference, beyond marginal irritation to those in power? Across all three countries this has been a largely a middle-class, urban debate. It is not an electoral issue, nor a focus for true mass mobilization, except through co-option and linking of issues. Instead, it is increasingly bound up with a much wider debate about new relations between contemporary capitalism and society: issues of sovereignty, inequality, rights, justice and so on – an attempt, perhaps, to create an alternative 'grand narrative', one counter-posed to the mainstream neoliberal worldview. These bigger issues are not going to be resolved around the GM debate of course. It is, however, as this study has shown, emblematic of wider struggles, with very powerful political interests at play. In all three countries there are relatively progressive alliance governments at the centre, all with professed commitments to social justice, development and poverty reduction. Yet the neoliberal economic agenda, the need to attract foreign investment and the pull of global politics are very strong. This makes in-roads into debates based on different framings of development very hard indeed.

The positioning of an 'anti-GM' stance has inevitably resulted in ambiguities and tensions within activist networks. Some groups feel more comfortable focusing narrowly on environmental issues; others focus on consumer food safety questions; while others frame the debate in terms of rights and justice. How these link to broader questions of agrarian change and politics is often not clear. Holding a broad front often means engineering strategic silences about some tough issues, with contradictions and tensions held in abeyance. But avoiding some of these deeper issues may also mean the unravelling of coalitions and alliances. When farmers are keen to engage with global markets to improve their livelihoods, how does this square with a positioning around food sovereignty pushed by some, for example? And, while farmers register real concerns about changes in the rural economy and agrarian conditions, they are often not averse to trying out new GM crops, whether 'pirated' or supplied by a multi-national company. These contradictions remain unresolved, and often unaddressed, among anti-GM groups.

That said, collectively, anti-GM activists have managed – sometimes only temporarily and often at the margins, and always with deep ambiguities and contradictions at play – to articulate a set of positions that shows how the debate about GM crops is actually a debate about a much wider set of issues: about the future of agriculture and small-scale farmers; about corporate control; about property rights; about global trade rules and so on. While pro-GM commentators argue that such activists are smuggling in debates that should not be part of the discussions, anti-GM activists argue forcefully that debates about values and politics must be seen as central, part of new negotiations around citizenship, knowledge and politics in an era of globalization (cf. Ellison 1997; Leach and Scoones 2007).

REFERENCES

ActionAid, 2000. *Indian Farmers Judge GM Crops*. ActionAid Citizens' Jury Initiative. London: ActionAid.

AgBioIndia, 2002. 'Demand for Monsanto's Ouster Picks Up: Karnataka Bans Monsanto's Bt Cotton Seeds'. 11 August. http://ngin.tripod.com/110802b.htm, Accessed 4 February 2008.

Appadurai, A., 2000. 'Grassroots Globalization and the Research Imagination'. *Public Culture*, 12 (1): 1–19.

Assadi, M., 2002. 'Globalisation and the State: Interrogating the Farmers' Movement in India'. *Journal of Social and Economic Development*, 4 (1): 42–54.

Bond, P., 2000. *Elite Transition: From Apartheid to Neoliberalism in South Africa*. Durban: Natal University Press.

Brass, T., ed., 1995. *New Farmers' Movements in India*. London: Routledge.

Business Day, 2003. 'Biowatch SA Says Modified Organisms Need Regulation'. 20 August. Rosebank, Johannesburg, South Africa.

Business Day, 2004. 'Open Wide for GM Bite'. 6 February. Rosebank, Johannesburg, South Africa.

Business Line, 2000. 'MNC Seed Interest vs Farmers' Plight'. 26 September. Chennai, India.

Castells, M., 1997. *The Information Age: Economy, Society and Culture; Volume 2: The Power of Identity*. Malden, MA: Blackwell.

Da Silveira, J.M. and I. De Carvalho Borges, 2005. 'An Overview of the Current State of Agricultural Biotechnology in Brazil'. Paper presented at Science, Technology and Globalization Project, Agricultural Biotechnology for Development – Socioeconomic Issues and Institutional Challenges, Bella Villagio, Bellagio Italy, May–June 2005. Belfer Center, Kennedy School of Government, Harvard.

Deccan Herald, 1998. 'KRRS to File Criminal Case Against Monsanto'. 18 December. Bangalore, India.

Dow Jones, 2000. 'Brazil Government Fights Back Against Anti-GMO Court Ruling'. 10 July. New York, USA.

Dow Jones, 2001. 'GMO Soy Planting in Brazil Said to Be Spreading North'. 18 December. New York, USA.

Economic Times, 1999. 'Monsanto Proposal to Genetically Alter Rice Draws Flak'. 6 August. New Delhi, India.

Economic Times, 2001. 'KRRS Opposed Commercial Release of Bt Cotton Seeds'. 20 June. New Delhi, India.

Edelman, M., 2001. 'Social Movements: Changing Paradigms and Forms of Politics'. *Annual Review of Anthropology*, 30: 285–317.

Ellison, N., 1997. 'Towards a New Social Politics: Citizenship and Reflexivity in Late Modernity'. *Sociology*, 31 (4): 697–717.

Financial Express, 1999. 'Farmers Support Biotech Use for Increasing Yield – Monsanto Study'. 31 August. New Delhi, India.

Financial Express, 2000. 'Farmer Leader Pleads for Immediate Release of GM Crops in India'. 24 April. New Delhi, India.

Financial Times, 2001. 'Brazil Police Add to Bové's Radical Status – Anti-Globalisation Activist'. 31 January. London, UK.

Financial Times, 2004. 'Monsanto Takes GM Crusade to Brazil'. 6 February. New York, USA.

Friedberg, S. and L. Horowitz, 2004. 'Converging Networks and Clashing Stories: South Africa's Agricultural Biotechnology Debate'. *Africa Today*, 51 (2): 3–25.

Glover, D., 2007. 'The Role of the Private Sector in Modern Biotechnology and Rural Development: The Case of the Monsanto Smallholder Programme'. DPhil thesis, Institute of Development Studies, University of Sussex.

Habib, A. and V. Padayachee, 2000. 'Economic Policy and Power Relations in South Africa's Transition to Democracy'. *World Development*, 28 (2): 245–63.

Hajer, M., 1995. *The Politics of Environmental Discourse: Ecological Modernization and the Policy Process*. Oxford: Oxford University Press.

Heller, P., 2001. 'Moving the State: the Politics of Democratic Decentralization in Kerala, South Africa and Porto Alegre'. *Politics and Society*, 29 (1): 131–63.

Herring, R., 2005. 'Miracle Seeds, Suicide Seeds, and the Poor: Mobilizing Around Genetically Modified Organisms in India'. In *Social Movements and Poverty in India*, eds R. Ray and M.F. Katzenstein, 203–32. Lanham, MD: Rowman and Littlefield Publishing Group.

Herring, R., 2007. 'Stealth Seeds: Bioproperty, Biosafety, Biopolitics'. *Journal of Development Studies*, 43 (1): 130–57.

Huang, J., S. Rozelle, C. Pray and W. Qinfang, 2002. 'Plant Biotechnology in China'. *Science*, 295: 674–77.

Jepson, W., 2002. 'Globalization and Brazilian Biosafety: the Politics of Scale over Biotechnology Governance'. *Political Geography*, 21 (7): 905–25.

Lahiff, E., 2003. 'Land and Livelihoods: The Politics of Land Reform in Southern Africa'. *IDS Bulletin*, 34 (3): 54–63.

Leach, M. and I. Scoones, 2007. 'Mobilising Citizens. Social Movements and the Politics of Knowledge'. IDS Working Paper 276. Brighton: Institute of Development Studies.

Lodge, T., 2003. *Politics in South Africa: From Mandela to Mbeki*. Bloomington, IN: Indiana University Press.

Mail and Guardian, 2000. 'SA Receives Bulk Shipments of Frankenfoods'. 25 February. Rosebank, Johannesburg, South Africa.

McAdam, D., S. Tarrow and C. Tilly, 2003. 'Dynamics of Contention'. *Social Movement Studies*, 2 (1): 99–102.

Mercury, 2003. 'Government Hands Out Free GE Seeds in South Africa'. 27 August. Durban, KwaZulu, South Africa.

Nature, 2000. 'Smugglers Aim to Circumvent GM Court Ban in Brazil'. 25 November. London, UK.

New York Times, 2003. 'Monsanto Pursues Seed Pirates'. 13 June. New York, USA.

Omvedt, G., 1998. 'Terminating Choice'. *The Hindu*, December 26. Chennai, India.

Paarlberg, R., 2000. 'The Global Food Fight'. *Foreign Affairs*, 79 (3): 24–38.

Paarlberg, R., 2001. *The Politics of Precaution: Genetically Modified Crops in Developing Countries*. Baltimore, MD: John Hopkins Press.

Patel, R. and A. Muller, 2004. 'Shining India? Economic Liberalisation and Rural Poverty in the 1990s'. Food First Policy Brief 10. Oakland: Food First.

Pimbert, M. and T. Wakeford, 2002. 'Prajateerpu: A Citizens' Jury/Scenario Workshop on Food and Farming Futures for Andhra Pradesh, India'. Brighton: Institute of Development Studies and London: International Institute for Environment and Development. http://www.diversefoodsystems.org/lfs_docs/Prajateerpu.pdf, Accessed 4 February 2008.

Ray, R. and M.F. Katzenstein, eds, 2005. *Social Movements and Poverty in India*. Lanham, MD: Rowan and Littlefield Publishing Group.

Reuters, 2000a. 'Brazil Turns Away GM Argentine Corn'. 6 June. New York, USA.

Reuters, 2000b. 'Greenpeace Protest for Argentine GM Food Labels'. 20 July. New York, USA.

Reuters, 2003a. 'Brazil Lula Says Legalizing GM Soy Was Best Option'. 8 October. New York, USA.

Reuters, 2003b. 'Monsanto Invests to Improve GMOs' Image in Brazil'. 8 December. New York, USA.

RFSTE (Research Foundation of Science, Technology and Ecology), 1998. 'Indians Fight Biotechnology Giants: Implement "Operate Cremate Monsanto"'. Monsanto Quit India Campaign Press Release. RDFTE: New Delhi.

Rudolph, L. and S. Rudolph, 2001. 'Redoing the Constitutional Design: From an Interventionist to a Regulatory State'. In *The Success of India's Democracy*, ed. A. Kohli, 127–62. Cambridge: Cambridge University Press.

Samykta Karnataka, 1998. 'Police Protection to all American Companies in Bangalore City'. 25 November. Bangalore, India.

Scoones, I., 2005a. *Science, Agriculture and the Politics of Policy. The Case of Biotechnology in India*. Hyderabad: Orient Longman.

Scoones, I., 2005b. 'Contentious Politics, Contentious Knowledges: Mobilising Against GM Crops in India, Brazil and South Africa'. IDS Working Paper 256. Brighton: Institute of Development Studies.

Scoones, I., 2007. 'The Contested Politics of Technology: Biotech in Bangalore'. *Science and Public Policy*, 34 (4): 261–71.
Shah, E., 2005. 'Local and Global Elites Join Hands: Development and Diffusion of Bt Cotton Technology in Gujarat'. *Economic and Political Weekly* (Special Articles), 40 (43): 4629–39.
Shiva, V., 1998. 'Terminating Freedom'. *The Hindu*, 26 December. Chennai, India.
Shiva, V., A. Emani and A. Jafri, 1999. 'Globalization and Threat to Seed Security: Case of Transgenic Cotton Trials in India'. *Economic and Political Weekly*, 34 (10): 601–13.
Shiva, V., A. Jafri, A. Emani and M. Pande, 2000. *Seeds of Suicide: The Ecological and Human Costs of Globalization of Agriculture*. Delhi: Research Foundation for Science, Technology and Ecology.
Stone, G., 2007. 'Agricultural Deskilling and the Spread of Genetically Modified Cotton in Warangal'. *Current Anthropology*, 48: 67–103.
Suhai, S., 2003. 'Monsanto's Claims are Uninformed'. *Times of India*, 12 March. New Delhi, India.
The Guardian, 2001. 'Jury Delivers "No" Verdict to GM Crops in Brazil'. 23 May. London, UK.
The Guardian, 2002. 'Brazil's Farms Minister Supports Gene-Modified Crops'. 17 December. London, UK.
The Hindu, 1999. 'Monsanto Turns to Food Crops'. 6 August. Chennai, India.
Thirtle, C., L. Beyers, Y. Ismael and J. Piesse, 2003. 'Can GM Technologies Help the Poor? The Impact of BT Cotton in Makhathini Flats, KawZulu-Natal'. *World Development*, 31: 717–32.
Times of India, 1998. 'Monsanto Trials Illegal, Says Environmentalist'. 21 December. New Delhi, India.
Toni, A. and J. von Braun, 2001. 'Poor Citizens Decide on the Introduction of GMOs in Brazil'. *Biotechnology and Development Monitor*, 47: 7–9.
Xue, D., 2002. *A Summary of Research on the Environmental Impact of Bt cotton in China*. Nanjing: Greenpeace/Nanjing Institute of Environmental Sciences.

7 Trade and Biotechnology in Latin America: Democratization, Contestation and the Politics of Mobilization

PETER NEWELL

INTRODUCTION

A range of social movements have mobilized to contest the role of biotechnology in agricultural development in Latin America. The adoption of GM (genetically modified) crops by large global agricultural players in the region such as Brazil and Argentina raises a series of key questions about the industrialization of agriculture (Pengue 2005), issues of access and control of technology and thorny dilemmas about land distribution and its sustainable use. Ties to trade and aid also place the technology at the centre of the political economy of development in the region. Alongside contestation around biotechnology in the Mercosur bloc and the negotiations towards a Free Trade Area of the Americas accord, bilateral trade agreements have been used to introduce GM crops into countries that are a centre of origin for crops crucial to the livelihoods of millions of resource-poor farmers (such as potato growers in Peru). Meanwhile countries adopting postures critical of the technology such as Bolivia have found themselves subject to immense political pressure from GM exporters such as the United States on behalf of their agribusiness lobbies.

Against this background of change in the global and regional structure of agricultural production and investment, an interesting mix of civil society organizations and social movements have been evolving repertoires of resistance to the corporate control of agriculture (Jansen and Vellema 2004; Glover and Newell 2004). These have taken the form of battles over the legitimacy, reach and application of intellectual property rights. Fundamental questions of land ownership and distribution have been raised by movements such as Movimento dos Trabalhadores Rurais Sem Terra (MST) and Vía Campesina and the sustainability of land use patterns required by the technology placed in doubt by a wide range of social forces concerned about the social and ecological effects of mono-crop plantations (Peleaz and Schmidt 2004). From a development and rural livelihoods angle, the interface between biotechnology and food security has been a central concern amid calls for food sovereignty, explored elsewhere in the journal in more detail.

Though intimate connections exist between each of these issues and agendas, the focus here will be on contestations by agrarian and environmental groups around the trade-related aspects of biotechnology in Latin America. This implies a focus on transnational biotech-related struggles that seek to contest the promotion of biotechnology in regional and bilateral trade agreements. The foci of these contests are issues of intellectual property rights (IPRs), agricultural liberalization and protection, and biosafety. Given the closed nature of many of these processes, one of the key challenges for movements has been how to democratize policy-making around trade (Newell and Tussie 2006). For some, this has meant working within restricted spaces for institutionalized participation that exist within formal, principally inter-state, decision-making processes. For others, critical of the treatment of agricultural issues within the principal trade agreements in the region, attempts have been made to resist the treaties and to open up trade policy debates to fundamental questions about which commodities and services should be subject (and which should not be subject) to international trade and, critically, who benefits from such policies and who is expected to bear the inevitable costs of adjustment (Newell 2007).

The study is focused on the region's two key players in the biotech sector, Argentina and Brazil, which have had very different experiences of engagement with biotechnology, reflecting contrasting levels of civil society mobilization around the issue, distinct rural political traditions and divergent state strategies towards the technology. It raises key questions about who mobilizes and how, and about the strategic dilemmas that arise when movements with different histories, membership bases and cultures of protest attempt to work together. Issues of accountability, representation and participation run through the analysis of strategies of organization and claim-making adopted by an eclectic range of groups seeking to contest the role of biotechnology in the structure of agricultural production, the institutions that manage that relationship and the discourses which sustain it. In particular, analysis will centre on the responsiveness of those mobilizing to the concerns and agendas of poorer groups in the front line of the 'gene revolution' as it plays out in the Latin American countryside.

It argues that though activists have made important gains, opening up the debate about biotechnology to a plurality of voices, challenging the regulatory structures set up to manage the technology and constructing alternative arenas to debate its risks and benefits, the close alignment of state strategies, which embrace biotechnology as key to their global economic competitiveness, with powerful elements of national and foreign capital severely restricts the space available to contest biotechnology. Moreover, while organized environmental NGOs have made some gains in advancing regulatory reforms and using legal processes to stall the technology's adoption, rural social movements, despite their broader social base and strong international connections, have not been able to shift the debate about biotechnology from one about bio-safety and responsible handling to one about land ownership, property rights and the unequal relations of power which sustain them.

AGRICULTURAL BIOTECHNOLOGY IN LATIN AMERICA

Despite claims by its advocates that agricultural biotechnology has the potential to boost yields through the use of drought and disease resistant crops for example, reduce pesticide use and address both nutritional deficiencies and hunger (Conway 1999; Lipton 2001), it has generated an unprecedented degree of controversy in Europe and parts of Asia and, as we will see below, in Latin America. Critics have focused on concerns about bio-safety (the potential for loss of biodiversity and contamination of non-GM varieties), impacts upon human health and the level of control the technology affords the companies that provide it over the livelihoods of the rural poor (Warwick 2000; FoEI 2007).

In Latin America, modern agricultural biotechnology was adopted against a backdrop of structural economic reform from the early 1990s. For countries across Latin America, neo-liberal reforms have produced transformations in the structure of agrarian production, concentrations in land ownership and the removal of parastatal support to poorer farmers, all aimed at the intensification of production to meet export markets (Kay 2002; Bellisario 2007; Thrupp 1996; Oya 2005; Murray 2006). The uptake of biotechnology with its packages of seeds, fertilizers and the political technologies of contract farming and monoculture cultivation sat comfortably with the agribusiness-oriented organization of agricultural production in the region's key players. Agribusiness has a leading role in the capital accumulation strategies of the largest countries in the region. Brazil's agribusiness sector still accounts for more than 40 per cent of total GNP. Brazil is the world's third largest *exporter* of agricultural products and after the United States is the largest soybean (though not the largest GM soy) producer, accounting for 7.5 per cent of the country's exports. Moreover, according to the US's own Department of Agriculture, Brazil has sufficient land resources to expand its soybean area and production to challenge the US's leading position (Peleaz and Schmidt 2004). As of 2006, Brazil is the third largest cultivator of GM crops in the world, now cultivating 11.5 million hectares of GM crops. These statistics explain why 'in Brazil, agricultural policy is always high politics' (Paarlberg, 2001, 68). Argentina meanwhile, with 18 million hectares of land under GM cultivation, is now the world's second largest producer and exporter of GM crops, accounting for 23 per cent of global production (James 2006).

What is interesting about Brazil's position, however, is that it also continues to be the primary source for non-biotech soybeans and soybean meal. This explains why Brazil has become such a significant site in the wider social and political struggle over biotechnology, or as Paarlberg puts it 'an important battleground' in the 'global contest over GM crops' (2001, 67). Biotech companies have said of Brazil: 'we are very hopeful that last domino will fall'.[1] For this reason, environmentalists and rural groups opposed to GM development, according to the same industry spokesperson, have been 'putting up a stink down there in Brazil. They know if that goes, it's all gone'. We will see below how this in fact accurately describes the current situation. This also explains both why companies such as Monsanto have represented the large-scale illegal cultivation of GM crops as evidence of farmer demand for their products whilst

---

[1] Bob Callanan, spokesman for the American Soybean association, cited in Peleaz and Schmidt (2004, 237).

sending out a clear message about the futility of claims about keeping Brazil free of GMOs. Even back in 2001, one estimate from the seed industry itself was that of the 2 million hectares planted as soybeans in Rio Grande do Sul, between 400,000 and 750,000 hectares were already transgenic (cited in Paarlberg 2001, 81). Evidence uncovered as part of this research of firms giving out unauthorized seeds at rural fairs to farmers in Paraguay, suggests penetration of markets by default represents an important tool in the arsenal of companies wanting to undermine resistance to the technology and bypass formal decision-making processes.

We will see below how a wave of resistance to the technology dramatically slowed the rate of commercial approvals of GM crops, but how these have once again accelerated more recently in the wake of the approval of the biosafety law in Brazil in 2005 and the resolution, for the time being at least, of the question of which government body has the authority to approve their commercialization. In this sense, activists may have won the battle but lost the war, whereas in Argentina it is possible to argue that activists never even made it to the battlefield.

The take-up of the technology has been uneven across the region. While Brazil has at times been hesitant about adopting the technology as a central agricultural strategy for reasons discussed below, Argentina saw the potential of the technology to boost the countries growth and consolidate for itself a strong position in global markets. Seven GM crop types have been approved for commercialization in Argentina, all of them in response to evaluations requested by multinational companies. GM soya is Argentina's most extensive GM crop comprising almost 90 per cent of the 12 million hectares planted in 2001/2002 and nearly half of all Argentina's agricultural production by 2002/3. Underscoring the export-driven nature of this model, in 2003, 98 per cent of soya was exported in Argentina as beans, soy meal for animal feed and soy oil, representing about 20 per cent of Argentina's total exports by 2004 (Galli 2005). This explains Argentina's sensitivity to disruptions in trade flows, as borne out by its participation in a US-led WTO trade case against the EU's *de facto* moratorium on GM crops.

More importantly, GM exports acquired renewed importance after Argentina defaulted on its US$140 billion national debt in December 2001, and an enormous devaluation took place. As an Argentine agriculture trader (chief executive of Cresud) commented: 'the I.M.F. should be very happy with us. Without agribusiness and oil, Argentina would never meet the surplus they are demanding' (Elsztain, cited in Vara 2005, 8). The government decided that income derived from exports would help to increase foreign earnings that would in turn help the poor. Revenues earned from taxes imposed on exports of GM soya, currently 6 per cent of all government revenue, have been used to subsidize the internal market (*The Economist* 2006).[2] Other sought effects have been the savings

---

[2] Such taxes (at different rates) have been in place since 1991, but the claim that government revenue derived from this source is used to combat poverty gained salience in the wake of the financial crisis in 2001/2002.

from reduced pesticide use and reduced soil erosion from less intensive tilling. While there is some evidence of these effects, more critical accounts suggest that the record to date has been less positive if a wider range of issues are taken into account, such as evidence of increased chemical imports (such as glyphosate from China), of deforestation associated with land clearing for GM production (Greenpeace 2006a), as well as concentration of land tenure and decreasing employment amongst labourers lower down the agricultural supply chain. Nevertheless, the politics of agbiotech is often referred to as a 'non-issue' in Argentina. With a strong alignment of the interests of state and national and foreign capital about the value of biotechnology, the key contestations are around access and ownership of the technology, rather than around its desirability in social and environmental terms. The percentage of agriculture devoted to biotech and the percentage of exports based on agriculture make it abundantly clear that the material contribution of the biotech sector to the Argentine economy is immense. Indeed, the very nature of the approval system is structured around the export potential of the technology.[3]

MOBILIZATION

Against this background there has been a highly uneven degree of mobilization across the region at the national level and in relation to transnational policy-making arenas. The question of who is mobilizing is a source of dispute among commentators. In general, while environment-oriented urban-based groups have focused their campaigns on issues of biosafety and adopted legal-based strategies to engage with the regulatory regime as currently constituted, rural-based social movements have incorporated biotechnology into their repertoire of campaigning not as a unique technology which may generate new risks, new politics and require new forms of campaigning. Rather, it has been quickly absorbed within existing campaign priorities, ideological frames and modes of collective action.

Along these lines, Scoones, for example, draws a contrast between 'urban centres with significant middle-class populations and a strong NGO presence. They are part of the metropolitan, relatively elite circuits where the higher profile activist networks are centred' and 'the deprived rural hinterlands where a different activist discourse may be present' (2005, 3). For Peleaz and Schmidt, on the contrary, 'it would be wrong to conclude, from the lawsuits brought by IDEC (*Instituto Brasileiro de Defesa do Consumidor*) and Greenpeace, that resistance to GMOs in Brazil is recent and mainly fomented by urban organizations of consumers and environmentalists'. Instead, they argue, 'this resistance is a direct outcome of a movement that challenged the Green Revolution model of agricultural

---

[3] The DNMA (Dirección Nacional de Mercados Agroalimentarios) makes an assessment of the export potential of a crop being considered for commercialization. This was known as the 'mirror policy': not commercializing a crop that had not been already been approved in Argentina's key export markets.

modernization at the end of the 1970s' (2004, 239). It was this latter emphasis that gave rise to claims of farmers' rights and autonomy in relation to the production of seeds and consequent efforts to resist the extension of property rights over genetic material to corporations. In practice, it is probable that groups present different parts of the campaign to different audiences depending on their perceived receptivity to that element of the concern. For example, issues of labelling and bans play well with an anxious urban middle class, while socio-economic, livelihood and trade concerns resonate more powerfully for marginalized rural groups who are more directly affected by these. It is clearly also the case that some discourses and modes of engagement are required for influencing state, as opposed to corporate actors, for example. This also reflects ideological choices about whether campaign goals are achievable through engagement with government in formal institutional arenas, as opposed to maintaining oppositional stances through mass actions.

These distinct repertoires of protest can be observed in the way many ENGOs (Environmental NGOs) engage in the economics of the issue and in the technical debate about crop biosafety. They, to some extent, accept the debate on terms bounded by regulatory discourse and parameters. As Scoones notes, the danger here is that 'the types of science deployed and the nature of the argument used are in essence responses to and so framed by the pro-GM position' (2005, 39). The emphasis for many is upon improved mechanisms of oversight and control, drawing on the expertise they bring to bear and the concerns of their membership. This is in contrast to movements that seek to challenge the very structure of production and ownership which creates biotechnology. Even within the state the targets are distinct. While ENGO lobbying tends to centre around Environment ministries with responsibility for biosafety measures, Ministries of Agriculture and Economy are often the targets of rural movements, those with responsibility for making and implementing policies that impact their livelihoods most directly. Nevertheless, when it comes to formulating trade policy, agrarian and environmental groups rarely have strong ties with the ministries of trade and finance that lead policy deliberations, a distance from policy influence which is exacerbated within regional institutions such as Mercosur, which continue to function on a strongly inter-governmental basis.

Despite these distinct backgrounds and strategies, increasing links between urban NGOs and rural social movements have served the needs of each. Though resistance to GMOs has traditionally come from urban-based environmental and consumer groups, the main peasants associations, such as the landless movement (MST) and the National Confederation of Agriculture Workers (CONTAG) in Brazil, have also stated their opposition to GMOs. The CONTAG, for instance, in its 2002 Agenda, campaigned against the production and commercialization of GMOs, and for the labelling of GMOs that have been commercialized. Links to other groups have served both rural movements and the NGOs they align themselves with. Peleaz and Schmidt note: 'the emergence of . . . movements aiming at the reinforcement of small farmers' autonomy improved the bargaining power of social organizations in the rural sector and provided the necessary legitimization for NGOs to represent the political interests of the rural sector in congressional disputes' (2004, 242).

Vía Campesina was critical in bringing the MST into the struggle against GM agriculture, broadening the social base of the struggle and in so doing lending it greater perceived legitimacy. Having on board groups who represent smaller producers, leaving aside the questions about how effectively they do this, lends moral and political credibility to coalitions claiming to act on behalf of and defending the interests of smaller producers in the face of multinational biotechnology companies. Likewise, for the NGOs making up the campaign for a 'GM-free Brazil', a link to the MST provided the means to greatly strengthen their credibility:

> For the small group of NGOs, which to that point were the core network members, this was critical. They could no longer be criticized for just being unrepresentative, unaccountable, foreign-funded NGOs (which they were), but were linked to a mass movement with official and informal connections to the PT, allowing political clout well beyond what was possible before. (Scoones 2005, 22)

Some coalitions are more NGO-dominated, nevertheless. 'Brazil Free of GMOs', for example, is an NGO coalition made up by Greenpeace, the Institute for the Defence of Consumers (IDEC), Services and Advice to Alternative Agriculture Projects (ASPTA), ActionAid Brazil, Social Assistance and Education Federation (FASE), the Socio-Economic Studies Institute (INESC) and the Advisory and Research Centre (ESPLAR). Their main purpose is to promote public awareness concerning the environmental and health risks of the production and consumption of GMOs. There has, nevertheless, been enough common ground between NGOs and movements to articulate a common critique of the trade in GM products. Mobilizations around the production and trade in the products of agricultural biotechnology are anchored around a series of key mobilizing claims which in turn draw on broader critiques of prevailing models of agribusiness production and trade. There are differences among groups regarding the degree to which they would support all of these claims. Few, if any, would subscribe to all elements and certainly not all of them equally. What follows, therefore, are recurrent themes common in the discourse and claim-making of activists.

1. A key over-arching theme is the excessive, unchecked *power of agri-business* manifested most obviously in the concentration of land ownership and control of the market. In Brazil alone, for example, Monsanto is the largest foreign seed company, controlling 60 per cent of the national market of maize and 18 per cent of soybeans.[4] Monsanto purchased five different national seed

---

[4] It is alleged that in terms of GM seeds specifically, Monsanto is responsible for 88 per cent of a market (Igor Felipe de Santos, MST 'Multinationals seek to dominate the entire food chain', see Vía Campesina web site, http://www.viacampesina.org/main_sp).

companies in Brazil, including some of Brazil's best established seed companies such as Agroceres. In one year, Brazil's previously domestic hybrid seed industry became 82 per cent owned by Monsanto (Paarlberg 2001, 70). The agribusiness model is accused of concentrating wealth rather than distributing it, with one figure put forward by MST leader João Pedro Stedile suggesting that just ten multinational companies operating in Brazil receive more credit than four million smaller family farms (Stedile 2004). Trade liberalization is seen to have enabled and accelerated this concentration. Echoing what was said above about the suitability of existing systems of agricultural production for the cultivation of GM, activists in Brazil claim that the *latifundium* is a powerful ally of the biotech multinationals (Andrioli 2006). Specifically, there is an alleged link between *latifundium*, agribusiness and transnational firms such as Cargill, Monsanto, ADM and Bunge, seen as the 'main enemy of the MST' (Santos 2006).

2. Drawing strongly on dependency critiques which highlight the dangers of over-reliance on private capital for the development of countries on the periphery of the global economy, activist discourses invoke themes of autonomy and sovereignty around food. For MST and Vía Campesina, '*food sovereignty*' has emerged as a central theme in campaigning against agri-business in general and its role in biotechnology development in particular, generally taken to mean 'producing, exchanging and consuming food in and closely around one's territory'. Vía Campesina claims, for example, 'a nation will only be sovereign if it can control the production of its own seeds' (Vía Campesina n.d. b). This element is strongly related to trade: dumping, cheap seeds and the removal of quantitative restrictions on agricultural imports. Indeed, Vía Campesina issued an open letter to Brazil's President Lula arguing that the WTO has no role in regulating access to fundamental rights such as food and calling instead for a suite of measures aimed at promoting food sovereignty and genetic diversity (Vía Campesina n.d. a).

3. Beyond generic issues of sovereignty and autonomy, a specific set of concerns with access and control of technology and knowledge feature prominently in activist discourses. IPRs, a key instrument of trade policy in regional and bilateral trade agreements, have attracted particular fire. In Brazil there was significant mobilization around the Patent Bill proposed by President Fernando Collor, focused on the issue of patenting seeds. This included a mix of 'agro-ecology and environmental NGOs, as well as the Catholic Bishops Conference' (Peleaz and Schmidt 2004, 242). A campaign around the call of 'No patents on life' meanwhile brought together a Forum on the Free Use of Knowledge (Forum pela Liberdade do Uso do Conhecimento). This umbrella organization included over a thousand civil society organizations including trade unions, professional organizations, churches, scientific societies, NGOs and even elements of national capital benefiting from looser protection of intellectual property such as pharmaceutical companies. Bilateral trade agreements have also been a vehicle for strengthening the protection of intellectual property rights and hence become a focal point for activist campaigning. For

example, Peru's negotiation of a free trade agreement with the US included provisions to synchronize the country's levels of IPR protection. Activists have claimed that this will have a profoundly negative affect on peasant farmers cultivating seeds and crops that will now be subject to patent control (Third World Network 2006).

4. The campaign 'Terminate terminator' in Brazil was also centred around opposition to the use of GURTs (Genetic Use Restriction Technologies) in seeds which require farmers to buy new seeds each season rather than re-plant and exchange them with other farmers. The claims focused on the likely effect of such a technology in Brazil, where it is estimated that as many as 87 per cent of all farmers do not buy their seed (Radio Mundo Real, n.d.). On 22 March 2006, a day of action against terminator technology was declared in which, for example, 300 rural workers demonstrated in Curitiba against the suspension of the moratorium preventing the cultivation of terminator seeds (Santos n.d.). The decision of the 8th meeting of the Conference of the Parties of the Convention on Biological Diversity (CBD) held in Curitiba, Paraná to uphold the moratorium on the use of terminator seeds was used by Vía Campesina to lend credibility and support to this campaign (Vía Campesina n.d. c).

The trade in food aid has also drawn attention to the question of control over food amid claims about the double-standards employed by large biotech firms that use the developing world as a dumping ground for technologies rejected elsewhere. Clapp (2006) shows how GM food aid has been used to undermine resistance to the technology. Ecuador was the first known developing country to receive food aid containing GMOs in a shipment of soy from the US and channelled through the World Food Programme. It was eventually destroyed following complaints by Ecuador. When food aid maize was sent to Bolivia in 2002, in spite of the country's moratorium on the import of GM crops, the issue of double-standards once again came to the fore. The GMOs in the Bolivian aid contained StarLink corn in a modified form not approved in the US for human consumption (only as animal feed). NGOs claimed that despite the fact that when the StarLink was found in the US food supply it was immediately removed from the market, the US made no such effort to remove the maize from Bolivia, prompting accusations of double-standards (Clapp 2006). The use of food aid within Argentina at the time of the country's food crisis also generated a counter-discourse around the corporate response to the crisis that GM was being used as 'forrajeros para los pobres' (fodder for the poor) or worse still as an opportunity to access more potential consumers for GM products (Kossoy 2003).

5. The potential environmental impacts of biotechnology have been another key element in activists' critique of biotechnology. The cross-border trade in genetically-modified seeds is notoriously difficult to monitor and regulate and has hence heightened concerns about biosafety and the potential for cross-contamination of non-GM crops, such as occurred with GM maize in Mexico (Fitting 2006). Other mobilizing claims centre on the deforestation of land to

make way for monoculture GM crop cultivation. In Argentina, for example, groups such as Grupo Reflexión Rural and Greenpeace have protested the environmental consequences of the mono-cultivation of soya, including loss of soil fertility and deforestation in particular (Benbrook 2005).[5] Because of the extent of mono-crop cultivation of soya, they refer to Argentina as the 'Republica Unida de la Soja' (Boy 2006). The clearing of forests to make way for soya plantations in Argentina has served to globalize this campaign, which may follow the route of neighbouring Brazil in finding itself the focus of a transnational campaign. In Brazil, the clearing of Amazonian rainforest for soya cultivation has prompted environmental protests and attempts to elicit from leading firms promises not to buy illegally cultivated soya. In a full-page advertisement in *The Guardian*, Greenpeace described as 'A turning point in the fight to protect the Amazon' the pledge by food manufacturers and fast-food chains such as McDonalds not to use soya illegally grown in Brazil. The commodity MNCs Cargill, ADM and Bunge are named as those responsible for financing the forest destruction as a result of the soya farming (Greenpeace 2006b).

Some groups within Latin America have objected to the development of GM trees. ETC (Action Group about Erosion, Technology and Concentration)[6] Mexico has expressed concerns about the reach of the pollen GM trees can produce. The World Rainforest Movement (WRM) also mobilized around calls for a ban on the release of GM trees in support of a precautionary stance taken by CBD negotiators, an issue raised at COP8 (WRM 2006). Groups such as *Rede Alerta contra o Deserto Verde* meanwhile focus on the social and ecological consequences of mono-crop or tree plantations, particularly expulsions from land and the effects on indigenous Indians (RACDV n.d.).

Each of these discourses resonates with broader global debates and contestations around farmers' rights, food sovereignty, biodiversity protection and critiques developed by the anti-globalization movement. It is this broader framing of biotechnology's relationship to trade that connects anti-GM campaigns with the agenda of *campesino* groups. During negotiations towards the FTAA there have been explicit inter-governmental statements in support of the trade in genetically modified organisms, prompting concerns among activists that FTAA will provide a back door route to spreading the use of GMOs in the region (Global Exchange 2004). This would be against the expressed reservations of countries like Bolivia about the technology, and driven by the need for the US, Canada and Argentina, in particular, to find new markets for products rejected in Europe and parts of Asia. This issue has been raised by *campesino* groups in countries that serve as a centre of origin for key crops such as maize, like Mexico, a country which has already experienced contamination of non-GM crops by transgenic

---

[5] Critics of this position, including some environmentalists, suggest that biotechnology per se is not the main driver of mono-cultivation as some activist campaigns claim. Interview, ENGO activist, 9 November 2006.
[6] Formerly RAFI (Rural Advancement Foundation International) based in Canada.

varieties.[7] This has fed into a concern that proposed IPR provisions within the FTAA might continentalize North American patenting provisions, over-riding communal and indigenous peoples' rights (Acción Ecológica 2004).

Transnational Organizing

It is around these claims, and their links to broader concerns about the model of market integration for the region with which they are associated, that joint campaigning between transnational biotech activists has been possible. This section of the study, therefore, reviews and assesses the strategies adopted by activists in Latin America to contest biotechnology and its promotion through trade agreements in the region. It is difficult, nevertheless, to assess with any degree of accuracy the true extent of *transnational* mobilization per se as it pertains to the relationship between biotechnology and trade in Latin America. Transnational coalitions with a broad membership base often operate in a very loose, decentralized fashion, with the head of the organization functioning mainly as a coordination unit and public face of a coalition whose transnationality mainly derives from the fact that it has members in many different countries that subscribe to the basic position of the organization. For example, Vía Campesina has a very small staff consisting of an executive secretary and a handful of support staff relying heavily on yahoo list serves to distribute position papers and issue announcements (Edelman 2003; Borras 2004). Similarly, the HSA (Hemispheric Social Alliance), through bringing together a broad range of movements around opposition to the FTAA, exists on an incredibly small staff and budget. The outward appearance of mega-coalitions of transnationally organized activities often betrays the reality of small but tightly organized offices with exceptionally good global connections. Moreover, the ties that sustain networks such as this are more often than not personal relationships rather than formal or regular inter-organizational collaborations (Edelman 2003).

We have seen above how coalitions of environmentalists and peasant-based movements have adopted strident positions on the issue. Below I raise questions about the extent to which those concerns derive from their membership base or rather from strategic and opportune positioning in relation to a current focus for mobilizations. Here I briefly address the 'dynamics of inter-connectivity' alluded to in the Introduction, across levels from local to global via the regional and back again. As with the analysis above, such forms of organizing raise issues of representation and accountability in equal measure.

There is evidence both of generic anti-biotech coalitions within the region as well as alliances aimed at dealing with specific aspects of it. For example, the *Red por una América Latina Libre de Transgénicos* (RALLT) was formed in January 1999 at a workshop in Quito on the theme of GMOs and biosafety. The initial impetus came from 'the need for communities to develop global strategies against the

---

[7] Meeting with *Grupo de Estudios Ambientales*, Mexico City, August 2002.

introduction of transgenic organisms and to prevent their introduction into the region' (RALLT 2007; translated). The network states among its aims: the protection of local communities and national processes which seek to avoid the introduction of GMOs; the protection of governments' autonomy in deciding whether to accept GM agriculture and ensuring that civil society and affected groups are included in this decision-making; and, finally, to work towards a moratorium on the release and trade in GMOs until such a time as there is complete evidence of their safety and societies have had a chance to debate in a informed way the risks and impacts associated with the technology. Its membership is broad and includes *campesino* organizations, indigenous groups and environmentalists. Any group or individual that agrees with the declaration developed by the group on GMOs can be a member of the network. Alongside such broad coalitions whose position is against the technology in general, other coalitions target specific uses of it. The *Red Latinoamericana contra los Monocultivos de Arboles* is against the use of GM trees on grounds of the environmental damage they cause, noted above, and the Forum for Biological and Cultural Diversity was formed in 2001 in Chiapas, Mexico to 'defend native medicinal and crop plant varieties against threats posed by free trade'.[8]

Many such coalitions have a similar starting point to groups such as ETC (Action Group on Erosion, Technology and Concentration) but with strong international connections. ETI's focus has been on issues of IPRs and GURTs or 'terminator' technology. They have had a consistent presence at international meetings on biosafety, organizing workshops and side-events, lobbying governments and sharing research and position papers. An organization playing a similar role within Latin America is the *Red Interamericana de Agriculturas y Democracia* (RIAD) which operates as an information clearing house and source of analysis for NGOs and peasant organizations.[9]

Some pre-existing regional and international bodies working on rural issues and seeking to protect the interests of *campesinos* have adopted positions on the issue of biotechnology. For example, the *Coordinadora Latinamericana de Organizaciones del Campo* (CLOC), formed in 1994, brought together representatives from 84 organizations from 21 countries working on a common agenda of agrarian reform, food sovereignty and indigenous rights. Given this, it is closely aligned with Vía Campesina, and the overlap between their members is extensive. Almost all Vía Campesina member organizations in Latin America participate in CLOC and many CLOC organizations participate in Vía Campesina (Edelman 2003). As well as supporting one another's networks, both coalitions have, in turn, played instrumental roles in constructing other coalitions on specific issues such as the Forum for Biological and Cultural Diversity mentioned above or the IPC (International Planning Committee for Food Sovereignty). This is a global alliance of dozens of transnational agrarian

---

[8] For more details on the forum see http://www.laneta.apc.org/biodiversidad/.
[9] APM-Mondial played an important part in creating this network providing another illustration of the way in which existing networks tend to spawn new networks.

movements plus other civil society groups, involving about 500 organizations across the world.

There are clearly many biotech activists whose concerns are specific to the technology or particular aspects of it, whose critique is not underpinned by a broader preoccupation with resisting neo-liberalism. For anti-globalization activists on the other hand, with whom Vía Campesina is most closely aligned, biotech is one more symptom of an economic system which benefits agribusiness at the expense of smallholders. There has been a convergence of diverse and competing agendas around a common critique of neo-liberalism. Edelman notes: 'trade, phytosanitary measures, intellectual property rights, animal and human health, environment, human rights, biotechnology, gender equity and food sovereignty have, in everyday political contention, become inextricably bound up with one another' (2003, 212). This is true both of groups such as Vía Campesina as well as coalitions set up to contest particular trade accords such as HSA, whose alternative agreement for the region '*Alternativas para las Américas: Hacia la construcción de un acuerdo hemisférico de los pueblos*' reflects the diversity and interplay of these issues.

The intimate connection between biotech's development in Latin America and the evolution of trade agreements aimed at securing market access for the technology has placed agricultural biotechnology firmly on the radar screen of opponents of market-driven regional integration. For example, opposition to FTAA is re-framed as a broader struggle against the global industrialization and intensification of agriculture. Connections have been forged to international campaigns against GMOs, which also have a regional resonance given the centrality of Argentina and Brazil to the global GM debate. According to Teubal and Rodriguez, 'various campesino movements have successfully articulated in recent years an authentic global movement' (2002, 197). This is grounded in opposition to TNC control of agriculture (including patenting and biopiracy), free trade in agricultural produce (especially dumping), the use of hormones and transgenics and in favour of food security and food sovereignty. Coalition-building of this sort and a range of protest activities have been the main strategies adopted by *campesino* movements, often aided by the financial support of sympathetic groups in Europe and North America, funding the travel of *campesino* groups to major anti-FTAA demonstrations (Newell and Tussie 2006).

As a result of this coalition-building, it is possible to argue, as Edelman, does that 'participants in the peasant and farmer networks have also come to have a dynamic sense of themselves as political actors, empowered with new knowledge, conceptions of solidarity and tools of struggle' (2003, 214). Coalition-building has also taken place horizontally between movements. In Brazil, though the campaign for a GM-free Brazil has formed the epicentre of protest, links have evolved between peasant-based movements such as MST and Vía Campesina and a range of environmental and consumer-based groups including Greenpeace and ESPLAR, *Centro de Pesquisa e Assessoria*, development groups such as ActionAid and AS-PTA (*Assessoria e Servicios a Projetos em Agricultura*

Alternativa), education groups such as FASE (see below) and INESC (*Instituto de Estudios Socio-Economicos*). Critically important links were also forged with the PT (Workers' Party). Symbolically important relations were also built with the Church through the Basic Christian Base Communities (Boff n.d.). The Pastoral Commission of Land produced a declaration on GMOs, for example, listing a series of social and environmental objections to the use of the technology (PCL 2004), as well involving itself in a campaign to 'save creole seeds' by creating community seed banks in the face of intellectual property claims by multinational companies (PCL n.d.). Amid such diversity it was possible nevertheless for each group to pursue its preferred strategies of protest and use its resources where the returns are likely to be highest. Scoones notes: 'each was able to focus on different areas of advocacy and action – some more direct and protest oriented; some more focused on practical demonstration of alternatives; some through courts and the media' (2005, 12).

One strategy activists have adopted is to disrupt the actual trade in GM produce through legal and direct means. In 1997 Greenpeace went to court to prevent two container boats carrying GM maize from Argentina bound for Brazil and its poultry industry, in particular, from setting sail. The move was successful in prompting public debate as boats full of Argentine corn arrived but were turned away and held offshore while a decision was made about whether to allow the GM commodities into the country. MST also intervened physically to stop the import of GM produce into Brazil when in July 2000 an MST-led group of farmers attacked a ship in Recife containing GM maize imported from Argentina (Scoones 2005). A legal action was also brought against the Ministry of Justice in an effort to block further shipments of soybeans from the US on the basis that no labelling provisions yet existed in Brazil to protect consumers. The controversy led to mandatory testing of imports for GM content and, according to Paarlberg (2001, 84), significant GM imports from the US and Argentina were avoided to minimize controversy. Though, in the short term, such actions did serve to draw attention to a poorly regulated trade and its associated risks, forcing the issue into public arenas and soliciting a response from the government, attempts to stall such an economically lucrative trade have been rendered futile by the shifts in the Lula government's position towards biotech. On 22 March 2005, for example, CTNBio approved the importation of 370,000 tonnes of GM corn from Argentina to be used as chicken feed.

It is often through working alongside media-savvy transnational actors that national struggles receive international attention, potentially acting as a lever for change. The problem for groups such as MST is that the more *quotidian* struggles of rural movements, particularly where the same strategy is used repeatedly (the land invasion), are less newsworthy. The radical content of the politics behind the action and the fact that such actions often take place in remote rural settings makes it more likely that they will be beyond the radar of national journalists based in urban centres. The lack of novelty of the land invasion for media seeking a new angle, and rarely sympathetic to the movement's goals, makes it hard to publicize key actions in mainstream media (Andrioli 2006). By contrast, the presence of 'international anti-GM stars' (Scoones 2005, 15) at World Social Forum (WSF) meetings in Brazil, who in many ways serve as the 'organic intellectuals' of the anti-GM movement lending it visibility, weight and direction (in different proportions), has helped to garner media attention, useful to national struggles against biotech. Media interest for example was ignited by the involvement of renowned anti-GM activist and farmer José Bové, who was arrested for participating in an MST-led invasion of a Monsanto farm in Rio Grande, destroying five acres of GM crops (Scoones 2005). Other celebrities include Vandana Shiva, the Indian writer and activist, and Canadian farmer Percy Schmeiser, who was involved in a court battle with Monsanto, also claimed he was using their seed illegally against his claim that his crops had been 'contaminated' by GM crop trials in neighbouring fields (Glover and Newell 2004). Involvement in the WSF proved effective for domestic movements in centring attention on this issue and building coalitions with international activists. Such fora also provide an opportunity to draw attention to groups' own agendas such that the WSF in 2001 became the venue for a 'World Forum on Food Sovereignty' supported by CLOC and Vía Campesina among others. Indeed as Edelman suggests, 'high-profile participation in international protests and civil society gatherings continues to be hallmark of Via Campesina activity' (2003, 206).

It is important to recall, nevertheless, that even seemingly transnational mobilizations often derive their energy and strategy orientations from particular national settings. Brazil has served this role as an epicentre in the broader social struggle over biotechnology. Particular sites in the global contest over the future of agricultural biotechnology become transnationalized because of the symbolic value they generate and the claims projected onto them by activists elsewhere. The romanticism that surrounds Latin American social movements in general is certainly evident in relation to the MST, particularly among Western-based anti-biotech activists keen to deflect claims that their rejection of the technology is elitist and serves to deny poorer farmers access to a potentially welfare-enhancing technology.

As we have seen within Latin America and in other arenas of global economic policy, most notably the WTO, efforts to shape regional and transnational processes are also channelled through the state. Much effort is still geared towards shifting or sustaining the positions of national governments as a way of shaping international policy. Hence, while there may be evidence of what Tarrow (2005) calls the 'new transnational activism' responding to and scaling up to political opportunity structures available within international and regional fora, the majority of biotech-related activism continues to take place at the national level, even if it is linked to groups elsewhere and is also often pitted in opposition to transnational actors operating in domestic arenas (such as biotech TNCs). For most activists, the primary target remains the state and the primary audience a national one, even if international levers can be used to generate domestic reform. This is particularly understandable for those rural movements that have engaged with the issue such as MST, where biotech is viewed as another manifestation of an unjust division of land and wealth from agricultural production, which the state has the overriding responsibility to address. These are issues beyond the jurisdiction of regional and international bodies even if decisions they make, especially around trade, have profound national and local impacts.

National campaigns within the region may well be managed through bodies serving as coordination points and with a transnational base of representation. They serve to direct their claims towards regional and global arenas, such as the WTO or multilateral arenas dealing with issues which impact upon rural livelihoods, such as the CBD's deliberations about access and benefit sharing or the debates about restrictions on the use of GURT technologies. Activists make use of their ability to operate simultaneously in different political spaces and to transgress scales. This involves, for example, mobilizing nationally within 'transnational arenas' such as the negotiations of the Cartagena Protocol. With access to transnational audiences such arenas serve as platforms to advance national political objectives. For example, around 1,500 peasants organized by Vía Campesina demonstrated in Curitibia, where the MOP3 (Meeting of the Parties) was being held. They were protesting the length of time allowed before requirements regarding the identification of transgenics in food come into play, which is currently four years.

Despite the claims that global social movements are more than the sum of their parts or that, rather like claims made of globalization itself, they render boundaries irrelevant, collapsing, transgressing and merging spaces of activism, there is a need for caution about the extent to which activism is truly borderless. The case of activism around trade and biotechnology in Latin America would seem to lend weight to Tarrow's (2005) claim that much activism that is called transnational is often actually the global expression of demands articulated in a particular national setting and intended to shift the position of a specific nation state. What we seem to observe are the emergence of 'focal points', moments or spaces of political convergence in which diverse national priorities, campaigning emphases and strategic preferences can be expressed. These can take the form of a 'discourse coalition' (Hajer 1995) or serve as a narrative that unites eclectic movements across different territories, a malleable story-line or device that works in different settings and therefore allows for the political buy-in of groups that might not otherwise work together. The concept of 'food sovereignty' certainly works in this way with a generic ability to serve diverse (and sometimes conflicting) campaign needs simultaneously. Hence while for MST this provides a starting point for claims about land redistribution, for other rural groups it is a rallying cry for the defence of local economies while, at the same time, allowing many green groups to argue for ecologically sustainable (non-GM) food production.

What we may find more evidence of is high levels of transnational coordination *within* international NGOs such as Greenpeace. Groups such as this can play a bridging role between campaigns in different global sites, emulating tactics adopted by their groups in one part of the world in other locales. For example, there has been a strong tradition of using direct action in Europe by Greenpeace,

GenetiX snowball and other groups uprooting GM crops in field trial sites. Such strategies have been adopted in Brazil too, where there have been invasions of Monsanto and other research stations and the uprooting of trial crops by Greenpeace and other activists. In January 2001, as part of a broader anti-globalization protest, more than 1,000 workers from the MST invaded a Monsanto plant in Rio Grande do Sul and said they would stay there indefinitely to protest against GM crops being developed by the company. It seems likely that we will see heightened emphasis on strategies of direct action in the face of government moves to further open the market to GM development. For example, in March 2006 Vía Campesina occupied a field trial site of the biotech firm Syngenta to denounce the 'illegal' use of GM soy and maize in the area. In November 2006 state governor Roberto Requião signed a decree of intent to expropriate the Syngenta farm. According to Kenfield and Burbach (2007), 'the decree was a huge political victory for the rural and environmental movements, challenging the power of agribusiness in Brazil'. If it was, it was short lived. In July 2007 Syngenta and its allies succeeded in overturning the decree and Vía Campesina was evicted from the site. More recently (October 2007) such occupations have resulted in violence and even the death of a Vía Campesina activist in Paraná following an occupation by 150 members of the group (Vía Campesina 2007; Kenfield and Burbach 2007). This is certainly transnational activism around biotech, albeit within one organization with significant sharing of information and resources aimed at aligning national struggles with international campaign priorities.

In terms of peasant-based transnational movements per se, activism around biotechnology in Latin America can be located as part of the trend Edelman (2003) describes of a 'globalization from below', in which agriculturalists develop common agendas and protest repertoires on issues as diverse as trade and human rights issues. As a result he suggests 'farmers have achieved a prominence in international arenas that they rarely enjoyed in their own countries' (2003, 185). Despite the differences of material interests that exist between them, what unites poorer farmers and landless labourers is a common experience of exploitation, which can be described in class terms which transgress national borders. Hence although those involved in rural social movements invoke multiple identities such as ethnicity and gender in asserting their rights and explaining their exclusion, exploitation, not consciousness or common awareness are the hallmarks of class (Burnham 2002, 117), such that mobilizations around biotech can be usefully understood in class terms. De Ste. Croix claims that 'class (essentially a relationship) is the collective social expression of the fact of exploitation, the way in which exploitation is embodied in a social structure' (1981, 43, cited in Burham 2002, 117). The issue in this context cannot be reduced merely to control of the means of production, since many smaller farmers do have land titles, unlike the landless labourers they employ. Rather, the resistance that results from exploitation by larger commercial interests and the state creates a political tie between an affected class of farmers across Latin America and beyond.

194 *Peter Newell*

For example, the prospect of a trade agreement with far-reaching effects on rural economies serves to galvanize social action and provides a common threat for smaller producers. This was certainly the case with the NAFTA agreement that spawned a whole series of coalition-building amongst the probable losers of an integrated market and latterly the FTAA (Teubal and Rodríguez 2004). RMALC (Mexican Network Against Free Trade) was one manifestation of this, serving as an important source of analysis on trade policy for peasant movements (Edelman 2003; Icaza 2004). Issue-specific groups such as *En Defensa del Maíz* (EDM 2002), focused around free trade and GM corn, and national coalitions of *campesinos* such as *El Campo No Aguanta Más*,[10] drawing attention to the negative impacts of NAFTA upon the livelihoods of the rural poor, also sprung up in the wake of the agreement. Trade politics connect macro economic re-structuring with local livelihood concerns 'blurring' domestic and foreign policy in new ways. There is also a sense with activism around trade issues that nationality ceases to be the primary point of reference even if lobbying continues to be channelled through the state. Expressions of solidarity, encouraged through joint demonstrations, declarations, exposure tours and the like lend weight to Edelman's claim that trade policy 'increasingly divided people less along national lines than in relation to shared class, issue-based or sectoral interests' (2003, 198). With indigenous and *campesino* groups, identity politics which transgress state borders are certainly also key. The prospect of the FTAA served to stretch these ties across an entire continent embodied in the creation of the Hemispheric Social Alliance.

PARTICIPATION

Though many of the mobilizations described above seek to contest the relationship between biotechnology and trade through 'outsider' strategies of protest and resistance, some groups have sought to make use of available channels of institutional participation or 'invited spaces' made available to them by national governments and regional trade institutions dealing with the issue of biotechnology.

There is not space here to review the full array of spaces available to organized civil society inside the formal arenas of trade policy in Latin America associated with agreements such as NAFTA, FTAA, Mercosur, and in any case these have been reviewed elsewhere (Newell and Tussie 2006). It suffices to note that the participation of civil society actors has been aimed at generating public and political support for controversial trade agreements rather than opening them up to serious public scrutiny. Openings for activists to make their views heard have been further circumscribed by the overwhelming degree of support leant to the technology by leading players in the region such as Brazil and Argentina. Within the institutions of Mercosur, such as the SGT6[11] Working Group on the environment,

---

[10] Translated as: 'The Countryside Cannot Take Any More'. For an example, see http://www.grupochorlavi.org/php/doc/documentos/elcamponoaguanta.pdf.
[11] Sub-Grupo de Trabajo.

despite early framings regarding the potential risks associated with the technology, Argentina played a lead role in vetoing the biosafety clause of a proposed draft. The Framework Agreement accepted in 2001 has no section at all on biosafety issues (Hochstetler 2003). When Argentina called a meeting of Ministers of Agriculture in Mercosur in 2005 to generate support for its position against paying Monsanto royalties on soya crops (rather than seeds), initial support was forthcoming from Brazil and Paraguay. Intense pressure in the wake of the meeting, however, led to these governments retracting their positions on the basis that they were concluding their own agreements between the private sector and Monsanto.[12]

Beyond the regional trade politics of biotech in Latin America, many governments in the region are also signatories to the Cartagena Protocol on Biosafety, which in turn requires countries to create mechanisms for public participation and consultation around the design of their National Biosafety Frameworks (Glover et al. 2003). At national level, therefore, this had led to the establishment of national invited spaces of participation around biotechnology. Democratizing policy in this area has been extremely difficult, nevertheless. In Brazil, briefs of the CTNBIO meetings agendas are placed on their website before the meetings and decisions taken are officially communicated to local and regional authorities and to other federal institutions such as the Ministry of the Environment and Ministry of Health. The decisions are also posted on the Internet after each meeting. However, the meetings are held behind closed doors and the public cannot access the complete reports of the discussions. The government justifies this approach through reference to commercial confidentiality requirements. The requirements for experimental plantations are published in the official government journal and on the Internet one month in advance in order to allow every interested citizen to declare their position as part of this provision. However, despite increased access to information, the Council has received very few contributions from the public to date. The CTNBIO has supported events promoted by the Biosafety National Association (ANBIO), such as the first Brazilian Congress on Biosafety, on September 1999. Activists complain, however, that a body that is meant to operate as the regulatory organ on public health and environmental issues should have maintained its independence from the companies represented by ANBIO. The event received sponsorship from biotechnology companies including Monsanto, Novartis, Agrevo and Du Pont. As a result, according Pelaez and Schmidt (2004, 249–50) CTNBio faced a serious crisis of credibility in Brazilian society who suspected it was stimulating biotechnology to benefit the multinationals.

The conflictual and legal nature of the policy debate, described by Scoones in this book, is seen by some as a result of grievances by groups that felt they were not consulted and continue to be left out of policy-making processes on these issues. In this regard it is worth citing a report of the Brazilian MP Ronaldo Vasconcellos, concerning CTNBIO's activities:

---

[12] Interview at Ministry of Agriculture, Argentina, November 2007.

> We believe it desirable that the CTNBIO make its procedures more open to the Brazilian society, breaking down myths and versions that have arisen, in many cases, because of the closed, untransparent procedures that marked its activities. We know that a forum of scientists cannot become a popular assembly but, also, it must not be characterized by an atmosphere of gods above the claims of the civil society. The authoritarian style that marked the CTNBIO, especially its presidency up until the year 2001, did not effectively contribute to the development of a biosafety policy in the best interests of the whole Brazilian society. (cited in Glover et al. 2003, 9)

The MP ends his report recommending 'the definition of new criteria for the election of the members of CTNBIO and other measures to increase the transparency of its decisions, in order to bring it closer to the civil society'. In relation to the intensity of civil society opposition to GM crop development in Brazil, CTNBio has sought to broaden the range of stakeholders it consults within its decision-making. From 1999 onwards, it sought to conduct 'extensive' consultations in advance with non-biotechnologists in Brazil's National Academy of Science, responding to the criticism that it gives too great a voice to proponents of the technology. Nevertheless, its reputation as an essentially promotional rather than precautionary body has deterred critical civil society groups from getting involved to replace IDEC on the Commission once the group left in protest.

The degree of access and participation of civil society groups is also defined by their relation with the party in power. Activists note the difference in Brazil, for example, between the Cardoso administration for whom, in their view, agrarian questions were reduced to questions of police capacity to deal with dissident groups and the Lula administration on the other hand, which has invited MST to participate in cabinet councils dealing with issues such as hunger and agrarian reform. This is in spite of activist critiques about the limited nature of that reform process and the government's continued support for an export-oriented agribusiness model (Santos 2006).

Public discussion on the potential impacts of the GMO being assessed is required in Brazil even if some representatives of the CTNBIO believe that such public involvement is not necessary because biosafety issues are extremely 'technical'. The NGO coalition 'For a Brazil free of GMOs' struggled to make an environmental impact assessment with public involvement mandatory. The strategy has been to open up the 'black box' of decision-making to a plurality of voices and to contest justifications for excluding public inputs, namely that the technical nature of the issues prohibits all but narrowly defined expert participation. What Levidow (1998) refers to as the 'biotechnologizing of democracy' has served to restrict debate to a pre-defined set of scenarios about biotech futures rather than engage broader social and ethical concerns. 'Technical' problems amendable to neo-liberal risk-benefit analysis demand a privileged role for experts. As Newell puts it: 'with biotechnology we see the essential tenets of modernity and capitalism brought to bear to foreclose broader democratic engagements with what amount, in many settings, to key questions of rights, access and entitlements to food and livelihood security' (2006, 76). A common political device employed by state and corporate elites towards this end is to invoke the public as ignorant, ill-informed and only able to engage in issues of science and technology on 'emotional' grounds (Wynne 2001). The effect is that 'participation in decisions about GM crops for most publics takes the form of exercising consumer rights to buy, or refuse to buy, a product that has already been approved for market entry despite the efforts of activists to democratize decision-making through attempting to secure public rights to information, to expose approval processes to public scrutiny' (Newell 2006, 77).

In response to the limitations of formal channels of public participation, ActionAid Brazil, FASE (*Federacao dos Orgaos para Asistencia Social e Educacional*), MST, CUT (Confederation of Labour Unions), and Advisory and Research Centre (ESPLAR) have promoted citizen juries targeting small-scale farmers, landless people and poor urban consumers. The first one took place in Fortaleza, capital of the Northeast state of Ceará, in April 2001. The second was organized in Belem do Para, capital of the Amazonian state of Para, in September 2001. The jury was selected 'randomly' from lists provided by a representative range of community-based associations. Hundreds of small-scale farmers, landless people and poor urban consumers attended the events. Among the questions addressed by the jury was 'is there enough evidence that GMOs do not threaten the environment?' and 'is the process of testing and the commercial use of GMOs democratic, transparent and careful enough?' After hearing evidence from witnesses, the answer to both questions from the jury was 'no', in both events. The trial took place over two days.[13] A representative from ActionAid Brazil concluded that 'these people, always excluded from the process of policy-making in issues that affect them very much, had the opportunity to access all the information and to decide about it via members of the jury' (Campolina 2001, 29). Another citizen jury took place in Rio in August 2002. It followed the same procedures of the previous ones, but instead targeted urban consumers. In 2004 there was a citizen jury in Parana. As Scoones notes, 'this was more a mass rally than a jury process with 3000 MST activists attending' (2005, 36). The timing of the event also ensured political buy-in from state officials, coming as it did at a critical point in the electoral cycle for the governor.

In Argentina, there has been a high level of scepticism about the necessity of a National Biosafety Framework with provisions for public participation. Business groups, in particular, were fearful that such a process would generate doubts and scepticism about the technology where they claim none currently exist.[14] Within the formal decision-making process, access for NGOs critical of the technology has been difficult. CONABIA is, in many ways, the epicentre of the approval process for agricultural biotechnology applications. It is a multi-sectoral body, wherein private and public organizations are represented.

---

[13] For more on such deliberative policy experiments see Scoones and Thompson (2003) and Pimbert and Wakeford (2002).
[14] Interview, head of ArgenBio, November 2006.

Membership and coordination have been modified with increasing input from the private sector. Currently, CONABIA is made up of three public research institutions, four public universities, six private sector associations, one civil society (consumers') organization, four SAGPyA (Secretary of Agriculture, Fisheries and Food) representatives and two members of the Health Ministry, though this composition is subject to change over time. The key decisions about who sits on the committee are taken by the Secretary of Agriculture, though members of CONABIA claim that anyone with 'professional experience' can apply to participate. When asked about the absence of consumer or environmental groups, one official claimed to have invited Greenpeace and consumer organizations to participate 'but they have no one to propose with professional capacity in this area'.[15] This use of science-based criteria for determining participation in decision-making and as a mechanism for deciphering 'legitimate' from non-legitimate stakeholders is also common to many other regulatory systems, but it undoubtedly serves to entrench a less critical perspective regarding the risks associated with the technology (Newell 2006). A former CONABIA representative from the Secretary of Natural Resources and Sustainable Development, who attempted to raise critical questions about biosafety issues that had not been adequately studied, is reported to have been strongly outnumbered on the body by advocates of the technology.[16]

What we observe then is an absence of formal invited spaces that function as arenas for the effective deliberation, let alone contestation, of biotechnology for activists. Given this, attempts to construct alternative venues for articulating 'dissident' voices are unsurprising, though their inability to impact authoritative decision-making will continue to be a serious limitation upon their ability to effect lasting change.

REPRESENTATION

As the editors of this collection note: 'it is clear that while a few globalized movements have made impressive strides, far more groups of the rural poor are not represented at all, even nominally in these transnational agrarian movements' (Borras et al. 2008). This is true even in countries such as Brazil, 'purported to be strongholds of these movements'. In part, this results from the multiple roles farmers perform simultaneously as producer, activist and local and global citizen. Edelman spells this out concretely: 'the same individuals who mobilize for international conferences may also have to assemble a legal team to defend contested property titles, follow up on orders for a cooperative's rubber boots and harvest a field of cabbages before the rains arrive' (2003, 214). This may be particularly true of 'rural celebrities' such as Bove and Schmeiser in the case of GM. Pressed by constraints of time, priority and lack of understanding of seemingly distant and technical trade policy processes, the possibilities for direct representation are,

---

[15] Interview, CONABIA, November 2006.
[16] Interview, agronomist, November 2006.

realistically, few. This inevitably leaves tremendous scope for autonomous agenda-setting by the leadership of broad-based coalitions.

The rise and fall of ASOCODE (Association of Central American Peasant Organizations for Cooperation and Development) provides a cautionary tale for other transnational coalitions seeking to scale up to regional and international arenas while remaining responsive to their social base. As Edelman reflects: 'in ASOCODE's case, a top-heavy organisation, a preponderance of activities that responded to donor rather than peasant priorities and incessant internecine squabbling brought the association to a point where it still exists in name but enjoys little of its earlier support, dynamism or prestige' (2003, 191).

In relation both to 'mass' mobilizations as well as channelling influence through formal institutional channels, issues of who speaks for whom and about what inevitably arise. Such issues can also raise tensions between civil society groups as they attempt to work together and are faced with the question of who is being represented. In coalition-building among trade activists around NAFTA, trade unions raised concerns about the representative base of NGOs who they caricatured as non-governmental individuals, while labour movements were also accused of having hierarchical non-inclusive decision-making structures (Newell and Tussie 2006). When issues of funding and diverse protest cultures in which middle-class students rub shoulders with very poor *campesinos* are added to the mix, the potential for friction is immense. For example, in mobilizations around NAFTA, questions were asked of those groups working to secure an environmental agreement within the treaty about whether their support for the trade agreement bore any relation to the corporate funding they received from economic interests likely to benefit from the trade provisions.

There is also an interesting set of contradictions that emerge from the adoption of anti-GM stances by movements such as the MST when many of their rank and file members have adopted GM seeds, albeit in most cases illegally. Membership of transnational alliances such as Vía Campesina and participation in fora such as the World Social Forums, often hosted in Porto Alegre and, therefore, providing privileged access for high-profile Brazilian groups such as MST, encourage the leadership to adopt anti-GM stances, which are consistent with overall strategies aimed at challenging the power of agri-business in Brazil and beyond. But rejection of the technology by movement leaders sits uneasily with its widespread adoption by the members of the movement and perhaps points to an absence of internal debate over the movement's position on this particular issue. The personal reflections of one long-standing commentator on rural social movements in Brazil are worth quoting at length:

> We simply do not have social struggles against biotechnology in Brazil. At most we have some sporadic 'social reactions' against GMOs. There is a small group of NGOs that keep fighting the conservative prevailing trend on these themes in Brazil but with no social base. Their protests are marginal with no repercussion whatsoever in the media or elsewhere. The MST and Vía Campesina do not have any informed social groups backing

the political rhetoric of their leaders when dealing with GMOs or agricultural biotechnology per se. The fact is that when landless families and workers participate in any action sponsored by the movement, they can even carry banners attacking trade and so on, but when you talk to them there is deep ignorance about these issues. These differences suggest that the targets of the leadership are built by a small group and do not find any repercussions with the lower groups within the movement. One of the most bizarre political developments is, for example, the anti-GMO stance of the MST and the plain fact that farmers in rural settlements under the influence of the movement in grain producing areas are all using GMOs in the fields.[17]

Scoones is similarly sceptical about the level of comprehension and political buy-in of rural poorer groups to discourses articulated about biotechnology by elite or vanguard rural NGOs or movements. He notes:

Most rural dwellers . . . do not really have an idea of what GM crops are and are mobilized on other issues – fears about the unknown (terminator crops, Frankenstein foods etc.), concerns about patents, loss of local varieties and so on. Indeed, very often the very same farmers mobilized by organized farmer movements – whether KRRS in Karnataka or the MST in Brazil – are the same farmers planting pirated GM crops illegally, or would try them out if they could. (2005, 18)

Higher levels of direct engagement with those they claim to represent are possible around more central livelihood concerns. Questions of agrarian reform, access to land and the right to food tap into deeply-felt concerns by those whose livelihoods are tied to rural development. It might justifiably be argued that there is no need to consult a membership base of poor and often landless farmers about their desire for land. But 'discourse coalitions' that form around slogans such as 'food sovereignty' disguise the fact that many, even very small producers, are thoroughly dependent on an export-led system of agriculture and though they might benefit from land reform, localized economies divorced from global supply chains may well be disastrous for their existing livelihoods. In this sense Scoones might be right to suggest that biotech, rather than being an issue that bubbled up from below, with rank and file members demanding a strong leadership stance, represented a strategically useful opportunity for a movement like MST to raise its profile and consolidate a position within supportive international networks. He notes: 'as an increasingly internationalized movement – with website and support groups in the US and Europe . . . [and] the arrival of activists by the plane load to Porto Allegre for the WSF festivities each year, the MST were able to raise their international (and so local) profile and forge links with the hall of fame of international anti-GM activism' (2005, 22).

---

[17] Personal communication, commentator and activist on rural development in Brazil, 12 December 2006.

As the editors of this collection note in their introductory text, there is at best only a 'partial representation' at work when claims to speak on the basis of a mass social base are made without corresponding mechanisms within movements for soliciting, let alone acting upon, concerns that may percolate from the bottom up. As the editors note, 'effective representation of the social base's interests within their movements should not be assumed to be automatic or permanent and unproblematic. "Effective representation" is dynamically (re)negotiated within and between leadership and membership sections of movements over time.'

Here we encounter the politics of 'brokerage' performed by intermediaries 'speaking *for* not with', whose claim to legitimacy is alleged to be their social base. The danger is that maintaining a coalition, particularly when there is the prospect of continued funding, becomes an end in itself. In this context, the rationale of the organization, to serve a particular marginalized constituency, becomes subordinated to the need to sustain the jobs of those in the secretariat, keep up the momentum of a campaign or secure funding for the next international event. This does not leave much time for establishing whether supporters agree that this is a priority for the coalition, even if the means to find this out were clearly established. As Edelman puts it, 'network practices of representation – submitting proposals, organizing meetings, publishing newsletters or web sites, drafting "action platforms" – sometimes seek to demonstrate the effectiveness of a network with reference to its own self-description and activities rather than to tangible impacts on targeted constituencies, policies and institutions' (2003, 214). Particularly when, as happened with the MST in Brazil, movements are invited to participate in formal invited institutional spaces, participation and, therefore, representation is rarely direct. The dilemmas of representation are intensified by such engagements with what activists often call 'friendly governments'[18] or movements in power. As with MST, frustration with the pace of reform through institutional means and the careful politics of compromise and negotiation that this necessarily implies, can lead to frustrations and an intensification of direct action strategies, for which the movement is notorious (Petras and Veltmeyer 2005).

We have then to view even allegedly 'mass' movements within the broader context of 'disorganized majorities' that are neither mobilized nor transnationalized. There is also the phenomenon of well-organized minorities; vocal groups with good international connections but little in the way of social base or national influence. The Argentine *Grupo Reflexión Rural*, a small rural development NGO, has managed to cultivate links with groups within the region such as RALLT (*Red por una América Latina Libre de Transgénicos*) and internationally through World Social Forum events. Their rejection of biotech is part of a broader critique of the social and environmental impacts of a model of export-oriented

---

[18] This theme emerged as a key theme of discussion in a workshop we organized with trade activists at the 'Cumbre por la integración de los Pueblos', a counter summit held in Cochambamba, Bolivia in December 2006 at the same time as the meeting of the Community of South American Nations.

agri-business development, an emphasis that aligns them with anti-globalization activists who have been willing to offer them a platform at activist gatherings in Europe and elsewhere.[19] Adopting the media-savvy tactics of groups such as Greenpeace, including protests outside corporate conferences and expos on agricultural biotechnology, they present journalists with good photo opportunities, but according to some, do not have the research or positions behind their advocacy.[20]

CONCLUSIONS

Evaluating the impact and broader significance of these interventions in the politics of trade and agricultural development with respect to biotechnology in Latin America with any degree of precision is a fraught exercise. There are clearly many variables that explain the degree of impact that groups and movements have had. In classic terms, the impact of collective action is affected by 'political opportunity structures', 'resource mobilization' and 'framing' (McAdam et al. 2003). But we have also seen how the micro-politics of mobilization suggest a series of other determining factors which affect the extent to which the agenda of biotech activists makes any advances: the importance of party politics, the role of the media and the differential impact of strategies and protest cultures in diverse political settings.

There are also of course varying indicators of that impact. For most, change in formal government or corporate policy is the aim: successful demands for a new regulation or the overhaul of a policy that the movements were opposed to. Examples would include the provision of environmental impact assessments, tougher biosafety provisions, enhanced restrictions on the use of GURTs or declarations from firms that they will not source soya from areas being deforested to make way for the crop's cultivation. For other movements, the goal is raising public awareness about the technology and the way in which trade agreements that governments are signing undermine national autonomy with regard to the governance of the technology. For others still, including the MST and Vía Campesina, campaigns around biotechnology are of secondary importance to the real issues, which are land reform and an economy dependent on powerful agri-business investors. Achieving change in those arenas given the alliances between state managers and (trans)national capital that sustain this accumulation strategy make struggles over biotechnology seem insignificant by comparison.

Engineering direct and dramatic shifts in policy may be a distant aim, but the short-term challenge is to engage the public in the debate and contest current media and government framings of the issue. The fact that major governments in the region continue to support biotechnology (Argentina, Brazil) or in some cases have had their opposition or scepticism towards it overturned (Peru, Bolivia) is not evidence of movement failure. The strength, economic weight

---

[19] For example, Adolfo Boy was invited to speak at the counter-summit to the EU-Latin America trade meeting in Vienna, May 2005.
[20] Interview with a journalist, 15 November 2006.

and political resources that proponents of biotechnology and potential beneficiaries of trade liberalization agreements have vastly outweigh those of their opponents. Yet, through careful strategic positioning, extensive networking and alliance building, they have, on occasion, been able to outmanoeuvre their opponents. As Peleaz and Schmidt note in relation to Brazil, for example, for some time 'with all its privileged technical and scientific knowledge, its experience in approaching regulators in other countries and with its considerable financial resources to establish a large marketing network and invest in "economic pressure", Monsanto could not get authorization for the release of its RR soybeans' (2004, 254).

Biotech activists have been highly astute at forging connections between biotech and issues which resonate with diverse public audiences, around the impact of the technology upon the rural livelihoods of poorer farmers, around health and environmental concerns for urban consumers. They have, on occasion, been able to challenge the terrain of the debate, introducing alternative forms of expertise and highlighting the importance of cultural values (Fitting 2006). They have successfully constructed informal spaces to enhance citizen participation in policy debate to take it beyond an arena of deliberation among experts. They have also been adept at forging connections with international anti-GM activists who, in turn, exercise influence in multiple arenas. This can create 'boomerang' effects (Keck and Sikkink 1998) where a biotech company in Europe finds itself subject to scrutiny for its actions in rural Brazil, for example.

Nevertheless, despite temporary victories in skirmishes with (largely foreign) capital and the state, the war over biotechnology in the largest countries of the region has for the time being been won. Through a slow process of attrition, careful coalition-building and concerted lobbying in institutional and public arenas across scales, biotech firms have had their products accepted. What remains undecided is the future of trade liberalization in the region and the extent to which it will be advanced regionally or bilaterally. The challenge for activists, as we have seen, is to ensure that success in challenging regional and hemispheric initiatives does not result in bilateral agreements whose provisions go even further than the proposed contents of the regional treaty (Gallagher, 2008). The case of the FTA between the US and Peru, mentioned above, is a case in point.

Contesting biotechnology in national arenas will continue to be key and here impact has varied significantly across the region. In Argentina, there are very few spaces available to contest the technology. The alignment of material, institutional and discursive power produces an effect of 'agro-hegemony'. The same is true of Uruguay and Paraguay by default. In Brazil, we have witnessed increasing acceptance of the technology despite early and ongoing objections to the promotion of GMOs. When faced with opposition, the state has leant strong backing to agribusiness. Despite the great hopes vested in President Lula by anti-GM activists because of views expressed prior to his election and his ties to oppositional movements, his incoming Minister of Agriculture, Roberto Rodrigues, was quick to lend his support to his allies in the agribusiness sector. Hence in March 2003, when faced with the decision about whether to uphold the GM ban and destroy the crop harvest for 2002–03 or to allow its sale, the government allowed the marketing of the GM crops, much to activists' dismay. A further Presidential decree again allowed the sale of GM soya from the 2003–04 seasons, despite it still not having been approved for planting (Scoones 2005). The Biosafety Law approved by the Senate in 2005 has paved the way for the consolidation of Brazil's position as a global leader in biotechnology, closely following the path set by neighbouring Argentina.

The country's dependence on agri-business circumscribes the possibilities for and effectiveness of political challenges. As one commentator said to me: 'it would be difficult to create political opposition to agri-business or problematize modern agricultural technology at the moment when the financial surplus of Brazilian exports are formed especially from export of grains, mainly produced by agri-business'.[21] In a further correspondence this informant noted: 'especially if we are thinking in terms of GMOs for agricultural production, it appears the battle has already been won by commercial interests. The reason is not political but practical. GMOs in grain production, especially soybean, are now spread throughout Brazil and I cannot see how to avoid its use anymore. Some state governments are still talking about restrictions and so on, but this is for the public because no real barrier can be imposed'.[22]

An article that appeared in the opinion pages of the *Folha de São Paulo* newspaper supports this reasoning, pointing out the scale of illegal planting of crops such as soya and cotton and the virtual impossibility of effectively regulating it, producing an 'institutional paralysis'.[23] Perhaps the battle for biotech has already been won. We certainly have to be cognizant of the power wielded by those in favour of the technology, supportive of the regional trade integration that ensures its diffusion and protection throughout Latin America, which includes some of the most influential lobbies in the contemporary global political economy from the agribusiness sector. There is a great deal of pressure being brought to bear upon governments throughout the region into accepting agricultural biotechnology. Whether it is through threats of trade retaliation or the forced acceptance of GM food aid as in the case of Bolivia and Ecuador, the use of bilateral trade agreements that synchronize regulatory systems with regard to biotechnology as in the case of the agreement between Peru and the US or the legal action taken by Monsanto against its former ally Argentina, countries in the region operate in a context of 'bounded autonomy' (Newell 2006). Many governments, it seems, believe they cannot afford to debate the future of the technology. The challenge for activists is to demonstrate clearly that they cannot afford *not* to debate the technology, given its ability to transform systems of agricultural production upon which many of region's poorest people depend.

---

[21] Personal communication, commentator and activist on rural development in Brazil, 12 December 2006.
[22] Personal communication, commentator and activist on rural development in Brazil, 12 December 2006.
[23] 'Barreira transgênica', *Folha de São Paulo* 24/12/06.

REFERENCES

Acción Ecológica, 2004. 'Area de Libre Comercio de las Américas'. http://www.accionecologica.org/index.php?option=com_content&task=view&id=165&Itemid=242, Accessed 2 February 2007.

Andrioli, A.I., 2006. 'Soja Orgânica versus Soja Transgênica'. *Movimento Dos Trabalhadores Rurais sem Terra*, October 30. http://www.mst.org.br/mst/pagina.php?cd=2334, Accessed 21 February 2007.

Bellisario, A., 2007. 'The Chilean Agrarian Transformation: Agrarian Reform and Capitalist "Partial" Counter-Agrarian Reform, 1964–1980: Part 1: Reformism, Socialism and Free-Market Neoliberalism'. *Journal of Agrarian Change*, 7 (1): 1–34.

Benbrook, C.M., 2005. 'Rust, Resistance, Run Down Soils and Rising Cost: Problems Facing Soybean Producers in Argentina'. AgBioTech InfoNet, Technical paper no. 8. http://www.earthscape.org/p1/ES16592/Greenpeace_rust.pdf, Accessed 12 October 2006.

Boff, L., n.d. 'Transgênicos? Não'. http://www.cebusai.org.br/transgenicos.htm, Accessed 12 May 2006.

Borras, S. Jr, 2004. 'La Vía Campesina: An Evolving Transnational Movement'. TNI Briefing No. 2004/6. Amsterdam: Transnational Institute.

Borras, S. Jr, M. Edelman and C. Kay, 2008. 'Transnational Agrarian Movements: Origins and Politics, Campaigns and Impact'. *Journal of Agrarian Change*, 8 (2/3): 169–204.

Boy, A., 2006. 'Grupo Reflexión Rural'. Presentation, *Enzando Alternativas*, Vienna, 10–13 May 2006.

Burnham, P., 2002. 'Class Struggle, States and Circuits of Capital'. In *Historical Materialism and Globalisation*, eds M. Rupert and H. Smith, 113–29. London: Routledge.

Campolina, A., 2001. 'Brazilian Small-Scale Farmers and Poor Consumers Reject GMOs'. *LEISA Magazine*, 29 December.

Clapp, J., 2006. 'The Political Economy of Food Aid in an Era of Agricultural Biotechnology'. In *The International Politics of Genetically Modified Food: Diplomacy, Trade and Law*, ed. R. Falkner, 85–101. Basingstoke: Palgrave.

Conway, G., 1999. *The Doubly Green Revolution: Food for All in the Twenty-First Century*. New York: Cornell University Press.

Edelman, M., 2003. 'Transnational Peasant and Farmer Movements and Networks'. In *Global Civil Society*, eds H. Anheier, M. Glasius and M. Kaldor, 185–221. Oxford: Oxford University Press.

EDM (En Defensa del Maíz), 2002. Conclusiones del seminario, Ciudad de México, 23 and 24 January. http://www.ceccam.org.mx/ConclusionesDefensa.htm, Accessed 4 December 2007.

Fitting, E., 2006. 'The Political Uses of Culture: Maize Production and the GM Corn Debates in Mexico'. *Focaal: European Journal of Anthropology*, 48: 17–34.

FoEI (Friends of the Earth International), 2007. *Who Benefits from GM Crops? An Analysis of the Global Performance of GM Crops (1996–2006)*. Amsterdam: Friends of the Earth International.

Gallagher, A., 2008. 'Trading away the Ladder? Trade Politics and Economic Development in The Americas'. *New Political Economy*, 13 (1): 37–59.

Galli, E., 2005. 'De la Chaucha de Soja al Reactor Nuclear de Investigación'. *La Nación*, 4 January, 2.

Global Exchange, 2004. 'Top Ten Reasons to Oppose the Free Trade of the Americas'. http://www.globalexchange.org/campaigns/ftaa/topten.html, Accessed 2 August 2004.

Glover, D. and P. Newell, 2004. 'Business and Biotechnology: Regulation of GM Crops and the Politics of Influence'. In *Agribusiness and Society: Corporate Responses to Environmentalism, Market Opportunities and Public Regulation*, eds K. Jansen and S. Vellema, 200–31. London: Zed Books.

Glover, D., J. Keeley and P. Newell, 2003. *Public Participation and the Cartagena Protocol on Biosafety*. A Review for DfiD and UNEP-GEF. Brighton: Institute of Development Studies.

Greenpeace, 2006a. 'Desmontes S.A: Quiénes Están Detrás de la Destrucción de los Ultimos Bosques Nativos de la Argentina'. Greenpeace Argentina Report. http://www.greenpeace.org/raw/content/argentina/bosques/desmontes-s-a.pdf, Accessed 15 October 2006.

Greenpeace, 2006b. 'A Turning Point in the Fight to Protect the Amazon'. *The Guardian*, 28 September 17.

Hajer, M.A., 1995. *The Politics of Environmental Discourse: Ecological Modernization and the Policy Process*. Oxford: Oxford University Press.

Hochstetler, K., 2003. 'Fading Green? Environmental Politics in the Mercosur Free Trade Agreement'. *Latin American Politics and Society*, 45 (4): 1–33.

Icaza, R., 2004. *Civil Society and Regionalisation. Exploring the Contours of Mexican Transnational Activism*. CSGR Working Paper No. 150/04, Warwick University.

James, C., 2006. *Global Status of Commercialized Biotech/GM Crops*. ISAAA Briefs No. 35. Ithaca, NY: ISAAA.

Jansen, K. and S. Vellema, eds, 2004. *Agribusiness and Society: Corporate Responses to Environmentalism, Market Opportunities and Public Regulation*. London: Zed Books.

Kay, C., 2002. 'Chile's Neo-Liberal Agrarian Transformation and the Peasantry'. *Journal of Agrarian Change*, 2 (4): 464–501.

Keck, M. and K. Sikkink, 1998. *Activists Beyond Borders: Advocacy Networks in International Politics*. Ithaca, NY: Cornell University Press.

Kenfield, I. and R. Burbach, 2007. 'Landless Rural Worker Shot by Security Company Hired by Multinational Syngenta'. Posted on the 'Brazil Network' list serve, 21 October 2007.

Kossoy, A., 2003. 'Iniciativas de Asociaciones de Productores Agropecuarios: La Incorporación de la Soja en la Emergencia Alimentaria'. In *Respuestas de la Sociedad Civil a la Emergencia Social*, eds I. González Bombal, 89–119. Buenos Aires: CEDES.

Levidow, L., 1998. 'Democratizing Technology or Technologizing Democracy? Regulating Agricultural Biotechnology in Europe'. *Technology in Society*, 20 (2): 211–26.

Lipton, M., 2001. 'Reviving Global Poverty Reduction: What Role for Genetically Modified Plants?'. *Journal of International Development*, 13 (7): 823–46.

McAdam, D., S. Tarrow and C. Tilly, 2003. 'Dynamics of Contention'. *Social Movement Studies*, 2 (1): 99–102.

Murray, W., 2006. 'Neo-Feudalism in Latin America? Globalisation, Agribusiness and Land Re-Concentration in Chile'. *The Journal of Peasant Studies*, 33 (4): 646–77.

Newell, P., 2006. 'Corporate Power and Bounded Autonomy in the Global Politics of Biotechnology'. In *The International Politics of Genetically Modified Food*, eds R. Falkner, 67–85. Basingstoke: Palgrave.

Newell, P., 2007. 'Trade and Environmental Justice in Latin America'. *New Political Economy*, 12 (2): 237–59.

Newell, P. and D. Tussie, 2006. 'Civil Society Participation in Trade Policy in Latin America: Lessons and Reflections'. IDS Working Paper, 267. Brighton: Institute for Development Studies.

Oya, C., 2005. 'Sticks and Carrots for Farmers in Developing Countries: Agrarian Neo-Liberalism in Theory and Practice'. In *Neo-liberalism: A Critical Reader*, eds S. Saad-Filho and D. Johnston, 127–35. London: Pluto Books.

Paarlberg, R., 2001. *The Politics of Precaution: Genetically-Modified Crops in Developing Countries*. Washington, DC: IFPRI and Baltimore, MD: John Hopkins Press.

PCL (Pastoral Commission on Land), 2004. 'Declaration on GMOs'. http://www.cptnac.com.br/?system=news&action=read&id=1230&eid=88, Accessed 21 August 2006.

PCL (Pastoral Commission on Land), n.d. 'The Solution is to Preserve'. http://www.cptnac.com.br/?system=news&eid=87, Accessed 21 August 2006.

Peleaz, V. and W. Schmidt, 2004. 'Social Struggles and the Regulation of Transgenic Crops in Brazil'. In *Agribusiness and Society: Corporate Responses to Environmentalism, Market Opportunities and Public Regulation*, eds K. Jansen and S. Vellema, 232–61. London: Zed Books.

Pengue, W.A., 2005. *Agricultura Industrial y Transnacionalización en América Latina: ¿La Transgénesis de un Continente?* Mexico City: PNUMA and Buenos Aires: GEPAMA.

Petras, J. and H. Veltmeyer, 2005. *Social Movements and State Power*. London: Pluto Press.

Pimbert, M. and T. Wakeford, 2002. *Prajateerpu: A Citizen's Jury/Scenario Workshop on Food and Farming Futures for Andhra Pradesh India*. Brighton: Institute for Development Studies and London: International Institute for Environment and Development.

RACDV (Rede Alerta Contra o Deserto Verde), n.d. 'Comunidades e Monocultures de Arvores'. http://www.wrm.org/uy/inicio.html, Accessed 12 April 2006.

Radio Mundo Real, n.d. 'Terminator Seeds are Killer Seeds, of Other Crops and of Peasants'. http://www.viacampesina.org/main_sp, Accessed 12 December 2006.

RALLT (Red por una América Latina Libre de Transgenicos), 2007. 'Una Red de Resistencia a los Organismos Transgénicos en América Latina'. http://www.rallt.org.menus/mision.htm, Accessed 12 September 2007.

Santos, I.F., n.d. 'Action Day Against Terminator Mobilizes Vía Campesina and Environmentalists'. http://www.viacampesina.org/main_sp, Accessed 30 March 2006.

Santos, M., 2006. 'Inimigo é Parceria Entre Latifúndio, Agronegócio e Empresas Transnacionais'. *Movimento Dos Trabalhadores Rurais sem Terra*, 13 December. http://www.mst.org.br/mst/pagina.php?cd=2593, Accessed 9 February 2007.

Scoones, I., 2005. 'Contentious Politics, Contentious Knowledges: Mobilising Against GM Crops in India, South Africa and Brazil'. IDS Working Paper, 256. Brighton: Institute for Development Studies.

Scoones, I. and J. Thompson, eds, 2003. 'Participatory Processes for Policy Change: Learning from Experiments in Deliberative Democracy'. PLA Notes 4.

Stedile, J.P., 2004. 'Who Does the Agribusiness Model of Agriculture Serve?' *Movimento Dos Trabalhadores Rurais sem Terra*, 1 June. http://www.mst.org.br/mst/pagina.php?cd=500, Accessed 6 March 2007.

Tarrow, S., 2005. *The New Transnational Activism*. New York: Cambridge University Press.

Teubal, M. and J. Rodríguez, 2002. *Agro y Alimentos en la Globalización: Una Perspectiva Crítica*. Buenos Aires: La Colmena.

The Economist, 2006. 'Argentina's Government and its Farmers'. 26 October. http://www.economist.com/displayStory.cfm?Story_ID=E1_RDRGRTJ, Accessed 2 November 2006.

Third World Network, 2006. 'US FTA Likely to Open Peru to GMOs?'. 2 October. Kuala Lumpur, Malaysia: TWN Biosafety Information Service.

Thrupp, L.A., 1996. 'New Harvests, Old Problems: The Challenges Facing Latin America's Agro-Export Boom'. In *Green Guerrillas: Environmental Conflicts and Initiatives in*

Latin America and the Caribbean, ed. H. Collinson, 122–32. London: Latin America Bureau.

Vara, A.M., 2005. 'Argentina, GM Nation; Chances and Choices in Uncertain Times'. NYU Project on International GMO Regulatory Conflicts. (http://www.law.nyu.edu/centers/etc/profrans/Argentina%20Country%20Case%20 Sept%20%202005>vara.doc'), Accessed 10 October 2005.

Vía Campesina, 2007. 'Attack by Syngenta's Armed Militia Results in Deaths and Wounded in Brazil'. Press Release, 21 October.

Vía Campesina, n.d. a. 'Open Letter from Vía Campesina to President Lula'. http://www.viacampesina.org/main_sp, Accessed 12 January 2007.

Vía Campesina, n.d. b. 'Vía Campesina Brazil and the Question of Seeds'. http://www.cptnac.om.br/?system=news&action=read&id=1226&eid=87, Accessed 20 December 2006.

Vía Campesina, n.d. c. 'Victory of Peasants in Defense of Seeds and Against the Terminator'. http://www.mmcbrasil.comb.br/noticias/080306/060406_via_campe.htm, Accessed 12 April 2006.

Warwick, H., 2000. 'Sygenta: Switching off Farmers' Rights?' Report by Genewatch Action Aid, The Berne Declaration and the Swedish Society for Nature Conservation.

WRM (World Rainforest Movement), 2006. 'Petition to the CBD to Ban the Release of GM Trees'. http://www.wrm.org.uy/temas/AGM/cartaCBD.html, Accessed 2 April 2006.

Wynne, B., 2001. 'Creating Public Alienation: Expert Cultures of Risk and Ethics on GMOs'. *Science as Culture*, 10 (4): 445–81.

8 Claiming the Grounds for Reform: Agrarian and Environmental Movements in Indonesia

NANCY LEE PELUSO, SURAYA AFIFF AND NOER FAUZI RACHMAN

INTRODUCTION

This chapter argues that the trajectories and strategies of 'new' agrarian movements need to be understood in relation to those of environmental movements and the positioning and power relations of both sets of movements within shifting political economic conjunctures. We focus on the alliances and divergences between movements in Indonesia and how these have changed under transformed and transformative political economic circumstances since the 1970s when the modern 'environmental movement' began. Indonesia is an interesting case to reflect on broader trends, as it helps demonstrate that these shifting alliances and conflicts do not derive from transnational forces alone. Rather, we suggest that they depend as well on the temporal and political economic origins and histories of the respective movements, on the types of land contested, on the politics of access to those lands and on the emergence of what Hajer (1993, 1995) has called 'discourse coalitions', that dominate discursive spaces. Going beyond this view of discourse as discursive space and into the realm of practice and institutions, we also examine the articulations of discursive and institutional practices of government and non-government, transnational, national and grassroots organizations involved in Indonesian agrarian and environmental struggles and movements.

These shifting coalitions across environmental and agrarian movements, and indigenous peoples' organizations with interests that might articulate with either or both, help us further understand Christodolou's (1990, 112) thesis that 'agrarian reform is the offspring of agrarian conflict'. Specifically, we argue that the forms coalitions take in the contexts of particular agrarian conflicts have lingering effects, even when movement groups move on and follow new, more separate trajectories. The diverse effects of different conflicts are particularly evident in Indonesia, where a number of key agrarian conflicts generated very different kinds of coalitions and helped produce new political opportunities in subsequent periods. Agrarian conflict in the 1960s, for example, did not lead directly to agrarian reform (Husken and White 1989; Farid 2005).

An important issue here is how the realm of the agrarian is defined. This is particularly important in understanding resurgent agrarian movements in Indonesia today, as the primary landlord targeted is a state 'environmental management' institution – the Ministry of Forestry. The state's expropriation of millions of hectares of land with the creation of a national forest puts forestry right at the heart of most agrarian struggles in Indonesia, though these have taken different forms in different parts of the country (see e.g., Bachriadi and Lucas 2001; Li 2007). In this chapter, we include apparently environmental conflicts over access to and control over forests and certain anti-dam campaigns in our use of the term agrarian conflicts, because these have engaged issues of rights to and use of agrarian land. Contestations take place within more broadly defined 'agrarian environments' (Sivaramakrishnan and Agrawal 2003), a term which recognizes that such sites were not always separated into discrete discourses and domains of 'forestry' and 'agriculture'.

Indonesian agrarian movements and their transnational connections are of significance because of the particular historical and political moments from which they have emerged and the forms they have taken. The massive drive to repeasantization[1] through land occupations and the formation of new rural organizations have taken place after over 30 years of depoliticization and structural violence since the largely rural massacres of 1965–66 (Cribb 1990) and subsequent, systematic land expropriation by Indonesian state agencies and their cronies in the private sector (Fauzi 1999; Farid 2005). The new agrarian movements have been concurrent with the efforts to make decentralization work and the 'thickening' of civil society (Robison and Hadiz 2004; Hadiz 2004a, 2004b). Those particularities make Indonesian agrarian movements unique and yet important to understand as they were constrained in their ability to connect with transnational networks until the late 1990s. Under the repressive 'New Order' regime, the name by which the Suharto regime was known for 32 years (1966–98), Indonesian environmental and agrarian movement actors often expressed a compulsion to retain a face of solidarity, despite some critical ideological differences.

The Indonesian case also demonstrates that state power has remained important, though in new institutional forms and ways, in the transition from state-led (including but not only authoritarian) development to a political economy dominated by neoliberal policy (see also, Borras 2004). Indeed, it is the involvement of state actors and institutions within the various coalitions formed across and within environmental and agrarian movements that often has determined their relative ability to literally gain ground – spatial zones of influence recognized by local, national and international actors and institutions.

That state power has remained important under changing global regimes was most evident in the gains of the environmental movement, occurring slowly but surely under Suharto. The intersections of state power and the history of discourse coalitions are also reflected in many environmental justice groups' willingness to work within the confines of forest law in CBNRM (Community-Based Natural Resource Management) and social forestry projects. Here, 'access to' forest land rather than 'private or communal rights' to forests are seen as

---

[1] 'Peasant' here refers to small farmers and land occupiers who have no land outside occupied areas. The word, '*petani*' in Indonesian translates into either 'peasant' or 'farmer'. Agrarian activists in the 1990s explicitly chose to translate it into 'peasant' in English for its more radical connotation.

justice accomplishments, even though these were not enough for more radical agrarian reformers who have demanded full land rights, including excision of lands currently under the formal jurisdiction of state forestry institutions.² In ways reminiscent of earlier environmentalist strategies, since *Reformasi* some agrarian movement groups have started working with government, trying to form coalitions with sympathetic district and national parliament members and government land managers. This had been impossible earlier, as farmer/peasant organizations were criminalized and repressed in the wake of the anti-left campaign and the agrarian violence that brought Suharto to power in 1966.

Also affecting the histories of Indonesian agrarian and environmental movements and their mutual constitution were the ways each formed relations with a nascent Indigenous Peoples' movement. Like the environmental movement, a major motivating factor for this movement came from international activity and grabbed Indonesian activists' imaginations in the mid-1990s after the UN declared 1993 'The Year of the World's Indigenous People'.³ However, this movement has to be understood as not only an artefact of transnational movement politics but as a historically grounded set of institutions whose participants, practices and policies found ways of expressing their political positions through articulations with transnational and national discourses (Li 2000, 2001; Moniaga 2007). It has been alternately claimed by and allied with environmental and agrarian justice advocates.

Resurgent agrarian movements today show the effects of these historical tensions and the emergence of new ones. As centralized state power declines, and with it some of the gains of conservation, decentralization, democratization and the increasing hegemony of neoliberal policy and practice have generated new splits in the environmental movement. This is reflected in new alliances between Big Conservation and capital and changes in the alliances between environmental justice organizations and agrarian reformers. While these are changing the coalitions on the ground, a new state-led agrarian reform initiative launched by the National Land Agency (NLA), the Ministry of Forestry (MoF) and Indonesia's president has once again made the state a critical site of contestation over agrarian reform.

Contemporary agrarian movements aim to change state policies and their implementation (Webster 2004, 2; Moyo and Yeros 2005). Peasant associations and movements are shaped by state policy and practice and various forms of class formation and accumulation, as well as by other movements and the political fields they help to create (Buechler 2000, 78; McKeon et al. 2004; McMichael 2005). The transnational dimensions of campaigns and advocacy, of knowledge exchange and communication, and new types and goals of collective action have produced new political and cultural spaces nationally and transnationally (Routledge 2004; Edelman 2005; see also Keck and Sikkink 1998). Environmental movements, because many of them operate in the realm once thought of as

---

² Mia Siscawati, personal communication, 2006.
³ It was extended for a decade.

'agrarian', are particularly important to understand. While some work has begun on these relations, a great deal remains to be unpacked (Edelman 1999; Franco and Borras 2005, 2006; Kowalchuk 2005).

THE FORMATION AND TRAJECTORIES OF ENVIRONMENTAL AND AGRARIAN MOVEMENTS WITHIN INDONESIAN POLITICS

In Indonesia, many of the first environmental activists advocated environmental justice and worked at the grassroots (Tsing 1999; Lowe 2006), creating a situation where alliances could be made with groups fighting for other forms of agrarian justice. In this section, we discuss the political economic contexts within which early agrarian, environmental justice/indigenous peoples, strict conservation and resurgent agrarian movements emerged and changed in Indonesia, from 1945 when Indonesia declared its independence from the colonial Netherlands East Indies, through the New Order (1966–98). The practices and strategies of both state-led development and authoritarianism affected the emergence, room to manoeuvre and strategic approaches of both environmental and agrarian organizations and activists. The growing importance of forestry nationally, and of conservation internationally, radically transformed agrarian politics during this period. Changing political fields also affected movement relations to one another.

Agrarian organizations are not new to Indonesian history. The Indonesian Peasants' Front, BTI (*Barisan Tani Indonesia* or the Indonesian Peasants Front) was formed in 1945. It was the first of several peasant/farmer organizations operating across the archipelago, and had its strongest bases in Java, Sumatra and Bali. In the 1950s and 1960s, peasant organizations were closely allied with political parties.[4] Indonesia's Communist Party and the BTI, legal organizations at the time, became the most active proponents of the state's land reform programme, set in motion by the passing of the Basic Agrarian Law in 1960. They mobilized rural people to take 'unilateral action' by occupying the lands of large private owners and demanding redistribution (McVey 1965; Lyon 1970; Mortimer 1972). In 1965, when the military, led by then-General Suharto, violently seized power, all left and left-leaning parties, associations and organizations were criminalized and banned. Hundreds of thousands of peasants and farmers alleged to be supporters of PKI and their affiliates, including BTI, were killed (Mortimer 1972; Cribb 1990).

The violence constituted a critical moment of primitive accumulation that has underpinned all further phases of capitalist development and forms of state, corporate and private accumulation in Indonesia (Farid 2005). In addition, the massacres turned rural social movements and land reform agendas into history; they

---

[4] BTI (*Barisan Tani Indonesia* or The Indonesian Peasants Front) was affiliated with the Indonesian Communist Party (PKI), PERTANI (*Persatuan Tani Nasional Indonesia* or The Indonesian Peasants Union) with the Nationalist Party (PNI), and PERTANU (*Persatuan Tani Nadhatul Ulama* or the NU Peasants' Union) with one of the Islamic parties (*Nadhatul Ulama*) (Pelzer 1982).

were absolutely stopped from operating in their previous forms. Agrarian transformations during the New Order generally meant large-scale land dispossession by central state institutions and their corporate or other capitalist cronies (Aditjondro 1993; Fauzi 1999).

Passed under Indonesia's first president, Sukarno, The Basic Agrarian Law of 1960 had both eliminated legal pluralism based on racial or indigenous categories of rightful access to land, and established a single, unitary land law that represented a 'classic' form of agrarian reform legislation for the times. It promoted land to the tillers and ceilings on private landholdings based on quality and location of land (with different ceilings for irrigated land and dry fields or uplands). After the mass agrarian violence and Suharto's rise to power, the law's land reform tenets were largely ignored, though not struck from the books.

A variety of pro-capitalist agrarian programmes were implemented, including 'green revolution' rice and maize production programmes, mining and forest exploitation, and large state and corporate plantations (Fauzi 1999; Husken and White 1989). All of these programmes were tied in with sources of global capital and backed by the Suharto regime's military-bureaucratic-authoritarian state. In some cases, such as the forestry and mining sectors, Indonesian versions of state capitalism were developed (Mortimer 1973; Robison 1986; Barr 1997). The strong state aimed to guarantee political stability and maintain control.

In upland areas of Java and in the larger, more forested 'Outer Islands' of Sumatra, Sulawesi and Kalimantan, state and corporate agro-industry and forestry institutions concentrated landed power (Barr 1997). The 'sectoral' or natural resource management laws on forestry, mining and land acquisition for development projects legislated under the New Order in 1967 radically centralized resource management and enclosed significant tracts of land for forest reserves and industrial agriculture, to be managed by government agencies under a variety of financial arrangements. Moreover, these natural resource laws did not recognize the Basic Agrarian Law as one of the legal tenets that preceded it, harking back instead to a clause in the national constitution on state sovereignty over 'the national territory and all the land and resources within' (Moniaga 1997; Zerner 1992). This legislation constituted some 70 per cent of the nation's land as a national 'political forest' (permanent forest zoned and maintained by professional state foresters) and strengthened forest law for Java and Madura, dating from the late colonial period. Forest Law 5/1967 in particular was a critical move that constructed agrarian and forest environments as legally, institutionally and conceptually separate spaces (Peluso and Vandergeest 2001). On and off Java, these state-based resource management institutions had their own territorial hierarchies and jurisdictions that were inconsistent with the territories and authority of civil administration and separate from the urban and agricultural areas under the National Land Board (NLB) (Peluso 1992; Afiff et al. 2005). This legislation accompanied and facilitated critical changes that were transforming the political economy of Indonesia, the structure and perception of the countryside, and the political ecological contexts within which subsequent agrarian movements would take place (Robison 1986; Barber et al. 1995; Dauvergne 1994).

Under the New Order, institutional and legal controls were accompanied by 'de-politicization', particularly of the rural population (Mortimer 1972; Mas'oed 1983; Robison 1986). Independent peasant organizations were replaced by HKTI (*Himpunan Kerukunan Tani Indonesia*, Indonesian Peasant's Harmony Association), an organization managed by military or other government officials and formally affiliated with GOLKAR, the state's ruling party (Hikam 1995). Violent punishments and incarceration were inflicted on anyone who resisted.

Until the 1980s, the New Order state successfully maintained such overwhelming power and control that rural protest was almost unknown. Coalitions of rural and urban activists – NGOs, students and local leaders – began to break the state stranglehold in the early 1990s (Lucas and Warren 2000, 2003; Aspinall 2004). Much of their initial work had to be done underground and involved various configurations of students and other activists (Ganie-Rochman 2002). The students discussed the potential and strategy for agrarian movements and worked with upland villagers in Java and Sumatra who, like many indigenous people outside Java, had been forced to give up their lands to military, other state agencies or corporate enterprises.

The Light in the New Order's Dark Ages: Environmental Justice

The grassroots character of the early environmental movement in 1980s Indonesia would today be described as an environmental justice movement (Lowe 2003; Tsing 2005). During the New Order, where any political opposition to the state was intolerable, environmental law and advocacy seemed – and proved – to be a safe arena for concerned activists working both to help local people and to advance environmental agendas.

Perhaps because state capitalism under Suharto had reordered development priorities, focusing on large-scale extractive projects and transforming the distribution of agrarian resources, it is not surprising that grassroots environmentalists engaged in agrarian struggles. The environmental concerns of these grassroots activists were advocated simultaneously by transnational environmental institutions and environmental lawyers and policy advisors working with and through state institutions to make state policy and development practice more environmentally sustainable. This multi-scaled approach turned out to be critically important to both the resounding successes of environmental discourses in Indonesia and the constraints environmentalists later faced.

Within transnational environmental protection discourses, a great deal of variation lay beneath the surface. For example, professional foresters had claimed experience in the conservation realm since the colonial era's establishment of protection forests, while ecologists and other environmental activists critiqued foresters for being overly production-oriented. By the late 1980s, environmentalist critiques of traditional forestry had gained significant international and national purchase. International conservation NGOs with strong preservationist agendas began to set up offices in Indonesia. Other advocates of justice were interested in addressing the exclusions of people from their lands that had occurred when the political forests were reserved and reinforced with the establishment of internationally supported nature reserves (Djuweng 1997).[5]

Environmental and conservation power in legal and policy domains in Indonesia grew in part as a reaction to the rapid growth and destruction wrought by natural resource industries in the wake of modernization policies, and the massive amounts of state, corporate and individual profits being generated at the expense of the environment. Yet conservation advocates often joined government foresters in blaming deforestation on shifting cultivators and forest-dependent peoples, working against the agendas of environmental justice groups. At the same time, though many of these big conservation interests may not have agreed with the justice advocates on the appropriate solutions to environmental problems, they helped make the government aware of environmental problems; they also helped establish institutions and provided funding to deal with these problems in formal-legal ways at the same time that environmental justice advocates were working on the ground. For example, as we see below, the Suharto regime established a Ministry of Environment in addition to the Ministry of Forestry, and thereby contributed to the creation of a multiplicity of contexts for environmentalist agendas. But the former institution's budgetary and ideological power within the state was hardly a match for what foresters called their 'Golden Ministry' (Peluso 1992). Thus, a variety of alliances and divergences characterized this environmental 'movement' from early on, even when a veneer of common ground was deemed politically expedient.

The commonality of origin stories for grassroots-oriented and national or international environmental institutions is illustrated by the story of WALHI (*Wahana Lingkungan Hidup Indonesia*/The Indonesian Forum for the Environment). This umbrella institution was formed in 1980 and explicitly traces its origins and inspirations to the Stockholm meeting on Sustainable Development in 1972, a meeting which inspired many mainstream environmental and conservation organizations as well.

WALHI's establishment was during the peak of forestry power in Indonesia, when forest (timber) and oil extraction and trade accounted for most of Indonesia's GDP (Robison 1986). Forestry had been elevated to Ministry status in 1983, with jurisdiction over hundreds of millions of hectares of the national land base. Hundreds of timber concessions had been allocated all over the country – over 500 in Kalimantan alone (Barr 1997). With technical assistance and loans from the FAO (Food and Agriculture Organization) and the World Bank, the foresters had declared, allocated and mapped (in that order) national forests, dividing them into production and protection forests, as well as nature reserves and conversion forests (for conversion to industrial agriculture or development projects).

---

[5] At the time, dominant paradigms in ecological sciences tended not to include humans in their analyses of forest succession except as 'disturbances' – responsible for destroying the forest rather than creating it – a view that has continued among ecologists until the present. For a recent iteration of this argument in English, see Terborgh (1999).

Given the institutional power and wealth of the Ministry of Forestry and the strength of the authoritarian government, WALHI could not have survived without some kind of support from within the government. In response to international pressures, but in ways meant not to interfere with the tremendous accumulation from the forestry sector, Suharto created the Ministry of Environment (MoE), headed by Dr Emil Salim. The MoE had a meagre budget and little actual power inside the state. However, Salim's MoE provided a safe haven for environmental justice activists – its own marginal position in government creating a basis for alliance with these justice advocates. At the same time, the MoE was developing relations with international environmental lawyers and legal advisors – who were also marginalized by the dominant global interests in state-led, large-scale, development. In some ways, then, the MoE served as a unifying institution for the justice and darker green components of the environmental movement. At a time when the New Order government interpreted any criticism of its development policy and practice as subversive, environmental debates were the only public media through which farmers' rights or access to land lost to extractive enterprises could be discussed. Further, political context made it crucial for them to have connections with a government body.

Thus both the strategies of transnational actors and the practical constraints of NGOs working within Suharto's authoritarian government helped forge connections among international, government and non-government environmental advocates – both those who would continue along an environmental justice path and those who would pursue more coercive conservation policies later. Close connections between environmental movement activists working at national legal and local levels were further strengthened through justice advocates' connections with YLBHI (the Indonesian Legal Aid Foundation), which also worked at multiple scales, collaborating with NGOs and student activists and transnational legal aid groups. Representative of this on the ground was the appointment in 1989, a member of YLBHI as the head of WALHI's presidium. At various points, YLBHI was supported by NOVIB (a Dutch donor NGO) and CIDA (Canadian Aid).

A defining moment occurred fairly early in this collaboration, one that in part explains later divergences within the environmental movement and the opportunities for alliances between environmental justice and still-underground agrarian activists. This now iconic 'moment' was constituted by activist mobilization against the Kedung Ombo project, a World Bank-assisted dam project in Central Java. Lasting more than five years, and ultimately failing, like the Narmada campaign in India (Baviskar 1995), to prevent the submergence of a huge swath of rural Central Java, the movement was significant for how it brought together different activists, simultaneously demonstrating commonalities and potentials for future alliances. For environmental groups inside and outside Indonesia, Kedung Ombo was part of a global 'Anti-Big-Dam' campaign. For agrarian activists, the campaign was an opportunity to help farmers in Java forced off their land without fair compensation (Rumansara 1998). And as development refugees forced into transmigration projects off Java, these farmers provided an unintended connection to indigenous people there, near or onto whose land they were forced to move.

This campaign gained greater national attention when YLBHI and other NGO members of INGI[6] sent an *aide memoire*, a letter to the head of the World Bank, to protest the project's violations of villagers' human rights. This internationalization of the case, what Keck and Sikkink (1998) call a 'boomerang effect', backfired in some ways. It reduced Suharto's tolerance of NGO activism; he spoke of the campaign as insulting and anti-national. This forced Minister of Environment Salim, the environmental NGOs ally and safe haven, to state publicly that the Indonesian NGO representatives in INGI had gone too far, mixing 'political' and environmental objectives. Salim toed the government line, stating that the political aspects of environmental cases were government concerns, not within the purview of environmental NGOs.

This rather dangerous encounter with 'politics' was critical, as it brought into the open the fact that 'environmental problems' were not as benignly apolitical as environmental justice organizations had represented. Indeed, Kedung Ombo is an example of articulated environmental and agrarian discourses in Java (Aditjondro 2003), a conjunctural moment. Nevertheless, the campaign to stop the dam failed in part because Java's violent agrarian history remained an obstacle to rights-based agrarian movement activities through the 1980s and early 1990s. Further, this history tempered the extent and manner in which some environmental justice activists were willing to proceed.

Off Java, particularly in Sumatra, Sulawesi and Kalimantan, there was more room for manoeuvre, literally and figuratively, where, in provinces with forest, an average of two-thirds to three-quarters of the land base came under the jurisdiction of the Forestry Department. In these regions, environmental justice advocates tended to ally with local groups who eventually called themselves Indigenous Peoples or *Masyarakat Adat* (Li 2000). The origins and identities of these groups, and the national organization formed in the late 1990s, saw itself as both an agrarian movement organizations and environmental justice organizations.

In Indonesia, the Indigenous Peoples' organisation was first comprised of people who had been pushed off their land by large-scale development projects related to forestry, plantations, transmigration, dams or large-scale tourism. The expropriations they experienced put them up against the development and environmental discourses of the state and the expanding dark green conservation world. They differed from agrarian activists and farmers in Java and Bali and parts of Sumatra who were eager to reclaim land lost to government and corporate

---

[6] INGI (Inter-NGO Conference on Inter-Governmental Group of Indonesia (IGGI) Matters) was formed in 1985 by YLBHI and NOVIB (de Nederlands Organisatie voor Internationale Bijstand), both transnational and local actors providing input to the inter-governmental conference of donors providing aid to Indonesia. WALHI is a member of INGI. Later, IGGI changed to CGI (Consultative Group on Indonesia) and INGI became INFID (International NGOs Forum on Indonesian Development).

expropriations, but who rarely talked the environmental talk.[7] *Masyarakat Adat* were critical participants in early environmental justice struggles, in part because they were usually represented as having environmentally friendly 'customary' practices. This was beneficial in some senses, but raised flags in others: the definition their national organization, AMAN (*Aliansi Masyarakat Adat Nusantara* or the Alliance of Indigenous Peoples in the Archipelago) decided on for *Masyarakat Adat* was clearly based on an (anthropological) definition of 'tribe', with all the positive and negative baggage that went with that label (Tsing 1999; Li 2001). They were, however, a harbinger of the future: NGOs working with them sought to legitimize their territorial claims by tying their agrarian practices to environmental goals of sustainability.

Moreover, their histories were quite different from those of peasants involved in agrarian struggles during the 1950s and 1960s on Java, Bali and the parts of Sumatra mentioned above. At the time, groups identifying as *Masyarakat Adat* lived mostly outside Java, in areas that until the 1940s had been largely under indirect colonial rule. Javanese, Sundanese and Balinese ethnicities were not really considered 'tribal' in the ways many *Masyarakat Adat* had been characterized by colonial officials and observers, social scientists and Indonesian NGOs.[8] Indeed the terms of colonial legal pluralism had recognized the 'customary territories' of these groups as spaces where they exerted authority over land disposition, among other governance functions. For indigenous people in the forested regions of Indonesia, the most immediately threatening laws were the Forestry Laws (Ruwiastuti 2000). These defined many of their agroforestry holdings or reserved areas as 'empty' and 'abandoned' land, and criminalized their agricultural systems of swidden cultivation.

Back to Underground Agrarian Activism . . .

After the Kedung Ombo campaign, agrarian and environmental justice activists worked more closely with YLBHI to assist villagers in land rights disputes in Java and parts of Sumatra (Aspinal 2004). Despite the still very tangible risks of being accused as 'communist' supporters and arrested or worse, activists organized protests to draw attention to farmers' land struggles. Unlike WALHI activists who, by the early 1990s, were framing their advocacy in terms of indigenous people's rights (environmental justice) and the criticism of forestry and mining laws, YLBHI and student agrarian activists focused their critiques on

---

[7] See, e.g., Fidro and Fauzi (1998) who analyze 29 land dispute cases under the New Order and make practical and strategic suggestions to advance the nascent (still underground) agrarian movement. Nothing in any of these writings suggested a sensitivity to the strategic potential in environmentalist discourse, except a paper by Aditjondro, which argues that people's economies were 'poli-cultures' and better environmentally than the monocultures of the imposed plantations. This was not yet a rallying idea for the agrarian advocacy groups.

[8] This changed later, as Balinese, Baduy and other groups joined AMAN and broadened its national base and the working definition of indigenous people in Indonesia.

the implementation (or failure thereof) of The Basic Agrarian Law no. 5 1960 (BAL). Some environmental activists were invited to Bandung to take part in the discussions that were later the foundation of the KPA, the Consortium for Agrarian Reform (author interviews 2007).

The BAL was a national icon, as it formalized nationalist intentions to throw off the yoke of colonial differentiation based on race and ethnicity – realized in colonial legal pluralism – by establishing a unified national land law. A key component of the law was to be national land reform. The Suharto government, however, had implemented the law selectively, favouring the articles supporting the state's rights to acquire land for development projects 'in the national interest'. The parts that discussed the social functions of land for livelihoods and land reform were ignored. YLBHI and their student allies framed their movement and mobilization activities around land reform, advocating that the government implement the BAL more comprehensively. Although they worked with YLBHI on an embryonic multi-scaled approach as the environmental activists had, the land reform activists on Java and Sumatra lacked national cohesion and the extensive international funding available to environmental movement groups. They were also still politically sensitive and had to tread softly.

Several key historical and geographic differences affected the types of reform sought by these two types of agrarian movements – indigenous peoples' and peasant-based land reform – and the ways they engaged with or embraced environmentalist discourses and movements, national or international. The student activists who worked in Java, Bali and some parts of Sumatra, where rural class tensions had long been an issue, tended to separate agrarian reform agendas from environmental agendas, even though at times (such as in Kedung Ombo) they had seen fit to ally with environmental activists. They had to operate largely off the public radar screen, underground, until the very end of the Suharto regime – when he was forced to step down in 1998 during the economic crisis. By the early to mid-1990s, '*Masyarakat Adat*' types of organizations characterized a closely knit set of agrarian movements outside Java and Bali, most of which had connection to environmental justice concerns and were active more publicly.

Some activists and leaders moved easily among groups working on environmental, indigenous peoples and agrarian reform issues. In essence, this was facilitated by the groups' common dissatisfaction with the Suharto regime's policies of expropriating huge tracts of land and extracting resources for state or private accumulation. They found common ground in mobilizing against tenets of two laws: the Forestry Laws and the parts of the Basic Agrarian Law that enabled state land acquisition. Both of these laws were products of the 1960s, but came out of the different ideological frames that had animated the first two national regimes. Over time, their cumulative effect had been to eliminate or ignore communal and other customary rights in favour of Western private and state property rights (Djuweng and Moniaga 1994; Heruputri 1997).

REFORMASI AND NEOLIBERALISM: THE EXPLOSION, RADICALIZATION, ROUTINIZATION AND FRAGMENTATION OF MOVEMENTS

The lid of repression literally burst off with the fall of Suharto in 1998; major changes came with Reformasi. Immediately after Suharto's demise, tens of thousands of peasants and farmers, landless people and smallholders occupied state forest and plantation lands. They chopped down rubber, cocoa, teak, pine and many dipterocarp species in the rainforests. On plantation and state forest lands, they planted their own cassava, rice, banana, durian and oil palm. As early as September 2000, the Director General of the Department of Forestry and Plantations estimated that some 118,830 hectares of national estate land had been seized, along with 48,051 hectares of private estate lands (Kuswahyono 2003; in Fauzi 2003). For the first time in 35 years, peasant organizations formed, debated land politics, found allies in and outside government, and laid the basis for a new trajectory of mobilizing for agrarian reform (Lucas and Warren 2003).

Agrarian reform groups no longer had to work underground, but the long period of violent repression had affected the forms of oppositionist expression. Not until several years into 'reformed' Indonesia, after neoliberal policies had made tremendous headway and decentralized state power had become the norm, were large numbers of peasants and small farmers willing to openly join agrarian organizations. Even though the academic and activist mobilizers framed these movements differently, farmers must have initially seen the demonstrations, protests and demands as quite similar to those that had been violently repressed in 1965. Moreover, this time, the biggest landlord they had to oppose was the state itself: the powerful Ministry of Forestry that now controlled some two-thirds of the nation's land base. At the same time, under the rubric of decentralization, district level government – parliament members, regents (*Bupati*) and district sectoral agency bureaucrats – had gained more authority and administrative power. Some of them were willing to support the new agrarian organizations and their calls for land reform.

Critically, however, these resurgent agrarian movements came into a national political context that had been reshaped by the environmental movements and agendas that began to influence Indonesian land management and allocation in the 1980s. Massive investment in Indonesian forest extraction after 1970 was paralleled a few years later by the meteoric rise of the international environmental movement. While this movement did not really take organizational form in Indonesia until the early 1980s, its effects were already being felt when FAO held its annual Forestry Congress in Indonesia in 1978 with the theme, 'Forests for People'. This meeting also foreshadowed future struggles, as new forms of territorialized environmental power were being envisioned and realized through the work of professional foresters and other ecological scientists, as well as by budding environmental justice NGOs. By the time pro-poor agrarian movements re-emerged publicly in Indonesia in the late 1990s, environmental discourses had transformed national and local political fields, as well as international arenas of law and policy, activism and moral authority. These changes were critical to shaping the strategies, positioning and rhetorics of both agrarian and environmental movements after the turn of the twenty-first century. At the same time, some of their common ground began to erode.

In 2004, Vía Campesina, arguably the most influential transnational peasant organization in the world, set up global shop in Indonesia, moving its International Operative Secretariat from Honduras. Over the past decade, Vía Campesina, an active and vocal critic of neoliberal agricultural and land policies, had consolidated a transnational network of peasant organizations from Asia, America and Europe, coordinated global protests and campaigned for a specific vision of 'agrarian reform and food sovereignty' (Desmarais 2002; Edelman 2003; Rosset 2006). When they moved to Indonesia, they announced that Henry Saragih, Director of the Federation of Indonesian Peasant Unions (*Federasi Serikat Petani Indonesia* or FSPI), one of several peasant unions in Indonesia that had a national constituency, would serve as the new International Coordinator.

The largest single peasant organization in Java, *Serikat Petani Pasundan* (the Sundanese (West Java) Peasant Union or SPP), was a member of FSPI at that time and had some 30,000 members, most of whom were landless or extremely poor[9] (Wargadipura 2005, 19). Formally established in 2000 (but organized underground in the mid-1990s), SPP's 52 local chapters in 2005 were located at sites of agrarian conflict in three districts of West Java and occupied more than 15,000 hectares of state plantation and forest lands in West Java's uplands. They are among the peasant unions that have moved against state landlords controlling land in West Java, that is, the State Forestry Corporation (SFC) and State Plantation Corporations (SPC). SPP's operations illustrate the new form of agrarian movement organization, as it works not only through direct land occupations and other forms of collective action, but also by its leaders negotiating with politicians, officials in government land management agencies, and members of district and national parliaments. SPP leaders and associated NGOs (primarily the Consortium for Agrarian Reform, KPA and, more recently, KARSA-*Lingkar Pembaruan Pedesaan dan Agraria*/Circle for Village and Agrarian Reform), have also collaborated with NGOs working for human rights (YLBHI) and conservation (WALHI, LATIN-*Lembaga Alam Tropika Indonesia*/The Indonesian Tropical Institute). KPA also was formed in 1995 during the underground period of agrarian activism. Although technically an NGO, some members of KPA are leaders of peasant organizations.

SPP and its membership put into practice an idea brought to them by a pro-reform academic, Gunawan Wiradi (1997), the notion of 'land reform by leverage' (Powelson and Stock 1990). Their ability to occupy, hold and transform the use and vegetative cover of these occupied lands was significant, even though the areas seem small relative to the million hectares of forest lands and plantations in West Java. The occupations are most meaningful for their *duration* of

---

[9] Although a systematic class analysis of SPP's membership has not been done, the organization's leaders and affiliated NGO contacts make this assertion.

over three-quarters of a decade. Under Suharto, such land occupiers would have been forced off or worse. Indeed, the SFC and SPCs have hired hundreds of thugs to evict peasants from these occupied lands, but have ultimately failed to remove them.

More recently, agrarian reform groups have campaigned for and achieved participation in government decision-making. In their West Java working area, the SPP lobbied local governments and district parliaments to set up committees to resolve agrarian conflicts, primarily those in the significant areas of land currently allocated to the SFC and SPCs, which cover some 44 per cent of all Garut District (Fauzi 2003).[10] SPP has also organized an association of village heads and parliaments in those districts and won some 10 per cent of village head elections in their working areas in 2006.

We should step back a bit and examine the major political economic transformations that have made these new institutional activities possible: land occupations, the revival of criminalized organizational forms and movement groups 'negotiating' with government officials. In a word, *Reformasi*, the Indonesian version of Philippines' style 'people power', represents the changed and changing political context. *Reformasi* is the Indonesian term for the transformations of the political economy of Indonesia, causing radical changes in the contexts, programmes, strategies, positioning and alliances among agrarian and environmental organizations. Suharto's successor in 1998 almost immediately passed a Presidential Decree to decentralize many functions of government to districts, particularly budgetary functions and the management of resources (Resosudarmo 2005). An important anomaly affected the expected re-distributions of state lands and the ways that grassroots groups and NGOs had to manoeuvre subsequently. Sectoral central state institutions, most importantly the Ministry of Forestry,[11] retained jurisdiction over those lands, although they are required by the terms of decentralization to negotiate with regional governments over management of the above-ground resources (McCarthy 2000, 2006).

It was into this political space of opportunity and constraint that the peasant organizations and supporting NGOs and NGO consortiums emerged and began their activities. After the collapse of the New Order, with the decentralization law to take effect within two years, agrarian activists, NGOs and student groups helped organize dispossessed peasants outside the major cities of Java (Jakarta, Bandung, Semarang, Yogyakarta and Surabaya) and parts of Sumatra (Medan). They set up 'action committees' and engineered public protests and dialogues with government officials and members of parliament. AMAN – the National Indigenous Peoples' Association – was formally established in 1999 at a national meeting in Jakarta. Before Reformasi, most delegates had been from outside Java, but at this meeting Java and Bali were represented.

---

[10] Cf. Wargadiputra (2005, 2), who states that 50 per cent of the district is state forest or plantation land.
[11] And its regional institutions, including the State Forestry Corporation (SFC) in Java.

The new opportunities offered by decentralization and the establishment of more civil liberties exacerbated an underlying source of difference amongst rural reformers in civil society and in government. This had to do with whether – and in what forms – land or broader agrarian reforms were appropriate for state forest lands. Such a split could be viewed as between environmental and agrarian reform interests, but, as we show below, the alliances and conflict did not line up precisely in that manner.

Transnational Agrarian and Environmental Movement Effects

FSPI and KPA are technically categorized as different types of agrarian movement organizations, though in practice they have some similarities, as both are led by activists, not peasants. When it was first formed in the mid–late 1990s until late 2007, FSPI was a federation of peasant organizations led by Sumatra-based activists associated with the Synthesis Foundation (*Yayasan Sintesa*, YS), an NGO. KPA is a West Java-based consortium of Indonesian grassroots NGOs, peasant organizations and activists, and registered in Indonesia as an NGO. The founding leaders of both began activist work during the Suharto-era struggles, working underground in their native North Sumatra and West Java, respectively. Differences emerged early on over the best base-site for a national level organization, funding and other sources of competition (Lucas and Warren 2003; Afiff 2004).

While supportive of many of Vía Campesina's policies and actions, and a member of Vía Campesina's main Indonesian affiliate, FSPI, until the national meeting in 2007, SPP has always been more closely allied with KPA (the Consortium for Agrarian Reform). KPA, as an NGO, could not be a member of either FSPI or Vía Campesina. Perhaps for that reason, it established connections with the International Land Coalition (ILC), which is also a kind of consortium of diverse types of organizations and representing more diverse ideologies among its members than Vía Campesina. For example, ILC has a less confrontational approach to the World Bank and its market-oriented land policies – indeed the World Bank is a member organization.

However, as Borras (2004) points out, the political positions of and actual relations between organizations and individuals in ILC and VC are confusing and variable, as some groups are members of both. SPP, for example, is a member of ILC and was a member of VC through its membership in FSPI until late 2007 when SPP resigned (see below). NGOs such as KPA are much more sympathetic to VC concerns and ideologies than to those of ILC and many of its member groups, but are barred from formal membership in VC by their NGO status. This all gets even more complex as the critical tensions between Vía Campesina (VC) and ILC in the global arena (Edelman 2003; Borras 2004) have been reflected in and created new tensions – real or represented as such – between some of their Indonesian allies and various sub-national member groups.

Learning of the ban on NGO membership during Vía Campesina's international meeting in Mexico in 1996, the Indonesian activists were surprised. The leaders of KPA (NGO), SPP (peasant organization) and YS (NGO) all saw themselves as equally committed to radical land reform and these Indonesian activists had been deeply involved with peasant organizations. Subsequently, they agreed in 1999 to form FSPI exclusively as a peasant organization, but with a non-peasant activist – Henry Saragih of *Yayasan Sintesa* – as its first secretary general. Some activists recall an understanding that, as the organization and its membership matured, an actual peasant/small farmer would take on the leadership. When FSPI became the site of VC's international secretariat, it also became the major international portal into Indonesian agrarian movement activities. At the same time, the organization's national agenda was now being more determined by international campaign priorities than by those of Indonesian member organizations. The leadership of KPA and SPP, in contrast, while aware of international developments and drawing on the experiences and ideas of MST and other international agrarian movements, tends to be more Indonesia-focused, explicitly looking inward to the demands and needs of their West Java membership, and in other Indonesian areas. Until late 2007, SPP remained an important member of FSPI as well, sending a full three-quarters of the 10,000 peasants who marched under FSPI's flag to a June 2007 agrarian reform demonstration in Bandung (http://fspi.or.id/en/content/view/120/1).

In their late 2007 meeting, FSPI's leadership changed the terms of membership in FSPI, making individuals rather than peasant organizations the membership units. This discouraged some member organizations, as it changed significantly the organization's federated character. SPP and several other peasant organizations refused to dissolve their own organizations to become chapters or individual members of FSPI, and resigned from FSPI. Activist Saragih was re-elected director again, causing further dissatisfaction in some quarters that he had not yet stepped aside to let an actual peasant lead the organization.[12]

Tensions between groups or their leaders have also been exacerbated by campaigns and connections with environmental groups and issues. In particular, differences intensified over various moves to integrate discussions of agrarian reform as discussed in the Basic Agrarian Law and the reclaiming of forest lands and mines under the purview of the Ministries of Forestry and Mining. How these tensions have played out on the ground can be seen through two key legislative and policy initiatives that animated these fora early in the reform period and recently (2005–07). The initiatives are Parliamentary Decree no. IX/2001 (hereafter called TAP MPR IX/2001) and the agrarian reform initiative announced jointly by the President, the Ministry of Forestry and the National Land Board (NLB) in September 2006. In addition, different notions of the ultimate goals for reform have split groups and individuals, demonstrated by the debates over community forestry or CBNRM. These debates demonstrate the changing (once again) agrarian and political contexts within which reform organizations have been operating.

---

[12] Despite the assertions of a reporter in a recent article on FSPI in *The Guardian*, Henry Saragih did not start out as 'a small farmer'.

TAP MPR IX/2001 Debates

By late 2000 and 2001, it was becoming clear not only that Reformasi would change the Indonesian state, but that environmental initiatives had already, throughout the Suharto period, reconfigured state institutions, including law, and normalized new regimes of territorial control. The environmental movement – and here again, we mean land and forest conservation movements, as these affected patterns of land control and access most – had constituted different state and civil society arenas within which its separate but entwined components ('big' conservation and environmental justice) operated. This was a context – the context – within which agrarian movements would have to negotiate and struggle.

KPA had long been one of the agrarian groups willing to work with national and sub-national environmental justice NGOs, in part because some environmental justice groups (e.g., LATIN and WALHI) had occasionally provided funds to KPA and SPP to finance occupations and protests (author interviews 2007). In 2001 they set up a Working Group on Agrarian Reform and Natural Resource Management. To agrarian reformers and environmental justice groups, this represented the first time the two issue groups openly shared a public forum to jointly strategize and influence national policy and legislation, although in ways still limited by the nature of state and corporate power and their own differences (Rosser et al. 2005, 67).

This environmental–agrarian activist collaboration on TAP MPR IX/2001, and the participation of AMAN, the indigenous people's national organization, which, remember, claimed and was claimed to be both an agrarian and environmental justice organization, was enabled by the focus on tenurial issues in environmental conflicts, even though the word 'tenure' was not used consistently. Rather, it referred both to 'land tenure' or 'resource tenure', the latter not necessarily meaning formal land rights. At the same time, sympathetic academics and policy makers attempted to integrate the agro-ecological and social structural dimensions of the two approaches to reform. Among those who started out first as agrarian reformers, one strategy was to link agrarian structural inequities with ecological crises, and show how solving the first might help ameliorate the second (e.g. Sangkoyo 2000; Kartodihardjo 2002). It was argued that the new agrarian reform agenda should not only restructure land access and control or alter land use, production and consumption systems to guarantee the basic rights and welfare of poor people, but also to ensure ecosystem integrity and improve productivity (Fauzi and Zakaria 2001, 2002).

Many of these ideas had been integral to both global and local (Indonesian) social and community forestry discourses, and part of strategies for community-based natural resource management (CBNRM) (ARuPA et al. 2003). Here, they were repackaged by agrarian activists as part of their own and common environmental-agrarian agendas. Interestingly, what started out as a community-based ideology intended to wrest control of tightly controlled forest resources from the central government was soon being represented at an international level as both a neoliberal strategy for reducing big government (McCarthy 2005; cf. Belsky 2008) and a mobilizing/organizing strategy for agrarian reform.

Another contradictory representation of 'reform' issues characterized the debates over TAP MPR IX/2001. The alliance between some agrarian reform groups (notably KPA) and environmental justice groups in the interest of passing the TAP MPR IX/2001 was repackaged by other agrarian reform groups as a capitulation to neoliberalism. FSPI's leadership, for example, argued that the parliamentary resolution could be used by pro-market forces to change – or invalidate – the reform tenets of the 1960 BAL.[13] On FSPI's website, for example, it is stated that the TAP MPR IX/2001 has been 'used as a basis of various law drafts (RUU) that clearly opposed and depleted the spirit of UUPA No. 5/1960 and UUD 1945 chapter 33. This legislation also reveals flaws in agricultural laws in Indonesia that have resulted in an unjust agrarian structure and enhanced the process of liberalization of natural resources'. The website also states that former state-owned companies (parastatals called BUMN in Indonesia) have been taken over or strengthened by transnational companies (http://viacampesina.org/main_en/images/stories/lvcbooksonwto.pdf).

This view however, distorted the intentions of allied agrarian and environmental justice advocates. To them, the issuance of TAP MPR No. IX/2001 had great symbolic and strategic meaning. Local SPP leaders used the decree as a means of justifying land occupations and as a bargaining chip with central and regional government land management agencies (Afiff et al. 2005). Well aware of, and not in agreement with, the problems around Suharto-era natural resources legislation, reinforced in the reform period by the revised Forestry Law (41/1999) and a subsequent presidential decree (KepPress 34/2003), KPA and SPP and other proponents of the resolution have used the TAP MPR IX/2001 to bring issues of agrarian reform back to the negotiating table with government. The decree clearly states its support for the BAL. Even member organizations of FSPI had differing views on these new pieces of legislation, such as KepPres 34/2003. SPP saw this presidential decree as legitimating district parliaments and administrations' engagement in agrarian reform, putting 'resolution of conflicts closer to the source' (Lucas and Warren 2003, 35–6; Afiff et al. 2005, 5). While FSPI, SPP and KPA supported the BAL and agrarian reform, the former saw the TAP MPR IX/2001 as more beneficial to big conservation (dark greens) and corporate environmentalism than to environmental justice issues. For KPA and SPP, environmental justice often sought the same solutions as agrarian reform initiatives.

The darker green environmental groups were also not enthralled with TAP MPR IX/2001, because of the agrarian alliance and what they saw as a threat to their recently gained territory. The prospect of redistributing whole tracts of forest land to impoverished or indigenous peasants and farmers on Java, or elsewhere, was a line that some environmental groups, moderate or conservative, were unable to cross. At the time of these debates, the large land management

---

[13] See Lucas and Warren (2003) for a more extensive elaboration of these debates.

institutions within the Indonesian state were being framed by agrarian activists as landlords. This move changed the terms of the conflict over agrarian reform and in some ways differentiated Indonesia's struggle from those of other countries.[14] It also pitted the most radical agrarian activists against more moderate agrarianists and the darkest green environmental NGOs. Some of the latter were not willing to consider the implications of the very recent history of state forestry expropriations (1967) in forming the 'national forest', a process which also involved taking over customarily managed or owned lands (Peluso and Vandergeest 2001). TAP MPR IX/2001 thus drove stakes through some collaborations, just as it had engendered new ones.

The TAP MPR IX/2001 debates also illuminated the fact that the big national-level challenge had switched from the criminalization of civil society movements to whether and how activists would work with or within the state. The Indonesian state was making its own radical changes in its structures, practices and policies. For both radical and moderate agrarian activists, working within government was an entirely new concept, minimally thought about and hardly – if at all – planned for.

Yet government had to be dealt with, as decentralization had created a major contradiction: many district level officials who were now in charge of making their own decisions on matters they had been ordered to take care of during the Suharto regime did not always know how to make such decisions. Thus some activist groups, including KPA, developed training programmes for district parliament members in the early years of Reformasi. Through workshops and training sessions, they could discuss the implications of the new laws, and the possibilities for different scenarios of agrarian reform. In many of these discussions, district level foresters were involved, requiring an almost constant awareness that ecological integrity and environmental sustainability would be invoked. Even the most radical rights-oriented agrarianists had to learn to talk this talk if their strategy included changing laws and policies of government land management institutions.

The Effects of Environmental Discourses

Environmental activists had always seen a major part of their movement as operating on the legal front – seeking territories over which they could exert control or influence state control – in addition to grassroots work. As shown above, over time, they succeeded: even under authoritarian rule in Indonesia, environmental discourses had gained legal ground and territory. Conservation benefited from state forestry and natural resource-based accumulation strategies (corporate and state) in part by contesting the capacities of natural resource agencies such as the Ministry of Forestry to sustainably manage land.[15] Newly

---

[14] But see Borras (2006), for example, who addressed agrarian reform on forest lands in the Philippines.
[15] On the 'natural' alliances between capitalism and conservation, see Smith (1984).

gazetted national parks, conservation areas and nature reserves were physical proof of environmentalists' power: they had even modified the terms of forestry, extending the amount of forests reserved for protection or conservation rather than large-scale production or development.

Increased conservation power changed and was changed by the political economic context. On the dark green conservationist side, the territorial bases of conservation were perceived as threatened by both decentralization and agrarian reform – whether through direct land occupations or through government policy (McCarthy 2006; Resosudarmo 2005). If management of forest and conservation areas was decentralized, who would take responsibility to fund and enforce the nature reserves and other protected areas that represented conservation's successful years of mobilizing on transnational, national and local fronts (Jepson and Whittaker 2002)? And, if the MoF and the SFC were being framed as illegitimate 'landlords' on political forestlands, conservation areas established under the New Order could be, and were, similarly challenged, as happened in the case of Dongi-dongi in Sulawesi (Adiwibowo 2005; Li 2007). These new developments pushed moderate or flexible agrarian movements and environmental justice groups to mobilize against exclusive areas of 'nature protection' (Lowe 2003; Afiff and Lowe 2008; Stedile 2002). Perhaps for this reason, some environmental groups made some previously unheard of deals with big capital, reflecting a neoliberal-era follow-up to the coercive conservation alliances made by international conservation groups with military and authoritarian states of the developmentalist era (Peluso 1993; Chapin 2004).

Environmental justice groups also had won some battles under the New Order, including under the rubric of community or social forestry. Community forestry proponents in environmental justice groups had long pursued their objectives of increasing farmer access to government forest lands as part of broader struggles for tenure reform. In the early years of the new millennium, they had some small but highly symbolic achievements. For example, the Ministry of Forestry formally recognized a few autonomous community forests in Sumatra and Sulawesi (D'Andrea 2003; Li 2007). In Java, the SFC expanded villagers' access to forest land through forms of joint forest management in the 'social forestry' areas of Java, in part responding to the threats posed by forest occupations and the mob logging that took place in the early years of Reformasi.

The Indonesian state's consolidated control over so much of the national territory as state forest made it necessary for the agrarian movement groups to deal explicitly with community/social forestry programmes and concerns. As agrarian reform activists in the late 1990s focused their attention on upland areas, their work unavoidably overlapped and sometimes conflicted with that of community-based forest management (CBFM) activists.

On Java, where land was tight and political forests had played a much longer role in the island's land use history, and particularly during the New Order, CBFM activists could not always insist on excising land from the island's forests.[16] The first step was seen as increasing farmers' access to land and forest resources.

However, the problem with community and social forestry, in the views of some agrarian activists, was that resource tenure rights and access constituted neither absolute control nor community autonomy, rather, they were held jointly with the SFC or the MoF. Further complicating the rights picture, many agrarian activists had been hoping for more communal property alternatives to emerge organically or be attempted. Private property ran the risk of being sold, a bitter lesson for many after the success and perceived failure of the Sagara campaign (Fauzi 2003; Lukmanudin 2001).[17] In these ways and others, both working with government in community/social forestry programmes and 'winning' private property became major bones of contention.

Some community forestry and agrarian reform activists made efforts to connect and collaborate, enabling both to see their work from new perspectives. For KPA, for example, community forestry connections represented access to new resources, support and networks. The connections also provided opportunities for agrarian reform activists to radicalize community forestry and other forms of environmental justice. To this end, the environmental NGOs WALHI and ARuPA (*Aliansi Relawan untuk Penyelamatan Alam*/Volunteer Alliance to Save the Environment) led a campaign on Java to dissolve the SFC – which actually seemed possible in the wake of the government enterprise's announcement of its bankruptcy in 2002. Although in the end the SFC was sustained by support from the MoF, the pressures imposed by activists, academics and some foundations/funders led the SFC to dramatically increase community access rights to state forest land and to increase the amount of territory and trees under social forestry programmes – but not to redistribute or reclassify state forest land (Simon 1993, 1994; Awang 2004). While this was criticized by agrarian reformers, EJ activists defended this 'access' approach, seeing it as a potential wedge in a very tightly closed door to forest lands. The incremental gains for farmer access would be difficult to refute once the SFC or the MoF had conceded them (author interviews 2007). In fact, this latter scenario is what transpired.

It is not so surprising, then, that after Reformasi some agrarian movements have found it expedient to embrace environmentally friendly stances as explicit components of their political strategies. Some agrarian movement leaders have taken a page from the environmentalists' book, deploying multi-scaled approaches to gaining legitimacy, territory and moral authority in various arenas

---

[16] On Java, the form implemented within the SFC was called Social Forestry. Activists posited community forestry as an alternative framing, one that would not involve intervention or decision-making power by the state, once the state had recognized the community's forest as outside the *kawasan hutan* or state forest territory.

[17] Sagara villagers had been mobilized by SPP and KPA to occupy land also claimed by SFC. The case ended with the SFC losing, when the National Agrarian Ministry and the Head of the National Land Bureau determined that the state land in question could be subjected to land reform (Decree No. 35-VI/1997). Yet, three years after the settlement, most of the redistributed 580 hectares of the land had been sold, disillusioning agrarian activists.

of contestation, including state law and policy (Afiff et al. 2005). The fact is that failure to commit to ensuring environmental sustainability could be a basis for denial of access to land – especially in areas that are classified as forest or conservation areas.

Indeed, a great deal of the environmental justice movement and their international academic allies had already focused on demonstrating the misunderstandings and misrepresentations of indigenous people's sustainable agroforestry practices and claims by foresters and others.[18] Just as environmental justice advocates had recognized the need to translate the traditional practices of *Masyarakat Adat* into notions of sustainable resource management, KPA advocates soon moved in that direction. The potential for land occupiers to be seen as sustainable managers of fragile upland environments made its way into SPP and KPA leaders' speeches, and was also used to transform some land occupiers' land use practices in favour of agro-forestry. SPP leaders have decried the lack of 'traditional agroforestry' on SFC-controlled uplands and associated that forestry department's upland monocultures with recent natural disasters such as erosion and landslides on Java's forest lands (author interview 2007).

Off Java, through AMAN and on their own, indigenous peoples' groups continued to work with the environmental justice movement. They wanted legal recognition of 'ancestral territories', meaning formal state acknowledgement. One possibility for this was a reintroduction of legal pluralism into the Indonesian legal system, a move supported in part by some national, local and international activists and even funders. But such a broad move could alienate many agrarian reformers because some of the 'customary rights' systems included as members of AMAN are dominated by the 'feudal elites' that agrarian reform and nationalist activists had resisted in the anti-colonial, anti-feudal movements of the 1940s and 1950s. Most contemporary activists were not willing to facilitate these elites now claiming 'customary rights' through inclusion in agrarian reform initiatives or strategic alliances. They have, however, maintained ties to AMAN member-groups with more democratic, communal or egalitarian connections.

Debates on Recent Government Land Reform Initiatives

The differences between the fundamental principles underlying agrarian organizations and how they would work (or not) with government were highlighted further in September 2006, when the President formally announced that land reform would be implemented in Indonesia after a hiatus of more than 40 years, with 8.15 million hectares of state land under the jurisdiction of the Ministry of Forestry and the National Land Board slated for redistribution (*Republika Online* 28 September 2006). Almost immediately, FSPI leaders campaigned about the danger of 'pseudo-agrarian reform' because the scheme could be implemented

---

[18] This also constitutes a huge literature, including in English, on Indonesia, most of the lifetime work of Michael Dove (see e.g., Dove 1983, 1985, 1996, and many others).

without substantial political-economic transformation (FSPI 2006). KPA, on the other hand, and some of the other peasant organizations, argued that although the political-economic structure would not radically change, the plan could be used to facilitate the legalization of hundreds of thousands of claims on occupied lands by dispossessed peasants and indigenous groups (KPA 2006). The confusion wrought by these two positions amongst supporters of agrarian reforms on state lands is exemplified further by the fact that leadership at SPP, then a member of both FSPI and KPA, engaged in formal and informal discussions with government actors about the implementation of the new initiative.

The agrarian reform initiative may be an indicator that it is now the turn of agrarian movements and their leaders to change the political-economic context. Moreover, this context in 2007, well into the neoliberal and decentralized post-Suharto period, is considerably different than when nascent underground movements first began in the late 1980s and early 1990s. In the place of the centralized New Order state is a mish-mash of diverse, decentralized districts, each dealing with shifting and uncertain politics in differently endowed agrarian environments with different institutional and social histories of land management. The MoF is still clinging to its claims to forest land. Parks as well as production forests look good to aspiring agrarian reform claimants, but they are off limits in the eyes of conservative conservationists and some environmental justice advocates. Despite these powerful forms of opposition, reform is in the air.

How all this will resolve is not entirely clear, as pilot project areas for the national agrarian reform initiative were being selected as of late 2007. Public positioning and private agitation and action by reform promoters and detractors will continue. As with other agrarian collaborations and reforms, the real stories are not yet public.

What is clear is that there will be some alliances between national and subnational environmental justice and agrarian reformers, just as these have been increasing in transnational campaigns. The common fight against GMO technology and the support of international organic farming are the most obvious examples of alliances (Stedile 2002, 103; Borras 2004; Biekart and Wood 2001). At the 2007 Climate Change conference in Bali and on their website, FSPI has come out in support of sustainable agriculture, organic farming efforts, anti-GMO campaigns and other 'initiatives' supported globally by Vía Campesina. In Bali, FSPI/Vía Campesina argued to the international community that 'small family farms make the earth cooler'. Again, it is interesting that FSPI leadership tends to use internationally oriented fora to make such claims, seeking legitimacy outward, while other national agrarian movement organizations such as KPA address the majority of their strategic initiatives to national and sub-national debates, institutions and actors.

CONCLUSION

Reformasi, decentralization and the resurgence of agrarian reform movements in Indonesia began at about the same time as the infamous global protests against

WTO took place in Seattle, suggesting that they were part of a common transnational movement. However, as Biekert and Wood have pointed out, 'Global protest existed long before the "battle of Seattle". What is new are the growing linkages between highly diverse campaigns' (2002, 1).

This chapter has shown that campaigns are constituted under different regimes and articulate within different conjunctures, with effects on each other arising from transnational, national and subnational sources. In a variety of ways that still need deeper exploration, the lines between environmental and agrarian justice movements are often blurred by their specific and common histories in opposing state expropriations. These have in part been hidden from view by the rhetorics and strategic moves of environmental and agrarian movements that suggest divergence rather than articulation and convergence. Using Indonesia as an example, we showed how agrarian and environmental movement rhetorics, strategies, alliances and other practices varied under neoliberalism and Reformasi, as well as under the previous authoritarian, state-led development regime. The outcomes and configurations of power have changed over time and have changed their times.

For observers of Indonesia's often violent agrarian and environmental politics, the extent of mass agrarian organizing, land occupations, public opposition to government policy and other critical actions since 1998 have been no less than revolutionary. Nevertheless, today's agrarian movements were forced to come to terms with the territorial and governance gains of yesterday's environmental movements. Similarly, and somewhat unexpectedly, as agrarian reform agendas have gained ground, at least in a symbolic or moral sense, environmental activists are being forced now to take account of their own campaigns, demands and claims. Early alliances with agrarian justice groups split Indonesia's environmental movement, between relatively coercive and justice-oriented environmental groups. Many agrarian movement groups, however, have found common grounds with environmental justice advocates.

Despite their common euphoria when Reformasi replaced authoritarian rule, the organizations that make up Indonesia's resurgent agrarian movements are not united in their struggles today. Among other reasons for the splits are the particularities and politics of movement groups' different associations with transnational agrarian and environmental organizations and the changes these and national or sub-national groups have generated both under state-led development and Reformasi.

As political fields have changed and alliances with government and transnational institutions formed and dissolved, formerly marginalized activist concerns have become normalized within state and development practices and ideologies – first those concerns of the environmental movement then of agrarian movement activists. Environmental activism and action have engendered legal, institutional and territorial changes in national land management and control. Under different political economic regimes, in different historical moments, the political opportunities open to environmental or agrarian movement groups have been shaped not only by international and national agendas but by strong grassroots organizing, and movement groups' perceived needs and capacities to ally with each other, with state and transnational actors, and with capital. Further complicating matters, the logic of what might be called a national grassroots – including migrants travelling around the nation seeking work in resource-based industries, or land they deem to be 'empty' – does not always work well with the logics of the local grassroots (Li 2007). Thus, partnerships across and within movements and transnationally have not been consistent.

What is interesting is that all of them have learned to talk the talk of environmental sustainability, which has become increasingly important to operationalize. The landscapes of occupation must appear sustainably managed in order to claim a piece of the moral high ground and create grounds for agrarian reform. Yet in the present conjuncture, environmentally sensitive agrarian reform has to be seen as having been pre-configured by environmental movements, conservative and radical.

REFERENCES

Aditjondro, George J., 1993. 'The Media as Development "Textbook": A Study on Information Distortion in the Debate about the Social Impact of an Indonesian Dam (Kedung Ombo Dam)'. PhD Dissertation, Cornell University.

Aditjondro, George J., 2003. 'Kebohongan–Kebohongan Negara Perihal Kondisi Obyektif Lingkungan Hidup di Nusantara'. Yogyakarta: Pustaka Pajar.

Adiwibowo, Suryo, 2005. 'Dongi-dongi – Culmination of a Multi-dimensional Ecological Crisis: A Political Ecology Perspective'. PhD Dissertation, Universität Kassel, Germany.

Afiff, Suraya, 2004. 'Land Reform or Customary Rights?: Contemporary Agrarian Struggles in South Tapanouli, Indonesia'. PhD Dissertation, University of California, Berkeley.

Afiff, Suraya, Noer Fauzi, Gillian Hart, Lungisile Ntsebeza and Nancy Peluso, 2005. 'Redefining Agrarian Power: Resurgent Agrarian Movements in West Java, Indonesia'. Center for Southeast Asia Studies Working Paper CSEASWP2-05.

Afiff, Suraya and Celia Lowe, 2008. 'Collaboration, Conservation, and Community: A Conversation between Suraya Afiff and Celia Lowe'. In *Biodiversity and Human Livelihoods in Protected Areas: Case Studies from the Malay Archipelago*, eds. Navjot S. Sodhi, Greg Acciaioli, Maribeth Erb and Alan Khee-Jin Tan, 153–64. Cambridge: Cambridge University Press.

Arupa, Koling, Yayasan and the Asian Forest Network, 2003. *Communities Transforming Forestland*. Java, Indonesia, Bohol, Philippines: Asia Forest Network.

Aspinall, Edward, 2004. 'Indonesia: Civil Society and Democratic Breakthrough'. In *Civil Society and Political Change in Asia. Expanding and Contracting Democratic Space*, ed. Muthiah Alagappa, 61–96. Stanford, CA: Stanford University Press.

Awang, San Afri, 2004. *Dekonstruksi Sosial Foresti: Reposisi Masyarakat dan Keadilan Lingkungan*. Yogyakarta: Bigraf Pub. & Program Pustaka.

Bachriadi, Dianto dan Anton Lucas, 2001. *Merampas Tanah Rakyat: Kasus Tapos dan Cimacan*, Jakarta, Kepustakaan Populer Gramedia.

Barber, Charles, Suraya Afiff and Agus Purnomo, 1995. 'Tiger by the Tail: Challenges to Biodiversity in Indonesia'. Washington, DC: World Resources Institute, WALHI, Pelangi.

Barr, Christopher, 1997. 'Discipline and Accumulate: State Practice and Elite Consolidation in the Indonesian Timber Industry'. Master's Thesis, unpublished, Cornell University, Ithaca, NY.

Baviskar, Amita, 1995. *In the Belly of the River: Tribal Conflicts Over Development in the Narmada Valley*. Delhi: Oxford University Press.

Belsky, Jill M., 2008. 'Creating Community Forests'. In *Forest Community Connections: Continuity and Change*, eds. Ellen Donoghue and Victoria Sturtevant, 219–242. Washington, DC: Resources for the Future.

Biekart, Kees and Angela Wood, 2001. 'Ten Reflections on the Emerging Global Protest Movement'. *The Ecologist*, 31: 8 (October).

Borras, Saturnino M., 2004. 'La Vía Campesina: An Evolving Transnational Social Movement'. Briefing Series No. 6, Transnational Institute, Amsterdam. http://www.tni.org/reports/newpol/campesina.pdf.

Borras, Saturnino M., 2006. Redistributive Land Reform in Public (Forest) Lands? Rethinking Theory and Practice with Evidence from the Philippines. *Progress in Development Studies*, 6 (2): 123–45.

Buechler, Steven M., 2000. *Social Movements in Advanced Capitalism: The Political Economy and Cultural Construction of Social Activism*. New York: Oxford University Press.

Chapin, Mac, 2004. *A Challenge to Conservationist*. World Watch magazine, November/December 2004.

Christodoulou, Demetrios, 1990. *The Unpromised Land: Agrarian Reform and Conflict Worldwide*. London: Zed Books.

Contreras-Hermosilla, Arnoldo and Chip Fay, 2005. *Strengthening Forest Management in Indonesia Through Land Tenure Reform: Issues and Framework for Action*. Bogor and Washington, DC: Forest Trends, ICRAF, World Bank.

Cribb, Robert, 1990. 'Problems in the Historiography of the Killings in Indonesia'. In *The Indonesian Killings of 1965–1966: Studies from Java and Bali*, ed. Robert Cribb, 1–43. Clayton, Victoria, Monash University Centre of Southeast Asian Studies.

D'Andrea, Claudia, 2003. 'Coffee, Capitalism and Culture in Katu, Sulawesi, Indonesia'. PhD dissertation, University of California, Berkeley.

Dauvergne, Peter, 1994. *Shadows in the Forest*. Cambridge, MA: MIT Press.

Desmarais, Annette Aurélie, 2002. 'The Vía Campesina: Consolidating an International Peasant and Farm Movement'. *Journal of Peasant Studies*, 29: 91–124.

Djuweng, Stepanus, 1997. 'Asal-usul Global dari Konflik Lokal versus Korban Lokal dari Masalah-masalah Global'. dalam *Tanah dan Pembangunan*, ed. Noer Fauzi, 253–81. Jakarta: Penerbit Sinar Harapan.

Djuweng, Stepanus and Sandra Moniaga, 1994. 'Kebudayaan dan Manusia yang Majemuk di Indonesia. Masihkah Punya Tempat?' In *ILO Convention 169*. Jakarta: ELSAM dan LBBT.

Dove, Michael, 1983. 'Theories of Swidden Agriculture and the Political Economy of Ignorance'. *Agroforestry Systems*, 1 (2): 85–99.

Dove, Michael, 1985. 'The Agroecological Mythology of the Javanese, and the Political-Economy of Indonesia'. *Indonesia*, 39: 1–36.

Dove, Michael, 1996. 'So Far from Power, So Near to the Forest: A Structural Analysis of Gain and Blame in Tropical Forest Development'. In *Borneo in Transition: People, Forests, Conservation, and Development*, eds. Christine Padoch and Nancy Lee Peluso, 41–58. Kuala Lumpur: Oxford University Press.

Edelman, Marc, 1999. *Peasants Against Globalization: Rural Social Movements in Costa Rica*. Stanford, CA: Stanford University Press.

Edelman, Marc, 2003. 'Transnational Peasant and Farmer Movements and Networks'. In *Global Civil Society Yearbook 2003*, eds. M. Kaldor, H. Anheier and M. Glasius, 185–220. Oxford: Oxford University Press.

Edelman, Marc, 2005. 'Bringing the Moral Economy Back In . . . to the Study of Twenty-first Century Transnational Peasant Movements'. *American Anthropologist*, 107: 331–45.

Farid, Hilmar, 2005. 'Indonesia's Original Sin: Mass Killings and Capitalist Expansion, 1965–66'. *Inter-Asia Cultural Studies*, 6 (1): 3–16.

Fauzi, Noer, 1999. *Petani dan Penguasa, Dinamika Perjalanan Politik Agraria Indonesia*. Yogyakarta: Insist Press.

Fauzi, Noer, 2003. 'The New Sundanese Peasants' Union: Peasant Movements, Changes in Land Control, and Agrarian Questions in Garut, West Java'. Paper prepared for workshop New and Resurgent Agrarian Questions in Indonesia and South Africa. Center for Southeast Asia Studies and Center for African Studies – Crossing Borders Program 2003–2004, 24 October 2003. Institute for International Studies – Moses Hall University of California, Berkeley, CA. http://repositories.cdlib.org/cseas/CSEASWP1-03.

Fauzi, Noer and R. Yando Zakaria, 2001. *Mensiasati Otonomi Daerah, Panduan Fasilitasi Pengakuan dan Pemulihan Hak-hak Rakyat*. Yogyakarta: Insist Press.

Fauzi, Noer and R. Yando Zakaria, 2002. 'Democratizing Decentralization: Local Initiatives from Indonesia'. Paper submitted for the International Association for the Study of Common Property 9th Biennial Conference, Zimbabwe.

Fidro, Boy and Noer Fauzi, 1995. *Pembangunan Berbuah Sengketa: 29 Tulisan Pengalaman Advokasi Tanah*. Bandung: Lembaga Pendidikan dan Pengembangan Pedesaan.

Franco, Jennifer C. and Saturnino M. Borras, 2005. 'Changing Patterns of Peasant Mobilizations for Land and Democracy in the Philippines'. In *On Just Grounds, Struggling for Agrarian Justice and Citizenship Rights in the Rural Philippines*, eds. J.C. Franco and S.M. Borras. Quezon City: Institute for Popular Democracy; Amsterdam: Transnational Institute.

Franco, Jennifer C. and Saturnino M. Borras, 2006. 'Pandangan dan Sikap FSPI Tentang Program Pembaruan Agraria Nasional'. http://www.fspi.or.id/index.php?option=com_content&task=view&id=366&Itemid=1, Accessed 26 January 2007.

Ganie-Rochman, Meuthia, 2002. *An Uphill Struggle: Advocacy NGOs under Soeharto's New Order*. Jakarta: LabSosio FISIP-UI.

Hadiz, Vedi, 2004a. 'Decentralisation and Democracy in Indonesia: A Critique of Neo-Institutionalist Perspectives'. *Development and Change*, 35 (4): 697–718.

Hadiz, Vedi, 2004b. 'Indonesian Local Party Politics: A Site of Resistance to Neo-Liberal Reform'. *Critical Asian Studies*, 36 (4): 615–36.

Hajer, Maarten, 1993. 'Discourse Coalitions and the Institutionalisation of Practice: The Case of Acid Rain in Great Britain'. In *The Argumentative Turn in Policy Analysis and Planning*, eds. Frank Fischer and John Forester, 43–67. London: Mobipocket.

Hajer, Maarten, 1995. *The Politics of Environmental Discourse*. Oxford: Oxford University Press.

Heruputri, Arimbi, 1997. 'Penghancuran Secara Sistematis Sistem-Sistem Adat oleh Kelompok Dominan'. Kertas Posisi WALHI No. 06-1997.

Hikam, Muhammad, 1995. 'The State, Grass-roots Politics and Civil Society: A Study of Social Movements under Indonesia's New Order 1989–1994'. PhD dissertation, the University of Hawaii at Manoa.

Husken, Frans and Benjamin White, 1989. 'Java: Social Differentiation, Food Production, and Agrarian Control'. In *Agrarian Transformations: Local Processes and the State in Southeast Asia*, eds. Gillian Hart, Andrew Turton and Benjamin White, 235–65. Berkeley, CA: University of California Press.

Jepson, Paul and Robert J. Whittaker, 2002. 'Histories of Protected Areas: Internationalisation of Conservationist Values and their Adoption in the Netherlands Indies (Indonesia)'. *Environment and History*, 8: 129–72.

Kartodihardjo, Hariadi, 2002. 'Hutan Kemasyarakatan Dalam Belenggu Penguasaan Sumber-sumber Agraria'. In *Menuju Keadilan Agraria: 70 Tahun Gunawan Wiradi*, 339–57. Bandung: Akatiga.

Keck, Margaret E. and Kathryn Sikkink, 1998. *Activists Beyond Borders: Advocacy Networks in International Politics*. Ithaca, NY: Cornell University Press.

Kowalchuk, Lisa, 2005. 'The Discourse of Demobilization: Shifts in Activist Priorities and the Framing of Political Opportunities in a Peasant Land Struggle'. *The Sociological Quarterly*, 46: 237–61.

KPA, 2006. 'Komunike Internal KPA No. 02/Desember/2006. Mari Rapihkan Barisan untuk Mengawal Program Pembaruan Agraria Nasional!' http://www.kpa.or.id/index.php?option=com_content&task=view&id=111&Itemid=53&PHPSESSID=a7a475d0ec77bd7c3eb633425f92b5de, Accessed 26 January 2007.

Kuswahyono, Imam, 2003. 'Mencari Format Hukum dalam Menuju Reforma Agraria dalam Kerangka Otonomi Daerah'. http://www.otoda.or.id/Artikel/Imam%20Koeswahyono.htm, Accessed 4 October 2003.

Li, Tania, 2000. 'Articulating Indigenous Identity in Indonesia: Resource Politics and the Tribal Slot'. *Comparative Studies in Society and History*, 42: 149–79.

Li, Tania, 2001. 'Masyarakat Adat, Difference, and the Limits of Recognition in Indonesia's Forest Zone'. *Modern Asian Studies*, 353: 645–76.

Li, Tania, 2007. *The Will to Improve: Governmentality, Development, and the Practice of Politics*. Durham, NC: Duke University Press.

Lowe, Celia, 2003. 'Sustainability and the Question of "Enforcement" in Integrated Coastal Management: The Case of Nain Island, Bunaken National Park'. *Pesisir & Laut*, No. 1: 49–63.

Lowe, Celia, 2006. *The Wild Profusion: Biodiversity in an Indonesian Archipelago*. Princeton, NJ: Princeton University Press.

Lucas, Anton and Carol Warren, 2000. 'Agrarian Reform in the Era of Reformasi'. In *Indonesia in Transition: Social Aspects of Reformasi and Crisis*, eds. Chris Manning and Peter van Diermen, 220–38. Indonesia Assessment Series. Singapore: Institute of Southeast Asian Studies.

Lucas, Anton and Carrol Warren, 2003. 'The State, the People and Their Mediators: the Struggle over Agrarian Law Reform in Post New Order Indonesia'. *Indonesia*, October, 76: 87–126.

Lukmanudin, Ibang, 2001. 'Mari Bung Rebut Kembali, Rakyat Segera Menuntut Hak Atas Tanah'. In *Mengubah Ketakutan Menjadi Kekuatan, Kumpulan Kasus-kasus Advokasi*, Yogyakarta: Insist Press.

Lyon, Margo L., 1970. *Bases of Conflict in Rural Java*. Berkeley, CA: Center for South and Southeast Asia Studies, University of California.

Mas'oed, Mochtar, 1983. 'The Indonesian Economy and Political Structure During the Early New Order 1966–1971'. PhD dissertation, Ohio State University.

McCarthy, James, 2005. 'Devolution in the Woods: Community Forestry as Hybrid Neoliberalism'. *Environment and Planning*, 37: 995–1014.

McCarthy, John F., 2000. 'The Changing Regime: Forest Property and Reformasi in Indonesia'. *Development and Change*, 31 (1): 91–129.

McCarthy, John F., 2006. *The Fourth Circle: A Political Ecology of Sumatra's Rainforest Frontier*. Stanford, CA: Stanford University Press.

McKeon, Nora, Wendy Wolford and Michael Watts, 2004. 'Peasant Associations in Theory and Practice'. Civil Society and Social Movements Programme Paper Number 8, United Nations Research Institute for Social Development.

McMichael, Philip, 2005. 'Globalization'. *The Handbook of Political Sociology: States, Civil Societies and Globalization*, eds. Thomas Janoski, R. Alford, A. Hicks and M. Schwartz, 587–606. Cambridge: Cambridge University Press.

McVey, Ruth, 1965. *The Indonesian Communist Party*. Ithaca, NY: Cornell University Press.

Moniaga, Sandra, 1997. 'Keaslian Suku dari Suku-suku Asli di Indonesia: Dilihat dari Perspektif Siapa?' In *Tanah dan Pembangunan*, ed. Noer Fauzi, 233–51. Jakarta: Penerbit Sinar Harapan.

Moniaga, Sandra, 2007. 'From Bumiputera to Masyarakat Adat: A Long and Confusing Journey'. In *The Revival of Tradition in Indonesian Politics. The Deployment of Adat From Colonialism to Indigenism*, eds. Jamie S. Davidson and David Henley, 275–94. London: Routledge.

Mortimer, Rex, 1972. *The Indonesian Communist Party and Land Reform, 1959–1965*. Clayton, Victoria: Centre of Southeast Asian Studies, Monash University.

Mortimer, Rex, 1973. *Showcase State: The Illusion of Indonesia's Accelerated Modernization*. Sydney: Angus and Robertson.

Moyo, Sam and Paris Yeros, 2005. 'The Resurgence of Rural Movements under Neoliberalism'. In *Reclaiming the Land: The Resurgence of Rural Movements in Africa, Asia and Latin America*, eds. S. Moyo and Paris Yeros, 8–64. London: Zed Books.

Peluso, Nancy Lee, 1992. *Rich Forests, Poor People: Resource Control and Resistance in Java*. Berkeley, CA: University of California Press.

Peluso, Nancy Lee, 1993. 'Coercing Conservation: The Politics of State Resource Control'. *Global Environmental Change*, 3: 199–217.

Peluso, Nancy Lee and Peter Vandergeest, 2001. 'Genealogies of the Political Forest and Customary Rights in Indonesia, Malaysia, and Thailand'. *The Journal of Asian Studies*, 60 (3): 761–812.

Pelzer, Karl J., 1982. *Planters Against Peasants: The Agrarian Struggle in East Sumatra, 1947–1958*. The Hague: Martinus Nijhoff.

Powelson, J.P. and R. Stock, 1990. *The Peasant Betrayed: Agriculture and Land Reform in the Third World*. Washington, DC: Cato Institute.

Resosudarmo, Budi, 2005. *The Politics and Economics of Indonesia's Resources*. Singapore: The Institute of Southeast Asian Studies.

Robison, Richard, 1986. *Indonesia: The Rise of Capital*. Sydney: Allen & Unwin.

Robison, Richard and Vedi Hadiz, 2004. *Reorganising Power in Indonesia: The Politics of Oligarchy in an Age of Markets*. London: Routledge Curzon.

Rosser, Andrew, Kurnya Roesad and Donni Edwin, 2005. 'Indonesia: The Politics of Inclusion'. *Journal of Contemporary Asia*, 351: 53–77.

Rosset, Peter, 2006. 'Moving Forward: Agrarian Reform as a Part of Food Sovereignty'. In *Promised Land. Competing Visions of Agrarian Reform*, eds. P. Rosset, Raj Patel and Michael Courville, 301–21. Oakland, CA: Food First.

Routledge, Paul, 2004. *Resisting and Shaping the Modern*. London: Routledge.

Rumansara, Augustinus, 1998. 'Indonesia: The Struggle of the People of Kedung Ombo'. In *The Struggle for Accountability – The World Bank, NGOs and Grassroots*, eds. Jonathan Fox and L. David Brown, 123–50. Cambridge, MA: MIT Press.

Ruwiastuti, Maria, 2000. *'Sesat Pikir' Politik Hukum Agraria. Membongkar Alas Penguasaan Negara atas Hak-hak Adat*. Yogyakarta: Insist Press, KPA dan Pustaka Pelajar.

Sangkoyo, Hendro, 2000. 'Pembaruan Agraria dan Pemenuhan Syarat-syarat Sosial dan Ekologis sebagai Agenda Pokok Pengurusan Masyarakat dan Wilayah'. Consortium for Agrarian Reform (KPA) Position Paper.

Simon, Hasanu, 1993. *Hutan Jati dan Kemakmuran. Problematika dan Strategi Pemecahannya.* Yogyakarta: Aditya Media.

Simon, Hasanu, 1994. *Merencanakan Pembangunan Hutan untuk Strategi Kehutanan Sosial.* Yogyakarta: Aditya Media.

Sivaramakrishnan, Kalyanakrishnan and Arun Agrawal, 2003. *Agrarian Environments: Resources, Representations, and Rule in India.* Durham, NC: Duke University Press.

Smith, Neil, 1984. *Uneven Development: Nature, Capital, and the Production of Space.* Cambridge: Blackwell.

Stedile, Joao Pedro, 2002. 'Landless Battalions: The Sem Terra Movement of Brazil'. [Interview by Francisco de Oliveira]. *New Left Review*, 15 (May–June): 77–104.

Terborgh, John, 1999. *Requiem for Nature.* Washington, DC: Island Press.

Tsing, Anna, 1999. 'Becoming a Tribal Elder and Other Green Development Fantasies'. In *Transforming Indonesia's Uplands*, ed. Tania Li, 159–202. Amsterdam: Harcourt.

Tsing, Anna, 2005. *Friction: An Ethnography of Global Connection.* Princeton, NJ: Princeton University Press.

Wargadipura, Nissa, 2005. 'Bekerja Bersama Anggota Serikat Petani Pasundan Dalam Mempengaruhi Kebijakan Reforma Agraria'. Paper presented at Forum Pengembangan Partisipasi Masyarakat – FPPM national meetings, 27–29 January 2005, Lombok Barat, Indonesia.

Webster, Neil, 2004. *Understanding the Evolving Diversity and Originalities in Rural Social Movements in the Age of Globalization.* Civil Society and Social Movements – Paper No. 7. Geneva: United Nation Research Institute for Social Development.

Wiradi, Gunawan, 1997. 'Pembaruan agraria: Sebuah Tanggapan'. In *Reformasi Agraria, Perubahan Politik, Sengketa dan Agenda Pembaruan Agraria di Indonesia*, eds. D. Bachriadi, B. Setiawan and E. Faryadi. Jakarta: Lembaga Penerbit FE-UI, and KPA.

Zerner, Charles, 1992. *Indigenous Forest-Dwelling Communities in Indonesia's Outer Islands: Livelihood, Rights and Environmental Management Institutions in the Era of Industrial Forest Exploitation.* A report commissioned by the World Bank, Forestry Sector Review.

9 Whose Rules Rule? Contested Projects to Certify 'Local Production for Distant Consumers'[1]

HARRIET FRIEDMANN AND AMBER McNAIR

INTRODUCTION

The new world of transnational agrifood supply chains cuts off farmers from local markets. Farmers accustomed to selling locally, regionally or nationally cannot continue as before. Some farmers, processors, distributors and retailers are incorporated into production contracts within transnational supply chains; others are marginalized; none can rely on stability in a context where supermarkets bring farmers from across the world into competition for contracts – even to supply nearby customers. Small farmers have regrouped in the North through a variety of direct marketing strategies: farmers' markets, Community Shared Agriculture, box schemes and the like are mushrooming but still account for miniscule market share. In the South, supermarkets are expanding so rapidly that it is difficult to track the changes entailed as they disconnect farmers from food purchasers. In Latin America, Reardon and Berdegué (2002, 371) report that already in the decade 1990 to 2000, supermarkets' share of retail purchases in Latin America rose from 10–20 per cent to 50–60 per cent. While resistance takes many forms, it is important to examine creative responses in response to new conditions. While some draw on longstanding cooperative traditions (Reardon and Berdegué 2002, 382), innovative entrepreneurial approaches by organized groups in the food system are seeking to re-embed agriculture and food consumption in socially and ecologically defined regions.

Regional projects draw on the European nineteenth-century innovation of *terroir*[2] and now find institutional support in provisions such as Geographical Indications at the World Trade Organization. While this antecedent is often implicit, it makes clear that alternative projects face the same challenge as large capitals, with different types of resources. To avoid marginalization, local producers of crop and livestock varieties and artisanal foods must improve quality, appeal to consumers and market to distant places (Fonte 2006). The challenge for

---

[1] This is a Slow Food phrase, used by Fonte (2006).
[2] Terroir refers to 'the combination of natural factors (soil, water, slope, height above sea level, vegetation, microclimate) and human ones (tradition and practice of cultivation) that gives a unique character to each small agricultural locality and the food grown, raised, made, and cooked there' (Petrini 2001, 8).

both capital and those seeking alternatives, therefore, is to *make known* the qualities, and increasingly the origins of foods. Old practices of trademarks, brands, seals of approval and certifications become central to supply chains of all kinds. They become an arena of contestation, multiplication, confusion, and therefore open opportunities for creative strategies.

In this chapter we examine local experiments in embedding agricultural markets in local ecosystems within a global context. In doing so, we take a specific angle on the 'food sovereignty' project. Most research on transnational social movements has understandably focused on *resistance* to the displacements of remaining peasantries of the world since the 1980s (Bernstein 2006; Araghi 2000; cf. Pietrykowski 2004).[3] This chapter instead focuses on 'concrete actions and feasible projects' (Petrini 2001, 110) to protect food and farming cultures under threat of extinction as agrifood supply chains incorporate some and marginalize others. While they fit under the big visionary tent of food sovereignty, these projects seek to create or shape *market relations*. Since they cannot remain 'local', a key to their viability – and transformative potential – lies in transnational coordination and facilitation. The projects we describe share with 'fair trade' and related initiatives 'inherent contradictions... between movement and market priorities' (Raynolds and Murray 2007, 223), but the global coordination of ecologically and culturally embedded agrifood systems shifts the emphasis from political mobilization to building of alternative institutions.

The argument falls into four parts. First, we show that two projects for *global* agrifood relations are emerging, both requiring some sort of certification. From above, supermarket-led agrifood capitals, exemplified by the GLOBALGAP consortium, are actively creating flexible and traceable supply chains based on standard norms of quality.[4] From below, Slow Food theorizes and coordinates an increasingly vibrant array of regional organizations experimenting with production, knowledge, innovation and marketing, all under the rubric of quality, which together anticipate an alternative constellation of 'supply chains' embedded in diverse cultural and natural regions. Second, we offer one example from the South (Michoacán, Mexico) of how the two projects interact on the ground. Third, we report on an innovative project in the North (Ontario, Canada), which has broken through some limits to increasing markets for local, sustainable farmers through institutional requirements that giant agrifood capitals purchase

---

[3] Resistance comes from many directions – marginalized farmers, farmers caught in oppressive new contract relations, and even from rural populations massively being displaced by what Araghi (2000) calls a global enclosure of the peasantry. And it has not been restricted to the South. The remarkable emergence within a mere decade and a half of the Vía Campesina (Patel 2007; Edelman 2003; McMichael 2007; Desmarais 2007), for example, expresses an extraordinary solidarity among small farmers in the North, who are willing to identify as 'people of the fields', with much smaller farmers, rural workers and indigenous peoples in the South. Remarkably, this North–South solidarity contrasts sharply with the conflicts between states of North and South in international organizations. It is fascinating that the conflict between states of North and South at the World Trade Organization contributes to an impasse in inter-state regulation, which invites private regulation by agrifood capitals, as we shall see, but also opens spaces for a creative politics of markets from below.
[4] This may be an instance of a wider shift towards 'green capitalism' (Friedmann 2005).

their products. Fourth, we consider possible ways that bottom-up experiments multiplying across the globe and fluently linked in global networks might transform agriculture and food on a world scale. We adopt the unconventional strategy, therefore, of putting theory at the end of the chapter, hoping to guide the reader along the inductive path we have followed. To anticipate, we indicate a fresh approach to Polanyi's concept of embeddedness, linking it to Wright's (forthcoming) updating of Marxian conceptions of social transformation.

CERTIFICATION RULES, BUT WHOSE?

The politics of 'quality' alter conflict and negotiation in the agrifood system. A growing literature explores how 'food audit processes are implicated in the neo-liberal governance of agriculture and food' (Campbell and LeHeron 2007, 133). Auditing and certification of specific qualities has a fascinating lineage, but always at the margins of the food system. Kosher labels in New York at the turn of the twentieth century (Glaswirt 1974, summarized in Campbell and LeHeron 2007, 136–38) now find a fascinating parallel in creation of Halal certification in North America.[5] Organic certifications are intensely contested with the rise of industrial organics (Campbell and LeHeron 2007; Lyons 2007; Guthman 2004). The emergence of global supply chains takes marginal cases to the centre, entailing 'complex cultural micro-politics of food auditing, which brings together important and powerful players like supermarkets and corporate producers in negotiation with public communities of interest who possess the ability to legitimize and delegitimize cultural values of food' (Campbell and LeHeron 2007, 147). As governments cede regulation of food to private organizations, the 'cross-hybridization of public-private standards' leaves all but minimal hygiene to private organizations (Barling and Lang 2005, 41). Without minimizing the power imbalance between the corporations on one side, and consumers, citizens and farmers on the other, the point is that conflicts and negotiations now occur at both ends of supply chains, linking producers and consumers into highly audited private systems whose rules are difficult to stabilize.

Two new regulatory phenomena are at the strategic centre of play between standard and diverse institutions of certification: 'Third Party Certification' (TPC) and 'traceability'. TPC arose from two changes, in addition to trade liberalization: (1) devolution of government food regulation; for instance, the UK government substituted auditing of retailers' documents for on-site inspections (Marsden et al. 2000a), opening the way for other non-governmental forms of 'audit'; and (2) social movements that pioneered special farm products, such as organic and fair trade, which required organizational labels to identify their

---

[5] A label reflecting specific rules for diasporic Muslims, e.g., acceptability of machine (in addition to hand) slaughter has been created by the Islamic Society of North America (see http://www.isna.net). Their rules may be challenged by others groups, but we have not been able to find references on other websites, e.g., the Canadian Council of Muslim Theologians.

provenance to consumers. Thus, new standards, such as United States Department of Agriculture organic standards, are implemented through third party certifiers, which are authorized by the government. This opened the way for governments of the global South to 'benchmark' their public standards to both intergovernmental ones (Codex Alimentarius) *and* to standards set by private capitals, especially when organized into consortia.

Traceability is the goal of TPC: to make it possible for capitals and (sometimes) consumers to know the qualities and the origins of foods. The idea and some of the techniques were pioneered by small producers seeking to differentiate their products and ensure quality. A legacy of the last century is Denomination of Origin (DOC), for wines and other regional products in Europe. This is being transformed through new trade and intellectual property laws, and is now enshrined in the World Trade Organization Agreement on Agriculture as Geographical Indications (GI). Yet it is also, as we shall see, the basis for global traceability based on characteristics of products specific to ecological and cultural places. More recently, organic movements created various certification organizations and marks to differentiate and guarantee products according to the various criteria of certifying bodies. Certification rules and organizations have proliferated, undermining the purpose of informing consumers and protecting producers. Governments are very much in the mix: they are facilitating traceability systems[6] and 'benchmarking' their governmental regulations to private standards. The European Union is advancing research that could potentially lay the basis for inter-governmental regulation through GI at the WTO, based on 'the European model of agriculture'.[7] However, the main play is between private capitals and social movements (Busch et al. 2005).[8]

Standard Standards: Uniform Varieties from Everywhere and Nowhere

Supermarkets rose to dominance in the agrifood sector in the 1990s in a pincer movement. On the supply side was an explosion of world trade in fresh fruits, vegetables and fish, which began as a direct result of the debt politics of the 1980s. 'Structural adjustment' imposed by the World Bank and the International Monetary Fund on indebted governments included mandatory exports in any

---

[6] For example, in Ontario, see http://www.ontraceagrifood.com. A deconstructionist could write at length about new relations between government and private sectors, which go much deeper than 'partnerships'.

[7] The EU sponsors the SINER-GI project to 'maximize GIs contribution as a driver for rural development and consumers' confidence'. It is directed to new and prospective members, but also to support their legitimacy in the framework of the World Trade Organization (WTO) negotiations (see http://www.origin-food.org/2005/base.php?cat=20).

[8] Faced with unfavourable decisions in trade disputes concerning food safety and quality regulations, at the moment when its moratorium on genetically modified crops was allowed to expire, the European Union announced the intention to require traceability. Whether or not this was a 'restraint on trade' as other governments argued, impasse at the WTO meant that the game moved to supermarkets. As safety scares from industrial foods, including livestock and monocultural field crops, came to impinge on supermarkets from customers, they moved quickly to implement technologies for tracking and auditing along supply chains.

way possible. Many governments of the global South pushed land and labour towards exports of 'non-traditional' crops (compared to 'traditional' mass produced tropical crops, such as bananas, coffee and sugar, established in earlier centuries). As a result export promotion came to trump 'food security' in the hierarchy of national goals (Friedmann 2004). Fresh fruits, vegetables and fish became established as new spheres of global accumulation. Trade agreements in the 1990s greased the channels supermarkets were already opening between consumers and producers of fruits, vegetables and fish.

Supermarkets created highly controlled networks of contractors and subcontractors across the world (Burch and Lawrence 2007). They thus created highly controlled 'transnational supply chains'. The non-traditional crops which constitute most of these transnational supply chains are now distinguished from the undifferentiated mass-produced crops, such as maize and soybeans, which oddly (and not theoretically) came to be called 'commodities'.[9] The latter are the main ingredients composing the illusory variety of processed foods (Pollan 2006), but they, too, are coming to be differentiated, controlled and monitored according to highly specified criteria and norms, under such rubrics as 'identity preservation'.

On the demand side, incomes became more unequal in all countries beginning in the 1980s, opening the way for niche markets for 'quality' foods. Differentiation into rich and poor food systems was exemplified most vividly by the rapid growth of two new types of supermarket, Whole Foods and Wal-Mart (Friedmann 2005). Other supermarkets introduced their 'own brands', for which they contracted to anonymous manufacturers (Burch and Lawrence 2007). Quality lines came to be defined by fresh (and freshly prepared) foods, including introduction of exotic and counter-seasonal products, which were sold at higher prices than 'durable foods' of the manufacturing companies, whose dominant ingredients are maize and soy, recomposed to create various starches, textures, flavours and proteins, and now oddly called 'commodities' in contrast to 'quality' or 'niche' foods. New Zealand was a pioneer in shifting towards niche products. The first to remove agricultural protection from its farmers (something still resisted in other 'rich' export countries such as the USA and EU), the government of New Zealand launched an export shift from 'commodities' (dairy and meat) to 'quality' products; an early success was to promote a fruit barely known in the West, renamed 'kiwi', followed by wines and the like. Fresh fish and seafood exports similarly grew in importance, as did aquaculture and capital-intensive fishing to supply international demand. As incomes became more unequal across the world, proportionally small but absolutely large groups of privileged consumers in the global South provided new investment opportunities for supermarkets, which extended their retail operations to the point where supermarket sales of fresh produce in the global South are larger in the aggregate than the

---

[9] This partly stemmed from need to segregate genetically modified crops, particularly for the European market. However, uses of grains as raw materials for many complex processed final goods, including plastics, paints and now fuels, are requiring increasingly specific varieties. On the peculiarity of this usage, see Bernstein and Campling (2006a).

quantity they export to the global North. As the distinction between high-quality exports and low-quality domestic agricultural products diminishes, supermarkets are increasingly interested in merging domestic and export standards (Reardon et al. 2005, 47–56).

These transnational supply chains changed the relationship between agrifood capitals and consumers. In a context where inter-governmental impasses limited the scope for governments to respond to citizen demands on food quality, supermarkets became the front line of guaranteeing safety and quality. As agrifood commodity chains became, in the language of commodity chain analysis, 'buyer-driven' (Gereffi and Korzeniewicz 1994; Busch and Bain 2004), issues of trust (Callon et al. 2002; Wilkinson 2005) impinged directly on supermarkets. Food safety scares in the 1990s were exacerbated by consumer and citizen concerns about genetically modified organisms. Ethical concerns (expressed, e.g., through fair trade) and environmental and health issues (e.g., organics) came to the fore. Supermarkets began to invite their customers – in both North and South – to trust their authority to define and guarantee safety and quality (Dixon 2007). Partly this was through their own labels, such as in Canada, the pioneering 'green' and 'cultural' lines of giant supermarket chain Loblaw's called 'President's Choice'. For fresh and minimally processed foods, which are less easily differentiated by supermarket brands, quality controls had to be instituted along the whole of the networks of contractors and subcontractors spanning the globe. In the power relations reaching from corporate offices to farms across the world, supermarkets moved to anticipating and shaping consumer demands (Barling and Lang 2005, 43).

Inter-governmental standards failed to keep up with transnational agrifood supply chains (Busch and Bain 2004; Barling and Lang 2005). Standards, today as in the past, bring uniformity to products and facilitate shipment and use across cultures, geographies and human practices (Tanaka and Busch 2003, 30, 32). The genetic base of avocados today, like that of bananas a century ago, is narrowed. The difference from the past is that no single national state and food culture dominates, as it did when Britain (and more widely Europe) was the overwhelming destination of imports in the nineteenth century. With the necessity of inter-governmental standards, and the stalemate among powerful governments about what they should be,[10] supermarkets needed to find their own ways to control at a distance.

This led to increasing efforts for private oversight of activity in distant locations through standardizing products and processes (Konefal et al. 2007, 295). The fluid, highly controlled networks of subcontractors reaching across the world needed to be monitored for adherence to increasingly specific and uniform protocols. And in the current context of contested qualities – from genetic modification to pesticides to sanitation – the penetration into farming systems is

---

[10] The possibility for inter-governmental standards was introduced in 1995 with the authorization of the Codex Alimentarius Commission as the arbiter of permitted restrictions on trade, by the Agreement on Agriculture of the World Trade Organization. These were composed of 'Sanitary and Phyto-Sanitary' (SPS) measures, one of several permitted 'Technical Barriers to Trade' (TBT).

deepened, with attention to a wider range of criteria related to production and products. The technical procedures for creating standard practices by producers in multiple, distant sites is a formidable challenge.

The 'market regulation of agrifood supply chains' (Konefal et al. 2007, 284) is not an easy task to accomplish.[11] It requires coordination of agrifood capitals within major markets, which differ according to definitions of quality (EU vs US on hormones in meat, on organic standards and more). It thus also requires a new relation to governments and inter-governmental regulation (McMichael and Friedmann 2007). These are beyond the scope of this chapter, but we note them as context for the earliest and most comprehensive private institution to design and implement auditing technologies into farms and ecosystems throughout the world, which was – and is – based in Europe.

In 1997, the Euro Retailer Producer Working Group Good Agricultural Practices (EurepGAP) – a consortium of private retailers, food services, manufacturers and other agrifood industries – was founded to develop standards to facilitate global sourcing. Its goal was to create a common standard for everything from fruit to fish to flowers, as a service to all its members. Supermarkets would use their own brands, as EurepGAP serves as an intra-industry organization. With the intense concentration of the retail sector in Europe (Harvey 2007), a small number of retail chains all adhering to a single standard would cover the continental market. It quickly began to train 'auditors' to monitor farm practices in all parts of the world, to qualify them for purchases for sale in the European market. As EurepGAP standards met or exceeded European Union standards, they could guarantee 'quality' in any other market as well. Farmers all over the world were invited to invest in the practices, both agronomic and documentary, to meet higher standards in hope of receiving higher prices.

EurepGAP was recently reborn as GLOBALGAP (GG), and with it a deepening of emphasis on technical, uniform standards. In 2004, the EurepGAP website proclaimed:

> In responding to the demands of consumers, retailers and their global suppliers have created and implemented a series of sector specific farm certification standards. The aim is to ensure integrity, transparency and harmonisation of global agricultural standards. This includes the requirements for safe food that is produced respecting worker health, safety and welfare, environmental and animal welfare issues. (Eurep, cited in Friedmann 2005, 255)

By 2007, GG is 'a private sector body that sets voluntary standards for the certification of agricultural products around the globe . . . primarily designed to reassure consumers'.[12] GG has created a system for instituting transnational standards governing a uniform vision of 'Good Agricultural Practices'. Its 'aim

---

[11] Private regulation by agrifood capitals seemed likely to emerge as a project in the context of disintegration of the food regime of 1947–73, but its shape only gradually emerged out of both capitalist reorganizations and inter-state politics (Friedmann 1993, 2005).
[12] See GLOBALGAP web site (http://www.globalgap.org).

is to establish ONE standard ... with different product applications capable of fitting to the whole of global agriculture'. GG certifies products from farm inputs until they leave the farm gate, but only as a 'business-to-business label'. GG 'consists of a set of normative documents ... General Regulations, ... Control Points and Compliance Criteria', used by 'independent and accredited certification bodies [TPC] in more than 80 countries'. To deal with problems of multiple certification schemes, each requiring an audit, GG encourages 'national or regional farm assurance schemes' to benchmark so that they can be recognized by GG. Thus, the Mexican government, which has benchmarked national regulations for its *Mexico Calidad Suprema* label to GG, can accredit its products and be recognized by GG.[13] 'The GLOBALGAP benchmarking process can be compared to a filter system, which qualifies and harmonises different standards around the globe'. Benchmarking is accomplished by both 'member peer review' and 'an independent witness assessment'.[14]

Diverse Standards? Certifying Foods, Certifying Regions

The advances of GLOBALGAP have been met in global ambition and organizational innovation, if not in scale, by a remarkable project spearheaded by Slow Food to 'raise quality' of regionally specific crops, livestock and prepared foods, and to assist in making their merits known to customers near and far. Any certification scheme faces complex and changing politics of defining 'quality' and guaranteeing origins and production processes (Campbell and LeHeron 2007). Yet Slow Food explicitly challenges standardization itself and understands the contest to be global in scale. To do so, it takes on the ambitious project to displace the 'double fetishism' in which 'knowledge' is 'sold' in addition to the commodity itself (Freidberg 2003, cited in Bernstein and Campling 2006b, 430); this 'knowledge' conveyed on supermarket labels has evolved into a literary form called 'supermarket pastoral', which gives a very different impression from what an intrepid investigator discovers in actual production sites (Pollan 2006). Slow Food qualifies not only the product or commodity alone, but its cultural and ecological context. Regions of the world are defined as unique cultural bioregions, worthy of appreciation, criticism and support, and in need of mutual aid in order to reverse loss of biological and cultural diversity, which it sees as intertwined threats to the survival of humanity. The products of these regions must become commodities to survive, but the effort is not simply to improve their characteristics but to guarantee their embeddedness within unique contexts. A Presidium[15]

---

[13] See What is GLOBALGAP (http://www.globalgap.org/cms/front_content.php?idcat=2).
[14] See GLOBALGAP – Benchmarking (http://www.globalgap.org/cms/front_content.php?idart=44&idcat=29&lang=1&client=1).
[15] The Presidia is the 'working arm' of the Ark of Taste of the Slow Food Foundation for Biodiversity (see footnote 19). Presidia are small projects which help groups of artisanal producers to 'establish quality and authenticity standards for their product[s] ... stabilize production techniques; to establish stringent production standards and, above all, to guarantee a viable future for traditional foods' (see http://www.slowfoodfoundation.com/eng/presidi/lista.lasso).

product is for sale at a distance – i.e. for export in international terms; however, in contrast to classical export crops such as coffee, its origins lie in the agronomy and food culture of its region, and it is the latter which is the intended object of regulation.[16]

The project is based in a different European experience from the supermarket sector that birthed GG, *terroir* and Controlled Denomination of Origin (DOC). It departs from the DOC model in two ways. First, it does not freeze production techniques. Slow Food grew out of the initial experiment of Carlo Petrini and his colleagues to 'build . . . a territory' in the region of Langhe, Italy, where the city of Bra, home of Slow Food, is located.[17] It consisted in bringing together wine producers, restaurants and journalists to plan for renewed commerce based on quality foods. The plan was implemented through catalogues of products, oral histories of individuals and production sites, such as vineyards and cellars, emphasizing distinctiveness of each within the distinctiveness of the region, and – most innovative – education of farmers, vintners, and others based on combinations of formal knowledge, e.g., of animal breeders, and practical knowledge of farmers, artisans and cooks.

Second, these regions, which have multiplied across Italy and later the world, rarely correspond to political jurisdictions. 'Every single product defines and shapes a space of its own, and together they make up a geographical mosaic with multiple intersections' (Petrini 2001, 40). DOC (and in France, *Appellation d'Origine Controllée*) has been guaranteed by national states as part of the protectionist legacy of the nineteenth century. Intellectual property has become complex when the denomination is claimed by those outside the country, for instance, through transplanted crop varieties. For this reason, and for others stemming from still unresolved disputes at the WTO, the EU is promoting international implementation of Geographical Indications (GI), which also seem promising to governments and producers in other countries. Whatever happens with GI, however, Presidia and their guarantor in the Slow Food Foundation for Biodiversity, work differently, at least so far. Learning from the success of Parmigiano Reggiano cheese – which in the 1980s came to be sold at premium prices throughout the world precisely because it could guarantee specific production protocols from breed of cow to pasture to ageing conditions – they have created ways to institute quality guarantees for specific products and build networks of appreciative customers, both local and distant.

In this view, regionally distinctive crop varieties and livestock breeds, as well as artisanal foods, can only survive by finding customers. Appreciative and knowledgeable customers, in turn, require not only 'educated taste' – something

---

[16] Whether or not this can succeed, it avoids some of the problems analyzed by Bernstein and Campling (2006b).
[17] The Slow Food Movement has its origins in the Italian left in the 1970s. A group of young radicals aimed to raise awareness of local products, drawing attention to the appreciation of food and wine – linking culture and gastronomy to issues of land and labour, production and consumption. Arcigola, Slow Food's predecessor, was founded in Bra, Italy in 1986. The International Slow Food movement was inaugurated in 1989 with the signing of the Slow Food Manifesto (Petrini 2001).

Slow Food initiated before considering crop, livestock and food diversity on a world scale. It also requires definitions of quality, distribution to customers and guarantees of maintaining defined quality standards. This in turn requires that farmers and artisans be valued in social esteem and money, and encouraged to innovate. All these requirements were addressed in the Manifesto of the Ark[18] in 1997, and instituted through the Presidium, an organization invented and coordinated by the Slow Food Foundation for Biodiversity (SFBD), founded in 1999. Since 2003, Presidia are created at the initiative of local groups (not only producers) to define criteria for specific quality products 'with commercial potential' (Petrini 2001, 92). Each Presidium defines a specific product, either a crop or animal variety, or an artisanal product made from it, or both; it defines a specific combination of the boundaries of the region from which it may come, the people who may produce it, and the specific techniques which may be used. In contrast to GLOBALGAP, a Presidium defines a unique product such that it encourages innovation by local producers (within negotiated parameters), encourages horizontal sharing of knowledge through international gatherings of Presidia (e.g., by goat cheese makers in different regions), and stretches the concept beyond 'terroir' to include crops and foods introduced to new ecosystems (e.g., unpasteurized cheeses across the US, which belong to a single Presidium).

Of course, markets are not easily bounded, so SFBD works with the tension within 'short food supply chains'. The defining feature of the latter is how much knowledge about the product is embedded within it rather than geographical distance or number of links in the supply chain (Marsden et al. 2000b). By emphasizing knowledge, finding a way to incorporate it in specific regional products, and guaranteeing that quality definitions are adhered to, SFBD overcomes some of the problems of standardization found in long distance shipments of 'organics' (Guthman 2004; Raynolds 2004). Fonte argues that 'what really matters in the local agrifood system is the embeddedness of the food network in the territorial context . . . The territory (geographical proximity) is the cement, the support of a common history and a common belonging, that is solidified in collective values, norms, regulations and in co-ordinated economic activities (socio-economic proximity)' (2006, 3). To this end, SFBD accepts applications from territorially based groups, including but not restricted to producers, to create a Presidium for specifically defined crops and livestock varieties and artisanal foods at risk of extinction. The Ark of Taste acts as a database to catalogue at-risk foods across the globe. Presidia assist producers by bringing them together, documenting their histories, establishing production techniques, sometimes helping to refine the quality of production, building capacity and even infrastructures such as farm houses or abattoirs. Presidia formed a network coordinated by SFBD.

By operating transnationally, and not allowing its snail logo to be used by Presidia, SFBD seeks to overcome the 'label fatigue' experienced by consumers

---

[18] The Ark of Taste is a project of the Slow Food Foundation for Biodiversity to catalogue 'forgotten flavours' – species, breeds and artisanal practices to help save them from extinction (see http://www.slowfoodfoundation.com/eng/arca/lista.lasso).

in a world of proliferating certifications (Goodman 2003, 10). At the same time, its founding attention to 'educated taste' adds bodily pleasure of eaters/drinkers to the charitable appeal of, for example, fair trade brands and their corporate imitators; the attempt by Slow Food to transfer through 'education' the embedded pleasures of *terroir*, which once required actually being there, seeks to move beyond the 'voyeuristic' knowledge proclaimed in appeals to 'ethical consumers' and their testimonials (Goodman 2004, 900). A big step in this evolution was an agreement signed between Slow Food and the Coop Italia, the largest supermarket in Italy, to market Presidium products. Coop tries to take on, in practice and not just image, closer ties to producers. This is mediated via Presidia. Slow Food, in another parallel to GG, does not allow its logo to be used on Presidium products; however, in this case the background certification supports territorial embeddedness of crops, livestock, farming systems, processing techniques and therefore, it is hoped, complex food cultures.

The agreement between Slow Food and Coop Italia helps stimulate demand for products but Slow Food assures that this will take place at a pace and within the viable limits of small-scale, artisanal production, thus guarding against the specific challenges that often greet efforts to 'scale-up' niche production and marketing (Goodman 2003). Coop Italia, for its part, educates consumers, trains its staff and hosts tasting events to help cultivate awareness of its adopted Presidia. They capture, too, the hybrid consumer who purchases typical, artisanal foods in addition to their regular purchases. This agreement offers an opportunity to large food retailers to stock local products, deal with small producers and, against the dominant trend of global expansion of supermarkets and their restructuring of supply chains, localize supply and distribution channels. It offers the possibility of 'change in logic of the big retailing firm, away from a mass market, towards a universe of niche markets, territorially defined' (Fonte 2006, 16).

Whether or not Slow Food can extend this marketing strategy to the rest of the world, it offers an example of how a world market of niches might emerge and be coordinated. SFBD is trying to do this. It has strategically expanded its vision and network to the South – home to most of the world's biodiversity. Only four years since the Presidium was invented, 41 Presidia exist in 21 countries of the South (Slow Food Foundation 2007). In November 2007 Slow Food held its international congress outside of Europe for the first time in Puebla, Mexico (Slow Food 2007). As has already happened in Italy, biodiversity, food and farmers' rights activists from the North and the South appeared together on a panel (V Slow Food International Congress Agenda). Vandana Shiva and Carlo Petrini earlier joined forces with other theorists and activists from across the world to issue a joint Manifesto on the Future of Food, and its 'world meeting of food communities' called Terra Madre ('mother earth') meets on alternate years, and brings producers from indigenous and other regional communities together for mutual exchange and exposure.[19]

---

[19] See http://www.slowfoodfoundation.com.

TWO STRATEGIES MEET IN THE MICHOACÁN HIGHLANDS, MEXICO

In the highlands of central Mexico, a bioregional label is designed to support the re-building of indigenous agriculture threatened by conversion to export farming.[20] The bioregional label has emerged in the context of the implementation of phytosanitary standards and Good Agricultural Practices (GAPs), which are rapidly and profoundly transforming a farming system toward uniform exports. Reorganization of farming for avocado exports has marginalized the multitude of historical varieties in favour of a standard one, and has changed both agronomic and social relations of production for all crops including maize. Avocado export production, by endangering water and forest resources, deepens the threat posed by maize imports to subsistence production by the region's indigenous communities.

The Global Avocado

Michoacán's avocado sector has been growing dynamically since the 1970s. The Hass variety of avocado developed in California was brought to central Mexico, where it adapted brilliantly to the region's climatic conditions and eventually replaced preferences for native varieties (Stanford 2002).[21] Through the 1970s, the Mexican government's food policy could not keep pace with fast-moving events connected with the oil and food crises. Mexico's oil boom increased incomes and demand for food, but government attempts to combat inflation by keeping food prices down encouraged commercial farmers to move away from basic foods toward higher-value crops (Scherr 1985, 13–14). Those with sufficient resources shifted to producing fruits, vegetables and livestock. In the 1980s, after Mexico's default precipitated a globally recognized 'debt crisis' and drastic contraction of domestic purchasing power, these farmers began to search for export opportunities. Mexican avocados reached Europe, Japan and Canada but were blocked from the nearby US market because of a phytosanitary ban in place since 1914.

Access to the US and later to more demanding world markets than those which first imported avocados in the 1980s was pursued through a phytosanitary campaign which transformed disorganized and variable quality production for the national market into the world's largest quality avocado-producing region (Stanford 2002).[22] Phytosanitary regulations over plant diseases, pest control and pesticide use are overseen by municipal plant health boards, in concert with US

---

[20] This section reports preliminary results from fieldwork conducted from February through June 2007.
[21] For a long time Mexicans still preferred the native *criollo* variety. The prominence of Hass was shaped by production and marketing channels, which preferred the Hass's transportability and longer shelf life made possible because of the fruit's thicker skin.
[22] Fieldwork, 2007.

Department of Agriculture inspectors.[23] The engineer in charge of the state phytosanitary campaign explained that it was intended to correct complete absence of regulations for avocados. 'Irregular production'

> provoked unstable prices because those producers who didn't put money into their orchards sold cheaply.... [S]o this situation provoked ... standardization – that there be a law. And also ... the harvest was really poor. They would harvest with a pole and the fruit would fall on the ground and within eight days they would be rotten, really black from the hits.... So, all of this induced the making of the standard. The standard addresses how to harvest, how to transport with the end of improving the quality, how to store them and also of pests and all of that. And so the standard speaks to the whole process: production, packing and transport.[24]

Increasingly, Good Agricultural Practices are also required in the fields. In 1999 Mexico's Secretary of the Economy, Secretary of Agriculture and Bancomext (a national bank whose objective is to promote Mexican business abroad) created *Mexico Calidad Suprema* (Mexico Supreme Quality) – a label to promote the country's highest quality produce in select higher-end venues at home (such as Wal-Mart) and abroad. In 2006, this label was benchmarked to GLOBAL/EurepGAP, i.e. aligned government standards to comply with the private consortium of retailers. Now certification has moved from packinghouse to farm. There are 11,000 ha of avocados currently in process and should be certified by late 2007.

The avocado sector is universally viewed as exemplary by Mexican government officials responsible for promoting high-value exports – for the organization of avocado producers, eradication of pests and diseases, implementation of GAPs and downstream Good Practices, increased price and volume of exports.[25] Such is the success of this 'green gold' that people are trying desperately to cash in on it. Growers are buying up land – a hectare here, two or three hectares there – even purchasing communal (*ejidal*) lands, as soon as that was permitted by a 1992 amendment to the Constitution. Forests are denuded to plant avocado trees. They are being planted in place of other crops, even on lands that cannot support avocado production for climatic or ecological reasons.[26]

This great success for some creates socio-economic and ecological dangers for others. The reliance on one variety, not only in Michoacán but through all avocado-growing regions of the world, increases susceptibility to plagues and

---

[23] Interview, Direccion General de Inocuidad Agoalimentaria, Acuícola y Pesquera. (General Direction of Food Safety, Fisheries and Aquaculture) 9 April 2007.
[24] Interview, MA, Uruapan, 14 March 2007.
[25] Interview, RG Uruapan, 4 May 2007.
[26] One day while visiting an *ejido* in the *Tierra Caliente* that produces blackberries, a grower explained how his neighbour across the road (not part of the *ejido*) had been trying to cultivate avocados despite the inappropriate environment and soil conditions. The soil was shallow and could not support a deep root system. So he purchased top soil and planted an avocado orchard anyway. The roots had now reached their limits and the owner was trying to sell the orchard before the inevitable became too apparent.

blights that could wipe out entire regions.[27] There is pervasive awareness of regional climate change that has followed the widespread planting of avocados. Clear-cutting of pine forest in favour of avocado production uses up underground aquifers at accelerated rates in contrast to the water-preserving qualities of pine forest.[28] Soil and water are polluted by agrochemicals – allowed by GAPs, which focus on consumer health – and dependence on export markets leads to new vulnerabilities.

Convergent Threats to Indigenous Subsistence

The area known as the Meseta Purépecha in Michoacán's western highlands is inhospitable to commercial farming. The Meseta is populated by indigenous *Purépecha* who live in agrarian communities throughout the region, mainly at altitudes higher than 2,000 metres, which do not suit avocados.[29] Because of the mountain and volcano-pocked territory of Michoacán, radically different ecosystems exist in close proximity, including not only avocado orchards, but also those below their altitude range, which include mangos, limes and berries. Thus, *Purépecha* communities and communities in lowlands are interwoven with some of the world's most productive avocado orchards.

In contrast to nearby avocado export areas, *Purépecha* communities engage in subsistence and small-scale farming of typical products, especially maize, squash and beans (in a farming style referred to as the *milpa*), using family labour. The varieties of maize in the region are part of the mosaic of genetic diversity in America, the centre of origin of maize. Variation is the result of millennia of adaptation by farmers to changing geography and cultural uses. Countless ecological niches, each with unique climatic conditions, have led farmers in every locality to continuously adapt local varieties (Nadal 2006, 34; Mijangos-Cortés et al. 2007). Communities have selected and preserved diverse types of maize for distinct uses (Mijangos-Cortés et al. 2007, 309). Due to scarcity of water in the region, crops are not irrigated, but use water retained in the soil from the rainy season. Few external inputs are used in traditional farming. Seeds are carefully selected by farmers, saved, traded and re-planted (Mijangos-Cortés et al. 2007).

Milpas are labour intensive but can be cultivated across rough landscapes providing food security in an environmentally sustainable manner (Nadal 2006, 35). Average yields in Mexico are low when compared to the heavily subsidized, highly mechanized monocultural corn production in the United States. But the concept of efficiency measured in terms of yield fails to capture the complex reality of maize production in rural Mexico. Structural adjustment reforms in the 1980s and the North American Free Trade Agreement in 1995 led to declining

[27] Interview, SMA Uruapan, 10 April 2007.
[28] Interview, SHD Morelia, 4 June 2007.
[29] In Mexico, indigenous communities have a legal status involving collective land tenure that is distinct from the *ejidal* system.

government support for domestic maize production. It is now cheaper to import US yellow corn despite its inferior nutritional value (de la Tejera 2007). This has led to fear of contamination by genetically modified US maize,[30] which can enter the country despite a Mexican moratorium on domestic planting. GM corn has been distributed as food to locals, and in 2000 native landraces were found to contain GM traces (Quist and Chapela 2001; Ribeiro 2004), possibly from local planting of distributed maize. Diversity in the centre of origin is at risk.

Michoacán is facing severe deforestation. It has lost more than half of its native forests since 1963. Nationally, rural policy has always favoured agricultural expansion over forest preservation (Jaffee 1997, 3). Forestry makes an important contribution to the country's biodiversity, to regional ecology, and especially to protecting water levels in underground aquifers. It is also a major economic activity for *Purépecha* communities, where an important source of employment is woodworking in family or community workshops, using forest wood to make, among other things, wooden crates used in packing avocados. In addition to deforestation that occurs in order to plant avocado orchards, in harsh economic times many turn to illegal logging and clear-cutting without reforestation (Jaffee 1997, 15; Klooster 2003; Works and Hadley 2004, 26). As forests are denuded wherever the possibility might exist to grow avocados, the maintenance of remaining forests becomes even more important. Since in Mexico most forests are held communally by *ejidal* or indigenous land-holding systems, these communities must be able to make a sustainable living on their lands so that they will not have to exploit natural resources in destructive ways.

Coyote Rojo: Protecting Indigenous Varieties Through Sale

Against this backdrop *Coyote Rojo* has emerged. *Coyote Rojo* ('Red Coyote', or *Jiuatsi Xarhapiti* in *Purépecha*) is an organic bioregional label, which began to certify producers in August 2007. Its purposes are to safeguard and promote biodiversity, uphold cultural practices of seed saving, methods of production of crops and typical foods, to protect natural resources (water and forest) and sustainable means of harvesting them (*Coyote Rojo Reglamento*). In contrast to *Mexico Calidad Suprema* – the government label benchmarked to GLOBALGAP – it is not the farm but the natural bioregion that is addressed.

The standard is elaborated by the producers who participate in the label. It was initiated and is orchestrated by a visionary named Fulvio Gioanetto, an Italian immigrant who has lived in the region for 11 years. Gioanetto heads the Mexican office of international organic certifier Bioagricert (based in Italy), and

---

[30] Thirty per cent of Maize consumed in Michoacán comes from the United States (see http://www.lajornadadeMichoacán.com.mx/2007/01/12/index.php?section-finanzas&article=009n2fin). In 2006, 61 per cent of all corn planted in the US was of a genetically modified variety (see http://www.csmonitor.com/2006/0831/p15s01-sten.html).

is himself a farmer of native maize. He writes for *Il Manifesto* – the Italian publication, which first published the ideas of the founders of Slow Food (Petrini 2001, 7), marking his intellectual connection to the international movement.

Producers are certified by third parties to comply with Mexico's national organic standard and must document production processes with *Coyote Rojo*. Since few inputs are used in traditional agriculture in this region, certifying for organic production is a way to express longstanding agrarian practices in contemporary terminology.

Coyote Rojo broadens the scope of *terroir* by shifting to the bioregion. Bioregional farming communities take advantage of renewable sources of energy, promoting and preserving organic agriculture. *Coyote Rojo* protects their ability to continue despite threats to the ecosystem through development of businesses based in local skill, knowledge and capacity.

The *Coyote Rojo* label is a way to confront some of the many challenges facing indigenous communities with farming systems centred on maize. Hundreds of maize varieties specific to localities, representing Mexico's single greatest cultural symbol, have a higher value to consumers if they can be identified. If indigenous varieties can be valorized, then people can remain on the land and continue to renew precious knowledge of how to work *milpas*. By promoting commercialization of local maize, the project hopes also to protect farming practices unique to the region, to defend and replenish dwindling water resources and to protect local animals and many crops other than maize. Sustainable harvesting practices may allow villagers to earn a living while protecting forests. 'Bioregionalism supports the fight to preserve, restore and improve the life of localities which constitute the planet' (Coyote Rojo n.d., 2).

Commercialization requires consumers, to whom the label offers a quality guarantee on traditional products, including aesthetic and organoleptic (sensory) qualities. 'We are creating a market and this label guarantees to the most sensible, aware and refined consumers, not just excellent production for their health, quality of life and the environment but at the same time to assure authentic flavours and original beauty' (Coyote Rojo n.d., 1). Perhaps echoing urban support for regional foods in Europe, it appeals to the 'market of nostalgia',[31] to many of those who have left their homes for other places. For them and others, it creates an emotional/cultural/historical connection between consumers, producers and the places of origin of native foods. Products are currently available on farms and in local markets (*tianguis*) in several states, and there is discussion of selling in larger venues. Wal-Mart of Mexico has expressed interest, but *Coyote Rojo* has put talks on hold because the label is not yet certifying sufficient quantities to supply the chain.[32]

These developments occur in a national context that has seen extreme contention over Mexico's shift to an export-led agricultural policy. The Chiapas uprising in 1994 (Barry 1995) initiated demands for revision to the North American

---

[31] Interview, SBU San Juan Nuevo, 13 June 2007.
[32] Personal communication, Gioanetto, 24 October 2007.

Free Trade Agreement's chapter on agriculture to exclude free trade in maize and beans. A tortilla crisis in early 2007 (Hamm 2007) attests to continuing peasant and indigenous opposition to national policies. The *Coyote Rojo* bioregional label, linked intellectually to Slow Food and concretely to Mexican agricultural and environmental politics, is a practical project by one community to help safeguard regional biodiversity and cultural practices, the ability of people to maintain themselves and their families on their lands and rebuild regional ecology by marketing their products locally but, importantly, also in more distant markets.

LOCAL FOOD PLUS: A REGIONAL LABEL IN ONTARIO, CANADA

A recent innovation in increasing the market for local, sustainable foods comes from the Toronto region. A characteristic of North American agriculture is that it is based entirely on crops and livestock introduced by immigrants from Europe within a little over 200 years. Immigrants have been arriving with new tastes and crops from other parts of the world for half a century now, but settling in cities. The challenge in this context is to create an evolving regional food system within a typically 'placeless foodscape', which lacks deeply rooted food cultures (Morgan et al. 2006, 196). On one side, cultural mixes continually change tastes and requirements for ingredients; on the other, ageing farmers and loss of prime farmland to urban sprawl make it difficult to renew farming to supply them. This example, therefore, speaks to ecological and cultural embeddedness, which requires both renewal of farming systems and adaptability of relationships, practices and foods to changing cultural mixes (Friedmann 2007a).

Local Food Plus (LFP), formerly Local Flavour Plus,[33] was incorporated as a non-profit organization during the process of negotiating its first contract with the University of Toronto (UofT), which has 60,000 students. It is an innovative Third Party Certifier, which found its identity, people, rules and practices in the course of working with a large public institution to design and implement a food services contract to govern the transnational food service corporations offering meals on two of its three campuses. The request for bids, which took shape over almost a year, required corporations 'to use local and sustainable farm products for a small but increasing portion of meals as certified by LFP' (Friedmann 2007b, 389).[34]

Key features of the constellation which formed over a few months represent creative responses to the tension between movement and market goals. One was the shift in priorities to *local* farmers as the basis for a sustainable food system. Local has long been part of the mix of issues, but is still not a priority for most.

---

[33] The name change was caused by discovery of intellectual property issues in other parts of Canada. It indicates how quickly contradictory forces can arise on every scale in a rapidly changing political economic and legal context.
[34] Most of what follows is taken from an article written by Friedmann (2007b), which uses the metaphor of ladders to capture the particular way that LFP has adapted 'continuous improvement' models from antagonistic to collaborative.

Organic farmers scrambled for any markets in the face of 'industrial organic' imports and the persistent practice of supermarkets to use long-distance supply chains, especially from California. Food security and anti-hunger organizations, which are the oldest and most established parts of the food movement, included sustainability as a secondary goal. One of these, FoodShare, nonetheless provided crucial institutional support for LFP in its formative stages, especially LFP's founding president Lori Stahlbrand. Stahlbrand (2003) herself came to prioritize local food through work and contacts from environmental organizations. Early formulations in conversation with US organization Food Alliance had defined 'eco-label' without reference to distance travelled, or ecological and cultural contexts. As a result, Food Alliance quickly came to include large, monocultural export firms such as Cascadian Farms, which actually displaced local crops, such as berries, from Ontario markets even in season! Stahlbrand therefore shifted the pivotal concept from 'ecological' to 'local'.

LFP created a set of six standards, including not only sustainable agronomy, but also labour standards, wildlife management, energy, animal welfare and, above all, proximity. Each of these has a minimal requirement, with a point system, which encourages individual improvement and allows LFP to continuously raise the standards. It includes experiences learned from integrated pest management and elsewhere, which substituted a collaborative, problem-solving relationship between producers and certifiers in place of the all-or-nothing certification/decertification powers of organics inspectors. Continuous improvement by the corporations that won the bid is encouraged by contractual incentives to increase the proportion of LFP-certified products, which translates into continuous growth in demand. At the same time, the continuous improvement model replaces hierarchical with collaborative relationships along its supply chains.

The UofT contract immediately led to institutional support. Foundation grants followed, particularly from the Greenbelt Foundation, itself created as part of an innovative (in North America) Ontario provincial programme begun in 2005. LFP fitted perfectly with the Greenbelt's mandate to protect farmland from conversion to other uses.[35] Availability of foundation grants in turn guided LFP to incorporate as a non-profit organization. The large UofT contract, together with the personal contacts of the principals who were deeply embedded in the organics movement and economic sector, and their attention to the burdens on farmers of forms and fees, attracted farmers and processors to certify. Inquiries from educational institutions throughout Ontario exceeded the supply capacities of LFP and even of certifiable farmers and processors, suggesting conditions for rapid growth. In autumn 2007 came a breakthrough into supermarkets, which had persistently eluded these activists, when Fiesta Farms, a small supermarket owned by Italian immigrants, announced it would feature LFP products.

---

[35] For early grants, including to LFP, see http://www.ourgreenbelt.ca/greenbelt-grants/grants-action/ontarios-greenbelt-focus-3-27-million-grants-awarded-greenbelt-founda.

Lacking a continuous agrifood culture, perhaps the meta-culture surrounding LFP can be understood as political. LFP grew out of deep roots in a 'community of food practice' involving social movements, non-governmental organizations and municipal government over two decades. In particular, the Toronto Food Policy Council (and its Coordinator, an official of the municipal Public Health Department) was the crucial bridge linking founder Lori Stahlbrand with key figures from the organics movement – advocates, organizers, TPC inspection and retail (Stahlbrand 2003, 395–6). First three and then four of the principals could work together quickly to create innovative rules and practices for certification of local and sustainable foods, because their paths had crossed and their understandings had evolved through environmental and food security non-governmental organizations, notably the World Wildlife Fund and FoodShare Toronto. Their links to social movements and to farmers were strong enough to allow them to depart from longstanding practices in the sectors from which they came, including organic, and inspire people to join them. Writing of protocols, education and recruitment of suppliers, and building an organization to certify and guarantee products, all proceeded at the same time as farmers and processors were being recruited to certify with LFP. Even as people were meeting for the first time, they built on trust made possible by the social networks and public, private and non-governmental institutions built over 20 years.

Although some of the social activists/entrepreneurs who guided its unfolding are involved with Slow Food, their understandings and practices have evolved independently over more than two decades. The parallel evolution of ideas and practices suggests a wider and deeper global context, perhaps a *zeitgeist*, of which Slow Food is a leading exemplar.

The culture was political rather than 'cultural' in the sense of Italy or Mexico. At the same time, SFBD has been creative in its own relation to Ontario. It has authorized a Presidium for Red Fife Wheat, probably descended from a Ukrainian variety, 'first grown in the Otonabee region of what is now central Ontario in the 1840s . . . [and] reintroduced to its original territory in Ontario in 2005'.[36] Slow Food is one of many institutions and networks seeking to embed food and agriculture in the ecological and cultural context of Toronto and Ontario.

REFLECTIONS: EMBEDDEDNESS AND INTERSTITIAL TRANSFORMATION

What do these examples tell us about agrarian social movements? We argue that they exemplify what might be called the Builder as opposed to the Warrior approach to social change. Slow Food founder Petrini contrasts the 'Slow Food way' with the 'guerrilla fighter' and 'direct action' of Jose Bové, a leading militant and theorist of Vía Campesina (Petrini 2001, 26). Through Presidia, the

---

[36] See http://www.slowfoodfoundation.org/eng/presidi/dettaglio.lasso?cod=267.

Slow Food Foundation for Biodiversity aims to provide institutional support for re-embedding agrifood systems in their cultural and ecological contexts, thus providing alternatives to the stark simplifications and marginalizations imposed by transnational capitals. This is a non-confrontational approach to social transformation, which nonetheless creates myriad nodes in a network. Just as plants growing through cracks in asphalt can eventually replace a roadway with a forest, tiny projects in the interstices of agrifood capitals might potentially – and eventually – become a new way of organizing food and agriculture, at once locally embedded and globally connected. It can be understood via Marx's idea of change emerging in the *interstices* of capitalist society.

Erik Olin Wright (forthcoming, Ch 3, 12) argues that sometimes such scattered and small projects can change the dominant direction of power and exploitation. His idea of *interstitial* social transformation suggests that we 'shift our efforts from building a theory of dynamic *trajectory* [in Marxist theory] to a theory of structural *possibility*':

> The history of the future – if it is to be a history of emancipatory social empowerment – will be a trajectory of victories and defeats, winners and losers, not simply of compromise and cooperation between differing interests and classes. The episodes of that trajectory will be marked by institutional innovations that will have to overcome opposition from those whose interests are threatened by democratic egalitarianism, and some of that opposition will be nasty, recalcitrant and destructive. So, to invoke metamorphosis is not to abjure struggle, but to see the strategic goals and effects of struggle in a particular way: as the incremental modifications of the underlying structures of a social system and its mechanisms of social reproduction that cumulatively transform the system. (Wright forthcoming, Ch 9, 1)

Militancy and prefiguring institutions are thus complementary. Nineteenth-century institutions such as workers' councils and agricultural cooperatives had to defend themselves against capitalist firms and governments, while resistance movements have built social institutions to support their members and ensure continuity across generations. Timing is everything in understanding the balance between the two. We propose that the recent cascade of small, local projects to re-embed food systems is emerging in the wake of anti-globalization, resistance struggles. They do not supplant these struggles, as our description of the massive capitalist transformations of the global agrifood system suggests. But they may reflect a growing acceptance that the old world of national regulation (Friedmann 1993) cannot be recovered. Indeed, new understandings and politics of indigeneity, threats to crop and livestock genetic diversity, and defence of specific agro-ecosystems, among other factors, do not make such a return even desirable.

In the present global context there is little prospect for returning to any 'tradition'. Farmers everywhere are subject to massive transformations of ecological and social conditions. Those engaged in mixed farming systems adapted to local soils, climates, waterways and food cultures find themselves lacking one or more of these conditions. A few may transform into contract farmers of highly specified niche commodities for global supply chains – even for those sold locally in supermarkets – while most are marginalized and unable to produce as before for lack of viable farms and markets. In a convergent trajectory, those engaged in specialized commodities of the postwar years, notably grains, oilseeds, and meat and dairy products, find their products devalued as ingredients to globally sourced and marketed manufactured foods – as well as changes in trade rules. Vía Campesina is a global network that both defends small farmers and diverse agroecosystems in the North and South, and fights specific threats to them, such as genetically modified crops, intellectual property and trade rules favouring access by capitals to land and other resources. The Presidium and the Slow Food Foundation for Biodiversity, as Petrini argues, represent a different balance between institution-building and militancy from the one that led to the founding of Vía Campesina. The Presidium is a flexible form that accommodates small farmers in the North and South, tropical and temperate zones, growing ancient or recently introduced crops and livestock. The Slow Food Foundation for Biodiversity, therefore, suggests several distinct trajectories, which originate from different locations in the dying food regime, which was based on national regulation (Friedmann 1993), and tend towards a new system in which each part is at once autonomous and connected, empowered and interdependent.

At the same time, institutional frameworks inherited from the dying food regime are unable to integrate food and agriculture with massive issues related to health, energy and ecosystem integrity, including the ability of the biosphere to continue to support human life. As public institutions rely on and defer to barely visible private organizations, such as GLOBALGAP and the larger International Organization for Standardization, government regulations are more often part of the problem, for instance, when hygiene rules militate against small traders and traditional markets, which have been the lifeline of culturally and ecologically embedded farmers and food artisans, such as growers of local maize varieties and tortilla makers in Mexico. Governments are caught, moreover, in anachronistic structures. Faced with new issues connecting food to health and environment, they are caught between old farm politics, often entrenched in strongly supported agricultural ministries, and new alliances between non-governmental organizations and weaker environmental ministries, which were created no more recently than the 1970s. As the Mexican government benchmarks its national standards for the *Mexico Calidad Suprema* label to GLOBALGAP, it has not only to alter older farm and food ministries but also to respond to grassroots efforts to deal with environmental, cultural and employment problems arising from conversion to export agriculture. Not the province of Ontario, which is responsible for health and agriculture, but the municipality of Toronto, in a notable departure from the old rural–urban divide, has deepened the inclusion of farmers and farm organizations in its policy and programmes. The Province of Ontario is at once locked into conventional farm lobbies through the Ministry of Agriculture, and – through ministries of economic development, regional planning and environment – struggling to respond to social movements which see in food and agriculture convergent solutions to multiple dilemmas.[37]

We do not, of course, hang a vision of global agrifood transformation on the frail threads of local projects in the western highlands of Michoacán, Mexico or Ontario, Canada, or on the Slow Food network. Instead, we suggest that thinking about them allows us to observe changes in interstices of the agrifood system, and join in the emerging, fluid apprehension of the eventual shape of a new constellation of agrifood relations and practices. The organizational and intellectual creativity of Slow Food lies in its paradoxical support for diverse, embedded and interconnected food systems. This organization will no doubt again transform the way it works and thinks, and it may well be superseded by other organizations, networks and experiments.

A cascading shift in consciousness is taking place, in which it is increasingly possible to see food and agriculture not as sideshows of 'growth', but as the seed around which can crystallize multiple, intractable and life-threatening issues. Health, climate chaos, energy shortages, water shortages, soil degradation and pollution are all increasingly clearly connected to food and agriculture. Yet the governing institutions of national states and intergovernmental organizations cannot connect them. New institutions and policies, such as 'multifunctionality' in Europe, have achieved some modest success in linking some of these issues, in this case attempting to turn farmers from polluters of soils and water into ecosystem managers by paying for environmental services. It matters little that the impetus for such creativity came from trade strategy; that is the nature of interstitial change. What matters is that it opens new horizons of possibility.

The emerging perspective suggested by *Cojote Rojo* and Local Food Plus, as well as Slow Food, fits within the wide and diverse food sovereignty movement. What distinguishes the Builder movement is that it avoids direct confrontation with capital (at least for the moment) by working with and through markets. Can something so apparently congenial to the dominant system, so apparently subject to appropriation by governments and corporations, as participation in markets – and pursuit of 'educated pleasure' – actually transform the agrifood system? These projects are on an edge of absorption, cooptation and the like, and many fall over. Yet others arise and recover. They thus require constant vigilance and self-correction, experimentation and mutual learning. Interstitial social

---

[37] Attempts to integrate them even conceptually must themselves innovate institutionally. For instance, the International Assessment of Agricultural Science and Technology for Development is the most recent in an evolving institution of 'scientific assessments', which began with the Intergovernmental Panel on Climate Change. These have no formal status, as is clear from the erratic trajectory of climate politics, yet have arguably because of this allowed for their use by individuals such as Al Gore, as well as civil society organizations and social movements, to undertake a massive public discussion and – despite the imbalance in power of participants – a cascading shift in awareness of the need for fundamental change, so fundamental that it is impossible to see how to do it (see http://www.agassessment.org).

transformation is an idea that invites us to depart from a polar divide between autonomous oppositional movements on one side, and cooptation by powerful corporations and states on the other. It is a muddy terrain into which one can sink at any time, yet perhaps also one from which one can renew and redirect the journey as swamps are mapped.

We suggest a less polar understanding also of the politics of embeddedness. Polanyi observed and compared three contending types of response to the destructions of society and nature by 'the self-regulating market system', and all were state-centred. In his classic work *The Great Transformation* (1944), Polanyi compared three state-centred experiments in redistribution during the Great Depression of the 1930s – fascism, Soviet Communism and US-style social democracy.[38] Even in that work, however, Polanyi theoretically identified historical forms of embedding other than state-centred redistribution, including reciprocity and 'householding'. After World War II, Polanyi and his collaborators focused on alternative forms of exchange embedded in non-instrumental relations (Polanyi et al. 1957). Using ethnography and pre-capitalist history, these authors discovered a myriad of geographically bounded markets and long-distance trade, which did not affect local prices or production, and which embedded market relations in substantively organized 'society'. We suggest that the era of 'globalization' – and the various ways in which governments have ceded authority and have contracted out administration and collective services – partly shifts Polanyi's 'protective' movements to re-embed land and labour away from the state. Our examples suggest that scattered spontaneous efforts, which are beginning to evolve a broadly shared ecological and cultural vision, operate less as demands on government than in Polanyi's time. Now they work creatively with and against governments at all scales, just as private agrifood capitals do. Builders and Warriors are in a shifting balance.

Cojote Rojo and Local Food Plus are suggestive examples from Mexico and Canada – or more accurately, from Western Michoacán and Southern Ontario. Each is unique and embedded in its cultural and agronomic context. Both are loosely connected to a global shift in consciousness and practices about food and agriculture. The institutions that express these shifts are emerging everywhere one looks, if one cultivates ways to see them. Slow Food in its various manifestations is evolving its own ways of working to support creative regional approaches to build diversity and embeddedness. The innovation of the Presidium flexibly supports farmers and artisans embedded in localities, helping them to create and sustain embedded practices and guaranteeing that they have done so to distant buyers. They are far too recent and scattered to institute something new, but each experience adds to a common prefiguring of a new type of global economy based on culturally and ecologically embedded foods.

---

[38] Roosevelt's New Deal. We use the term 'social democracy' loosely, as Britain was devastated economically, while most of the continent was ravaged by war, and only the US model of government intervention stood as an example when Polanyi wrote during World War II.

REFERENCES

Araghi, F., 2000. 'The Great Global Enclosure of our Times: Peasants and the Agrarian Question at the End of the Twentieth Century'. In *Hungry for Profit? The Agribusiness Threat to Farmers, Food and the Environment*, eds H. Magdoff, J. Foster and F. Buttel, 145–60. New York: Monthly Review Press.

Barling, D. and T. Lang, 2005. 'Trading on Health: Cross-Continental Production and Consumption Tensions and the Governance of International Food Standards'. In *Cross-Continental Food Chains*, eds N. Fold and B. Pritchard, 39–51. London: Routledge.

Barry, T., 1995. *Zapata's Revenge: Free Trade and the Farm Crisis in Mexico*. Boston, MA: South End Press.

Bernstein, H., 2006. 'Is There an Agrarian Question in the 21st Century?'. *Canadian Journal of Development Studies*, 27 (4): 449–60.

Bernstein, H. and L. Campling, 2006a. 'Commodity Studies and Commodity Fetishism I: Trading Down. On *Trading Down: Africa, Value Chains, and the Global Economy*, by P. Gibbon and S. Ponte'. *Journal of Agrarian Change*, 6 (3): 239–64.

Bernstein, H. and L. Campling, 2006b. 'Commodity Studies and Commodity Fetishism II: "Profits with Principles"?'. *Journal of Agrarian Change*, 6 (2): 414–47.

Burch, D. and G. Lawrence, 2007. 'Supermarket Own Brands, New Foods and the Reconfiguration of Agri-food Supply Chains'. In *Supermarkets and Agri-food Supply Chains. Transformations in the Production and Consumption of Foods*, eds D. Burch and G. Lawrence, 100–30. Cheltenham: Edward Elgar.

Busch, L. and C. Bain, 2004. 'New! Improved? The Transformation of the Global Agrifood System'. *Rural Sociology*, 69 (3): 321–46.

Busch, L. et al., 2005. 'The Relationship of Third Party Certification (TPC) to Sanitary/Phytosanitary (SPS) Measures and the International Agri-food Trade: Final Report. Prepared for USAID'. http://www.msu.edu/~ifas/downloads/The%20Relationship%20of%20TPC%20to%20SPS%20Measures--Final%20Report%20+%20Annexes.pdf, Accessed 30 January 2008.

Callon, M., C. Meadal and V. Rabeharisoa, 2002. 'The Economy of Qualities'. *Economy and Society*, 31 (2): 194–217.

Campbell, H. and R. LeHeron, 2007. 'Supermarkets, Producers and Audit Technologies: The Constitutive Micro-politics of Food, Legitimacy and Governance'. In *Supermarkets and Agri-food Supply Chains. Transformations in the Production and Consumption of Foods*, eds D. Burch and G. Lawrence, 131–53. Cheltenham: Edward Elgar.

Coyote Rojo, n.d. *Reglamento para la Certificación Territorial Bioregionalista JIUATSI*. Nurio, Michoacan, Mexico: Coyote Rojo.

De la Tejera, B., 2007. '¿Que Hay Detrás del Precio de la Tortilla?'. *La Jornada de Michoacán*. 26 January. http://www.lajornadamichoacan.com.mx/2007/01/20/index.php?section=opinion&article=002a1pol, Accessed 30 January 2008.

Desmarais, A.A., 2007. *La Vía Campesina: Globalization and the Power of Peasants*. Point Black, NS and London: Fernwood Books & Pluto Press.

Dixon, J., 2007. 'Supermarkets as New Food Authorities'. In *Supermarkets and Agri-Food Supply Chains: Transformations in the Production and Consumption of Foods*, eds D. Burch and G. Lawrence, 29–50. Cheltenham: Edward Elgar.

Edelman, M., 2003. 'Transnational Peasant and Farmer Movements and Networks'. In *Global Civil Society 2003*, eds M. Kaldor, H. Anheier and M. Glasius, 185–220. Oxford: Oxford University Press.

Fonte, M., 2006. 'Slow Food's Presidia: What Do Small Producers Do with Big Retailers?'. *Research in Rural Sociology and Development*, 12: 1–39.

Friedmann, H., 1993. 'The Political Economy of Food: A Global Crisis'. *New Left Review*, 197: 29–57.

Friedmann, H., 2004. 'Feeding the Empire: Pathologies of Globalized Agriculture'. In *Socialist Register: The Empire Reloaded*, eds L. Panitch and C. Leys, 124–43. London: Merlin.

Friedmann, H., 2005. 'From Colonialism to Green Capitalism: Social Movements and Emergence of Food Regimes'. In *New Directions in the Sociology of Global Development*, eds F.H. Buttel and P. McMichael, 227–64. Oxford: Elsevier.

Friedmann, H., 2007a. 'Seeds of the City'. In *Food*, eds J. Knechtel, 240–50. Cambridge, MA: Alphabet City and MIT Press.

Friedmann, H., 2007b. 'Scaling Up: Bringing Public Institutions and Food Service Corporations into the Project for a Local, Sustainable Food System in Ontario'. *Agriculture and Human Values*, 24 (3): 389–98.

Gereffi, G. and M. Korzeniewicz, eds, 1994. *Commodity Chains in Global Capitalism*. Westport, Connecticut: Praegar.

Goodman, D., 2003. 'The Quality "Turn" and Alternative Food Practices: Reflections and Agenda'. *Journal of Rural Studies*, 19 (1): 1–7.

Goodman, M.K., 2004. 'Reading Fair Trade: Political Ecological Imaginary and the Moral Economy of Fair Trade Foods'. *Political Geography*, 23: 891–915.

Guthman, J., 2004. *Agrarian Dreams: The Paradox of Organic Farming in California*. Berkeley, CA: University of California Press.

Hamm, G., 2007. 'Mexicans Protest as Tortilla Crisis Hurts Calderon'. *The Boston Globe*. 13 January. http://www.boston.com/news/world/latinamerica/articles/2007/02/01/mexicans_protest_as_tortilla_crisis_hurts_calderon, Accessed 28 December 2007.

Harvey, M., 2007. 'The Rise of Supermarkets and Asymmetries of Economic Power'. In *Supermarkets and Agri-Food Supply Chains: Transformations in the Production and Consumption of Foods*, eds D. Burch and G. Lawrence, 51–73. Cheltenham: Edward Elgar.

Jaffee, D., 1997. 'Confronting Globalization in the Community Forests of Michoacán, Mexico: Free Trade, Neoliberal Reforms and Resource Degradation'. Presented at the *LASA Conference*, Guadalajara, April 1997.

Klooster, D., 2003. 'Campesinos and Mexican Forest Policy During the Twentieth Century'. *Latin American Research Review*, 38 (2): 94–126.

Konefal, J. and M. Mascarenhas, 2005. 'The Shifting Political Economy of the Global Agrifood System: Consumption and the Treadmill of Production'. *Berkeley Journal of Sociology*, 49: 76–96.

Konefal, J., C. Bain, M. Mascarenhas and L. Busch, 2007. 'Supermarkets and Supply Chains in North America'. In *Supermarkets and Agri-Food Supply Chains: Transformations in the Production and Consumption of Foods*, eds D. Burch and G. Lawrence, 268–90. Cheltenham: Edward Elgar.

Lyons, K., 2007. 'Supermarkets as Organic Retailers: Impacts for the Australian Organic Sector'. In *Supermarkets and Agri-Food Supply Chains: Transformations in the Production and Consumption of Foods*, eds D. Burch and G. Lawrence, 154–72. Cheltenham: Edward Elgar.

Marsden, T., A. Flynn and M. Harrison, 2000a. *Consuming Interest: The Social Provision of Foods*. London: UCL Press.

Marsden, T., J. Banks and G. Bristow, 2000b. 'Food Supply Chain Approaches. Exploring Their Role in Rural Development'. *Sociologia Ruralis*, 40: 424–38.

McMichael, P., 2007. 'Sustainability and the Agrarian Question of Food'. Keynote Address *European Congress of Rural Sociology*, Wageningen, 20–24 August 2007.

McMichael, P. and H. Friedmann, 2007. 'Situating the "Retailing Revolution"'. In *Supermarkets and Agri-Food Supply Chains: Transformations in the Production and Consumption of Foods*, eds D. Burch and G. Lawrence, 154–72. Cheltenham: Edward Elgar.

Mijangos-Cortés, J.O. et al., 2007. 'Differentiation among Maize (*Zea mays* L.) Landraces from the Tarasca Mountain Chain, Michoacán, Mexico and the *Chalqueño* Complex'. *Genetic Resources and Crop Evalutaion*, 54: 309–25.

Morgan, K., T. Marsden and J. Murdoch, 2006. *Worlds of Food: Place, Power, and Provenance in the Food Chain*. Oxford: Oxford University Press.

Nadal, A., 2006. 'Mexico's Corn-producing Sector: A Commentary'. *Agriculture and Human Values*, 23: 33–6.

Patel, R., 2007. 'Transgressing Rights: La Vía Campesina's Call for Food Sovereignty'. *Feminist Economics*, 13 (1): 87–116.

Petrini, C., 2001. *Slow Food: The Case for Taste*. New York: Columbia University Press.

Pietrykowski, B., 2004. 'You Are What You Eat: The Social Economy of the Slow Food Movement'. *Review of Social Economy*, 42 (3): 307–21.

Polanyi, K., 1944. *The Great Transformation: The Political and Economic Origins of Our Times*. Boston, MA: Beacon Hill.

Polanyi, K., C.M. Arensbert and H.W. Pearson, eds, 1957. *Trade and Markets in the Early Empires*. Chicago, IL: Henry Regnery.

Pollan, M., 2006. *The Omnivore's Dilemma: A Natural History of Four Meals*. New York: Penguin Press.

Quist, D. and I.H. Chapela, 2001. 'Transgenic DNA Introgressed into Traditional Maize Landraces in Oaxaca, Mexico'. *Nature*, 29 (414): 541–3.

Raynolds, L.T., 2004. 'The Globalization of Organic Agro-Food Networks'. *World Development*, 32: 725–43.

Raynolds, L. and D. Murray, 2007. 'Fair Trade: Contemporary Challenges and Future Prospects'. In *Fair Trade. The Challenges of Transforming Globalization*, eds L. Raynolds, D. Murray and J. Wilkinson, 223–34. Abingdon: Routledge.

Reardon, T. and J.A. Berdegué, 2002. 'The Rapid Rise of Supermarkets in Latin America: Challenges and Opportunities for Development'. In *Development Policy Review*, 20 (4): 371–88.

Reardon, T. et al., 2005. 'Supermarket Expansion in Latin America and Asia: Implications for Food Marketing Systems'. In *New Directions in Global Food Markets/AIB-794*, eds A. Regmi and M. Gehler, 47–61. Economic Research Service/USDA. http://www.ers.usda.gov/publications/aib794/aib794.pdf, Accessed 30 January 2008.

Ribeiro, S., 2004. 'The Day the Sun Dies: Contamination and Resistance in Mexico'. *Seedling*. http://www.grain.org/seedling/?id=292, Accessed 30 January 2008.

Scherr, S.J., 1985. *Agriculture and the Oil Syndrome: Lessons from Tabasco, Mexico*. New York: Praeger.

Slow Food, 2007. Programme for Fifth International Slow Food Congress. Puebla, Mexico, 8–11 November.

Slow Food Foundation, 2007. *International Presidia*. http://www.slowfoodfoundation.org/pdf/Elenco%20EN%2017-05-07.pdf, Accessed 30 January 2008.

Stahlbrand, L., 2003. *Ecolabelling as a Marketing Tool to Support Sustainable Agriculture*. Toronto: World Wildlife Fund.

Stanford, L., 2002. 'Constructing "Quality": The Political Economy of Standards in Mexico's Avocado Industry'. *Agriculture and Human Values*, 19 (4): 293–310.

Tanaka, K. and L. Busch, 2003. 'Standardization as a Means for Globalizing a Commodity: The Case of Rapeseed in China'. *Rural Sociology*, 68 (1): 25–45.

Wilkinson, J., 2005. 'Global Agrifood Chains, Retail and Catering: The Cast of the Fish Sector'. *FAO Corporate Document Repository*. Rome: Food and Agriculture Organization. http://www.fao.org.docrep.007/y5767e/y5767e0j.htm, Accessed 21 December 2006.

Works, M. and K.S. Hadley, 2004. 'The Cultural Context of Forest Degradation in Adjacent Purépecha Communities, Michoacán, Mexico'. *Geographical Journal*, 170 (1): 22–38.

Wright, E.O., forthcoming. Envisioning Real Utopias (unpublished). http://www.ssc.wisc.edu/~wright, Accessed 27 October 2007.

10 Migrant Organization and Hometown Impacts in Rural Mexico

JONATHAN FOX AND XOCHITL BADA

CONTENDING RURAL FUTURES: EXIT OR VOICE?

Political economists counterpose exit and voice as conceptual shorthand for different actors' possible responses to diverse challenges.[1] In this context, migration is often understood as an exit option and therefore an alternative to voice. Migrants vote with their feet, in the commonsense phrase. If this exit-voice dichotomy holds, then the hundreds of thousands of Mexicans who leave their villages each year are indirectly weakening rural civil society's capacity for collective action and political representation. This chapter suggests that while exit might well substitute for voice in the short term, exit can also be followed by voice.

This proposition emerges from analysis of the cross-border social processes in which migrants come together in transnational communities, which in turn constitute the social foundations of an emerging 'migrant civil society'. Mexican migrants have demonstrated a growing capacity to form their own representative organizations. For more than a decade, hundreds of US-based Mexican migrant hometown associations have raised funds and campaigned for community development and public accountability in their villages of origin. Widespread practices of long-distance community membership are generating notions of bi-national citizenship.[2] A bi-national framework for understanding Mexico's emerging migrant civil society allows analysts to take into account the feedback effects of migrant organization on power relations within home communities.

These cross-border and multilevel forms of active membership represent one dimension of the broader process of the formation of transnational civil society. So far, the study of transnational civil society has been dominated by discussions of transnational advocacy campaigns, often involving more openly politicized public interest groups and/or militant social movements. Such campaigns are often described as transnational social movements, though in practice they usually involve networks or coalitions whose actual density would fall short of most

---

[1] See Hirschman's classic works (1970, 1981). He applied the notion to individuals, firms and peoples, in both economic and political arenas.
[2] See the related conceptual discussion in Fox (2005a).

definitions of social movement.³ Yet for the most broad-based social and political organizations that are engaged in cross-border networking or mobilization, very few of the participants actually cross borders (in any sense). In many sectors and issue areas, the transnational engagement and liaison is often limited to a small handful of leaders or professional staff, who serve as the intermediaries between the global and the local. Yet some specifically agrarian transnational movements are quite different, involving broad and deep direct contact between the rank and file across borders, as in the case of the Campesino to Campesino agroecology movement (e.g., Holt-Giménez 2006). In this context, when considering the range of possible forms of expression of agrarian transnational movements, the formerly rural migrants who reach out across borders to engage with their hometowns and villages also 'count' as part of transnational civil society – even if their terms of engagement are often confined to less overtly politicized civic and community development agendas. While these territorially-based migrant civic organizations are only occasionally openly confrontational in their stance towards the state, and only a few of these mass-based migrant organizations pursue transformative goals, those migrants who are organized to promote community development and democratization in their communities of origin may well have more transnational density and cohesion than many cross-border campaigns that are less deeply grounded in their respective societies.⁴

To assess some of the ways in which migrant collective action can encourage rural democratization, this chapter focuses on the patterns and impacts of organized migrant participation in the federal government's 'Three-for-One' community development matching fund programme.⁵ As context, the study begins with a brief overview of rural out-migration trends, followed by a discussion of the relationship between the concepts of exit, voice and loyalty.

Accelerating Migration

Migration to Mexico's cities and to the US has long been a pathway to escape the limits of smallholder agriculture, often as part of diversified family survival strategies. While migration to the US was historically concentrated in Mexico's centre-west region, in the 1980s and 1990s out-migration spread throughout the nation's countryside, as well as into large cities and across a broader mix of social classes. Yet while the urban share of Mexican migration to the US is growing, migrants continue to be disproportionately rural, often coming from outlying

---

³ For example, the campaigns against the North American Free Trade Agreement, involving labour, agrarian, human rights, environmental and civic groups, arguably constitute a 'paradigm case' for assessing the degree to which globalization from below is catching up with globalization from above. In almost all sectors, the transnational dimension of the networks and campaigns proved to be thin and/or transitory (Brooks and Fox 2002). For related perspectives based on studies of other campaigns, see Laxer and Halperin (2003).
⁴ One of the exceptions in the Mexico–US context, in the sense of a bi-national mass organization that does pursue transformative goals, is the Bi-national Front of Indigenous Organizations (FIOB). See Fox and Rivera-Salgado (2004).
⁵ On the dynamics of rural democratization more generally in Latin America, see Fox (1990).

villages in their municipalities of origin. For rural Mexico, consider the implications of the fact that the million Mexican farm-workers who gained US permanent residency under the 1986 immigration reform were equivalent to *one sixth* of the adult men in rural Mexico at that time (Martin 2005, 6). In increasing numbers of villages, from the northern border to the Mayan southeast, young men and women increasingly *expect* to migrate, rather than envisioning their future in rural Mexico.

While this cross-border migration process represents the current phase of a century-long structural process, its pace was accelerated by conscious policy choices. At a 1991 Harvard forum, Mexico's then-undersecretary of Agriculture, Luis Téllez, predicted dramatic changes in the place of agriculture in Mexican society. He estimated that, within the following decade, the share of Mexico's economically active population in agriculture would drop from 26 percent to 16 percent – thanks to the Salinas presidency's three main rural policy reforms – the North American Free Trade Agreement, the withdrawal of government-subsidized production supports for family farming and a Constitutional reform that encouraged individual titling of agrarian reform lands.[6]

According to Mexico's 2000 census, 25 percent of the population continued to live in localities with less than 2,500 inhabitants. This suggests that Téllez's prediction was off the mark, especially when one considers that this official threshold for defining rural is exceedingly low. Yet if one looks at the share of the population that is 'economically active' in agriculture, then Téllez's prediction was on target. According to the most recent National Employment Survey, agricultural employment fell from 24 percent in 1991 to under 15 percent at the end of 2005 (INEGI n.d.). A similar survey found a loss of 1.3 million agricultural jobs between 1993 and 2002 (Polaski 2003, 20). These data indicate a growing gap between the population that lives *in* the countryside and the population that lives *from* the countryside. The growth in the share of the rural population that does not live off of agriculture has major implications for the future of public life in the countryside.

By the year 2000, only six years after the implementation of NAFTA, national census data indicated an increase in international migration rates, with 96.2 percent of the country's municipalities reporting international labour 'expulsion'. Increased migration combined with falling birth rates led to widespread depopulation in so-called 'sending' regions: between 2000 and 2005, 33 percent of Mexican municipalities reported negative growth. For instance, in the state of Michoacán, between 1990 and 2000, 93.8 percent of the state's municipalities reported population decreases, with some municipalities losing more than 9 percent of their population in that decade. The state's population fell 0.1 percent

---

[6] For an analysis of that political turning point, which immediately preceded the Zapatista uprising, see Fox (1994). Note that the constitutional reform did not lead to widespread individual land privatization and sale. Most land reform communities followed the law insofar as they agreed to confirm both their collective and family land boundaries, and longstanding trends toward commodification and rental of these lands accelerated, but very few *ejidos* took the final step of complete privatization – a decision which the law left in community hands (e.g., Cornelius and Myhre 1998).

annually between 2000 and 2005, compared to an annual average population increase of 1.2 percent reported between 1995 and 2000, thus becoming the first state in the country that registered a population decrease since the end of the Mexican Revolution (INEGI 2006; CONAPO cited in Ramos 2007; SEDESO 2004).

As a result of the accelerated pace of out-migration, family remittances back to Mexico skyrocketed over the last decade, from a total of just under US$3.7 billion in 1995 to more than US$23 billion in 2006, increasing five times in just one decade and currently representing 2.7 percent of Mexico's GDP and 66 percent of oil exports. Most of these resources are spent on basic consumption (86 percent) and a modest percentage is invested in commercial operations or community improvement (0.6 percent) (Banco de México 2007). Investments with collective remittances amount to an average of US$14 million per year, representing far less than 1 percent of migrant remittances.

Mexico's remittances are disproportionately concentrated in few states. In 2006, 65.5 percent of total reported family remittances went to nine states, including Mexico City and the traditional 'sending' states in central Mexico (Michoacán, Guanajuato and Jalisco). The rest of remittance income was spread among 23 states. There is still no consensus regarding the exact amount of family remittances in Mexico. Some remittance experts believe that the official figures are underestimated because they don't include cash transfers, US pensions received by returned migrants, gifts, and in kind transfers from migrants to their families. Others strongly question the Central Bank calculations and believe that remittances have been grossly exaggerated and manipulated as a political tool, to enable political leaders to depict migrants as heroes in the national public discourse (Lozano Ascencio 2003; Lozano Ascencio and Olivera Lozano 2005; Leyva Reus 2005).

Exit and Voice: Dichotomous or Interactive[7]

Overall, in 2000, 14 percent of Mexican-born workers were in the US (Martin 2005, 10). The cumulative result of this exodus of working-age adults must affect the prospects for future social and political change in the countryside, but the patterns of this impact remain unclear. It is no coincidence that analysts in Mexico often refer to this issue as the 'migration problem', even though – for the migrants themselves – access to the US labour market represents a 'solution'.

It is worth recalling that during the post-NAFTA decade, Mexico experienced no protest movement of the rural poor that was both sustained and of national scope.[8] The most notable apparent exception to this generalization, the well-known Zapatista movement, generated widespread sympathy nationwide, but remained a regionally-bounded social actor. The broad-based but brief 'Countryside Won't Take Any More' 2003 march on Mexico City was the

---

[7] This section draws on portions of chapter 10 in Fox (2007).
[8] The well-known Barzón movement for debt relief reached national scope, but represented primarily small-to-medium-sized commercial producers. Only a minority of Mexican farmers were sufficiently well-off to have received bank credit in the first place.

decade's only peasant protest of national significance that focused on making family farming economically sustainable. Though the mobilization was much larger than even sympathetic observers expected, it ended up having virtually no impact on national agricultural trade and investment policies, which continued to be extremely biased in favour of better-off producers.[9] In January 2008, the final implementation of NAFTA's agricultural measures also provoked a national peasant protest march, though it appeared to be a classic case of 'too little, too late'.

Mexican and US political elites presented NAFTA to the US public as a job creation strategy that would reduce migration, but estimated annual rates of out-migration grew sharply during the 1990s (Passell and Suro 2005). In this context, it is useful to rethink Mexico's 1994 national elections. The public policies that are now widely associated with the increase in out-migration, notably the withdrawal of support prices, input subsidies and trade protection for basic grains, date primarily from the Salinas presidency (1988–1994). In this sense, the 1994 elections, had they been fully democratic for rural voters, might have served as a referendum on this package of public policies.

Thanks to the citizens' movement for independent election monitoring in 1994, led by the Civic Alliance, their reports show that at least half of the polling places in the countryside lacked guaranteed ballot secrecy (Fox 2007). The Civic Alliance also found vote-buying pressures in 35 percent of rural polling places. As Hirschman noted, the secret ballot is a key mechanism for 'making voice retaliation-proof' (1981, 241). To put this in Hirschman's terms, given the lack of political voice for most of the rural poor, many turned to exit. While this was certainly not the only migratory push factor, out-migration rates did rise substantially over the rest of the decade, perhaps suggesting some relationship between lack of voice and the exit option – at least at that political turning point.[10]

The clearest expression of rural political voice during this period came from Mexico's indigenous peoples, whose numerous and politically diverse local and

---

[9] On the class bias of the Mexican government's agricultural spending, see the little-known but nominally public analysis by the World Bank (2004). Most analysts would agree that the very modest agricultural policy concessions that the 2001 peasant protest had appeared to win were quickly subsumed by old-fashioned corporatist-style payments to organizations. Once the protesters returned home, the combination of technocratic diversions and the persistent intervention of traditional corporatist peasant groups overwhelmed the national representatives of participating independent organizations. The newly-governing PAN discovered the political convenience of providing funds directly to PRI-style peasant organizations (e.g., Fox and Haight forthcoming). For background on the '*El campo no aguanta más*' movement, see, among others, Schwentesius et al. (2004), a thematic issue of *El Cotidiano* (No. 124, March–April, 2004) and the extended 2003 debate between leading rural analysts Luis Hernández Navarro and Armando Bartra in the pages of the left-wing daily *La Jornada*.

[10] More recently, Goodman and Hiskey's (2008) large-scale statistical analysis of voter turnout rates and survey data finds that, at the municipal level, higher rates of out-migration are associated with lower voter turnout levels at national elections. At the same time, high rates of migration are associated with higher levels of reported civic engagement at the local level, which may be related to bi-national partnerships with organized migrants.

regional social organizations flowered from the 1970s through the 1990s (Fox 1996, 2007). In the wake of the 500 year anniversary of the European conquest and bolstered by the Zapatista movement, they began to come together nationally for the first time during the 1990s. Yet during the same decade, cross-border migration processes began to extend for the first time to almost all of Mexico's indigenous regions (Fox and Rivera-Salgado 2004). Looking back over the past decade and a half, Mexico's indigenous peoples have been exercising *both* voice *and* exit more than ever before.

While conventional discourse in Mexico refers to migrants as 'abandoning' their communities, a growing body of research on migrant collective action based on shared communities of origin suggests that many migrants bring their sense of community with them, and recreate it with their *paisanos* in the US. This sense of shared collective civic identity is broadened when hometown associations form home *state* federations in Chicago or Los Angeles, constructing a sense of regional belonging that the migrants may not have shared before they left.

At the same time as one recognizes the emergence and consolidation of transnational communities, to be discussed below, one must also recognize that many who migrate *do* abandon their communities. Some do not return. In spite of the widespread attention to the growing volume of migrant remittances, substantial minorities do not send resources to support their families. Plus, when an organizer migrates, their organization clearly suffers a loss – especially if the leadership has invested in their training, as in the case of coffee coop certifiers of organic production (e.g., Mutersbaugh 2008).

These patterns suggest that while exit may sometimes weaken voice, and at other times they may reinforce each other, perhaps exit can also reflect the *prior weakness* of voice. Many observers point to regions of long-term out-migration and see a very thin civil society, yet the cause and effect relationship is not so clear-cut. Many migrants leave regions where rural civil society was *already* thin. In addition, even in regions that had experienced autonomous collective action, few campaigns had produced lasting change, and even fewer could offer viable future options for young people. But if we extend the temporal and geographic frame for considering the interaction between exit and voice and take the bi-national arena into account, new ways of considering the relationship between exit and voice emerge, as well as the role of loyalty as a mediating factor.

This review of rural out-migration trends and the conceptual dilemmas posed by the relationship between exit and voice sets the stage for a discussion of Mexican migrant civil society and the impact of collective community development remittances on village governance. But first, a brief discussion of the relationship between out-migration and poverty rates is in order.

Out-Migration and Rural Poverty Rates

The available data on rural municipalities suggest that there is no direct correlation between poverty levels and out-migration rates. Of Mexico's 2,443 municipalities, 82 percent are considered rural. One-quarter of Mexico's municipalities are also

Table 1. Rural and indigenous municipalities: migration and poverty rates

Migration intensity	High and very high marginality levels (%)	Intermediate and low marginality levels (%)	Very low marginality levels (%)
Very high	**3.3**	4.7	0
High	**7.1**	8.4	0.2
Medium	8.1	7.5	0.9
Low	*11.9*	7.5	1.4
Very low	27.8	6.0	0.4
No data	4.4	0.1	0
Total	62.6	34.3	3.1

Source: Carral Dávila (2006, 99–100), based on 2000 data from the National Population Council.

considered indigenous (a category defined by language use). Government census data indicate that 62.6 percent of these rural and indigenous municipalities are in extreme poverty, with 'high' or 'very high' levels of marginality – an indicator that refers primarily to access to basic services (water, sanitation, education, housing, etc.). These rural, low-income municipalities account for 20 percent of the national population (CONAPO 2000).

The government demographic agency considers 20 percent of Mexico's rural municipalities to register high or very high levels of out-migration. If one reviews Mexico's rural municipalities in terms of the varying degrees of what government discourse refers to as 'migration intensity' and 'marginality', one can get a sense that the relationship between poverty and out-migration is not as direct as widely assumed, as indicated by the data in Table 1.

These data suggest three distinct patterns among rural municipalities, indicated by the typeface in Table 1.

1. Just over 10 percent of rural municipalities experience both high poverty and high migration rates. Here the impacts of both government anti-poverty programmes and remittances have been very limited.[11]
2. Approximately 20 percent of rural municipalities combine high poverty and a medium degree of 'migration intensity', and they are likely to increase their out-migration rates in the future, in the absence of substantially increased investment in family farming.
3. Another 13 percent of rural municipalities combine high levels of out-migration and 'intermediate' poverty levels. Here the potential impacts of government anti-poverty programmes and remittances appear to be more significant.

---

[11] The federal government's flagship welfare programme, the *Oportunidades* conditional cash transfer programme, appears to reach a much larger share of the poorest rural population than receives migrant remittances (e.g., Muñoz 2004).

This approach to understanding the interaction between poverty and migration rates is complicated by the fact that rural municipalities are not the 'most local' governmental jurisdiction. They refer to local districts rather than to specific villages, and therefore usually include both an 'urban centre' and numerous outlying villages and hamlets (also known as 'localities'), which tend to experience higher levels of both poverty and out-migration than the town centre. Official municipal data therefore average the poverty and migration trends in these different kinds of communities. Most official out-migration data are not sufficiently disaggregated to the level of 'locality' to allow more precise analysis of their relationship to poverty levels.

From the point of view of understanding the dynamics of rural democratization, it is crucial to recognize that in much of rural Mexico, outlying villages are politically subordinated to municipal centres, both formally and informally. Many rural municipalities are in the midst of a long-term 'regime transition', largely invisible to outsiders, in which outlying communities campaign for the right to resources and self-governance vis-à-vis the town centres (Fox 2007). This is the context within which organized migrant hometown associations, together with their communities of origin in outlying villages, pressure municipal and state authorities to gain standing, voice and representation.

MIGRANT CIVIL SOCIETY AND HOMETOWN ASSOCIATIONS[12]

As many as hundreds of thousands of Mexican migrants work together with their *paisanos* to promote 'philanthropy from below', funding hundreds of community development initiatives in their hometowns. Tens of thousands signed up to exercise their newly-won right to cast absentee ballots in Mexico's 2006 presidential election. Many more are engaged with their US communities – as organized workers, parents, members of religious congregations and naturalized voters. In addition, some Mexican migrants are working to become full members of *both* US and Mexican societies at the same time, constructing practices of 'civic bi-nationality'.

What are some of the implications of putting together three words: 'migrant civil society?' Simply put, migrant civil society refers to *migrant-led membership organizations and public institutions*. Specifically, this includes four very tangible arenas of collective action. Each arena is constituted by actors, while each set of actors also constitutes an arena. These arenas include autonomous public spaces (such as large-scale cultural or political gatherings), migrant-led NGOs, the migrant-led mass media, as well as migrant-led membership organizations.[13]

---

[12] The following text draws on Fox (2005b). Note that 'migrant civil society' emerges from but is distinct from transnational communities, since they may or may not be engaged with the public sphere.
[13] For detailed discussion of these arenas of migrant civil society, see Bada et al. (2006) and Fox (2007).

Membership organizations composed primarily of migrants can range from hometown associations (HTAs) to worker organizations and religious congregations. Because of the focus here on cross-border impacts on home communities, this discussion will be limited to the HTAs. Hometown Associations are migrant membership organizations formed by people from the same community of origin. Though many began as very informal groups, by the turn of the century hundreds had become formal organizations. HTAs function as social support networks, as well as transmitters of culture and values to the US-born generation. Often in response to Mexican government encouragement, many of these translocal clubs later joined with others from their home states to form federations. These scaled-up forms of representation increased migrant leverage with their home state governments. They become involved in social development projects on behalf of their communities of origin, as well as in the defence of migrant rights in their region of settlement.

The Mexican consulates have registered well over 600 such clubs (Rivera-Salgado et al. 2005). The federal Ministry of Social Development is also developing a database of Mexican HTAs in the United States and Canada and so far they have found 815 clubs. However, this figure has not yet been disaggregated geographically (SEDESOL 2006). Mexican HTAs are heavily concentrated in California and Illinois, with 86.5 percent of them concentrated in the metropolitan areas of Los Angeles and Chicago. They are also expanding their presence in New York City (Cordero-Guzmán and Quiróz-Becerra 2005; Smith 2006). While they are concentrated in large US cities, most have rural roots in Mexico.[14] Each has a core membership of perhaps an average of two dozen families, some with hundreds more. Many HTA members are relatively well-established in the US, and many of their leaders have relative economic stability and are either legal residents or US citizens (which allows them to travel back and forth frequently).

It is difficult to measure with any precision how many migrants participate, especially given the wide variation in the size and activities of each HTA and federation. In addition, the official consular registries include some clubs that exist only on paper, while some active associations choose not to register. An unusually large-scale survey of relatively recent Mexican migrants found that 14 percent of respondents belonged to some kind of hometown association (Suro 2005). However, a much smaller national survey found that only 6 percent of foreign-born Mexicans interviewed reported membership in an ethnic immigrant civic or social organization (Waldinger 2007).[15]

Today's Mexican HTAs have a long history, with the first Zacatecan club in California dating back to 1962 (Moctezuma 2005). But their numbers and

---

[14] One of the few federations of HTAs that is located primarily in rural areas of the US is Alianza LUDA (Latinos Unidos de América), which includes 16 mainly farmworker-based clubs in the small rural communities of California's Salinas Valley.

[15] Whether these numbers are considered high or low depends on one's comparative frame of reference.

membership boomed in the past 15 years, as the result of several converging factors. Within the US, the massive regularization of undocumented workers that followed the 1986 immigration reform facilitated both economic improvement and increased cross-border freedom of movement for millions of migrants. On the Mexican side, the government deployed the convening power of its extensive consular apparatus, bringing together people from the same communities of origin and offering community development matching funds to encourage collective social remittances, through the Three-for-One matching fund programme. Though this policy began as a response to pressures from organized Zacatecan migrants, it also served as a powerful inducement for other migrants to come together in formal organizations for the first time. Indeed, many transnational social and civic relationships unfold outside of the clubs and federations (Fitzgerald 2000). In addition, the Mexican state changed the tone of its relationship with the diaspora by formally permitting dual nationality for the first time.[16] While most clubs emerged autonomously, from below, many of the state-level federations were formed through engagement with the Mexican state (Goldring 2002; González Gutiérrez 1997).[17]

COLLECTIVE REMITTANCES AND HOME COMMUNITY IMPACTS

Over time, the academic and policy discussion of the impact of migration on sending communities has shifted from an earlier focus on the loss of human capital, to a debate over whether family remittances contribute to more than relatives' subsistence, and whether remittances can become a lever for job creation (Goldring 2004). In terms of the dichotomy often posed between the use of remittances for consumption vs investment, documented experiences with sustainable job-creating enterprises beyond a very small scale are very limited, at least so far.[18] There are many powerful reasons why the results of job-creating investment of remittances have been limited, including unequal distribution of

---

[16] Note that a full discussion of the potential for migrant home country political impacts in the electoral arena is beyond the scope of this study. Briefly, beginning in the late 1980s, migrant civic leaders began campaigning for the right to absentee ballots, eventually winning a partial victory that allowed migrants who had brought their voting cards with them to navigate a bureaucratically complex process to vote by mail in 2006. Less than 1 percent of the estimated eligible migrant electorate actually participated, and it did not appear to make a difference even in a very close national outcome. While many Mexican migrants are certainly politicized, that energy has yet to be fully expressed through cross-border partisan electoral processes. It is therefore safe to say that so far, Mexican migrants' greatest cross-border civic and political impact has been at the community level. On migrant voting rights campaigns, dilemmas and results, see the archives of the bi-national civic journal *MX Sin Fronteras* (http://www.mxsinfronteras.com).

[17] For more on Mexican HTAs, see also, among others, Bada (2004a, 2004b, 2004c, 2008), Escala Rabadán and Zabin (2002), Fitzgerald (2000, 2004), Lanly and Valenzuela (2004), Moctezuma (2003a, 2003b), Orozco et al. (2004), Rivera-Salgado and Escala Rabadán (2004), M Smith (2003), R Smith (2003, 2006), Smith and Bakker (2008) and Williams (2004).

[18] See García Zamora (2005a, 2005b, 2006). For a heterodox critique of the conventional discussion of remittances and development, see the *Declaración de Cuernavaca* from the Migration and Development Network, at http://www.migracionydesarrollo.org. For an English translation, see *Enlaces News*, No. 10, August 2005 at http://www.enlacesamerica.org.

land, supply and demand mismatches, lack of technical capacity, a less-than-hospitable policy environment, the greater attraction of public vs private goods (in the case of collective remittances) and very limited investment opportunities in many sending communities. In addition, the wages usually offered by migrant micro-investors are rarely better than the prevailing rural wage, which limits the incentives to stay home instead of leaving in search of higher wages in the United States.

In the state of Michoacán, several job-creating projects using collective and individual remittances have been implemented, with mixed results. Many are struggling to survive despite repeated financial contributions from the government to prevent bankruptcy, while others have failed after a few years of operation (Bada 2008). This state was the pioneer in implementing agricultural projects with collective remittances, using a cooperative model that requires participation from at least 10 migrant investors. However, peasant production cooperatives in Mexico have had an uneven track record, and the state of Michoacán is no exception, with limited results after years of state government efforts since the economic restructuring of the late 1980s (Gledhill 1995, 212).

So far, migrants' main impact on the productive structures of rural communities is through the withdrawal of their labour, rather than through productive investment that creates sustainable employment. Yet they often do influence political and civic life. Do they encourage local democratization? Do they affect women's opportunities for participation and representation?[19] Many participants and observers expect that HTAs do have democratizing impacts, though the evidence is not yet clear. Returned migrants clearly play key roles in hometown public life as individuals. According to a survey carried out by the Michoacán state government migrant support agency, 37 percent of the 113 mayors who governed in the state during 2002–2004 were former migrants (Bada 2004c).

But the fact that some migrants return to fill local leadership roles does not answer the question about the civic and political impacts of HTAs. More generally, to what degree do the hometown associations reproduce the political culture that dominated Mexico in the twentieth century? Optimists often suggest that organized civil society generates democratic values and practices, and this is sometimes the case. But civil society also carries the weight of history, and is cross-cut by hierarchies and inequality between genders, classes and ethnic groups, as well as the legacy of less-than-democratic political ideologies. After all, many of the federations, as well as some of the HTAs, came together in response to Mexican government initiatives. If one interprets this relationship through the lens of state–society relations in Mexico, then this government strategy represents both a response to real demands from below, while also serving as an institutional channel to regulate relationships with migrant civil society. In principle, in contrast to similar government efforts *in* Mexico,

---

[19] For bi-national analyses of Mexican migrant organizing and gender, see Goldring (2004), Stephen (2007) and Maldonado and Artía Rodríguez (2004).

migrants in the US are less vulnerable to clientelistic manipulation, but some recent reports indicate that old habits die hard.[20]

While in some cases the persistence of home community political cultures across borders sustains persistent clientelism, in other cases a strong sense of local community membership grounds long-distance social cohesion. Indeed, many indigenous communities have strong, explicit criteria for determining local citizenship, based on high expectations of unpaid community service and informal taxation (Fox 2006a). As these communities become more involved in migration, some have created flexible approaches to allow for long-distance membership, permitting migrants called back for service to spend less time than usual, or to pay others to cover their dues (Kearney and Besserer 2004). In one Oaxacan case, returned migrants doing community leadership service formed a de facto coalition with locally excluded women to dislodge entrenched local bosses (Maldonado and Artía Rodríguez 2004).

Nevertheless, high levels of migration directly undermine indigenous community traditions that rely heavily on a large fraction of the adult male population providing service at any one time. While communities cannot prevent out-migration, some have found ways of discouraging exit by making return more difficult. For example, village elders may decide not to be flexible about long-distance membership, insisting that if villagers do not return to provide their service, they risk losing their local citizenship status. This carries both tangible and symbolic weight, land rights can be lost, and migrants who do not return when called can lose their right to be buried in the village cemetery. This adds up to what is known as 'civic death' (Mutersbaugh 2002).

The broad question of home community civic-political impact needs to be unpacked in at least two ways. First, to what degree do the HTAs themselves generate democratic values and practices? So far, research that compares the internal practices of different state federations finds a wide range of practices, from more to less democratic (Rivera-Salgado and Esacala Rabadán 2004). The second question would focus on their impacts in home communities. These questions are distinct because, in principle, hometown clubs could be highly representative of their US-based constituencies, but not necessarily of the non-migrant population.

Why might one expect migrant clubs to encourage democratization in home communities? Those that send collective remittances for community investments are taxing themselves for the benefit of others. Historically, those who pay taxes are accustomed to demanding some form of representation, which recalls the metaphor of exit, voice and loyalty. In this view, collective remittances are possible thanks to migrants' exit, they exist because of their loyalty, and they then tend to encourage the exercise of voice.

---

[20] For example, Fitzgerald's (2004) study of the cross-border/home community politics within a migrant-led California trade union local suggests that 'old politics' can persist across borders and triangulate homewards to involve communities of origin.

Such civic practices suggest the hypothesis that HTAs tend to hold local governments accountable.[21] However, even if most clubs are internally democratic, and even if they hold local governments accountable (to the HTA 'donors'), this does not necessarily generate democratization within the home community. Accountability refers to a power relationship, checks and balances, in this case between a specific constituency and the local government – but not necessarily vis-à-vis the majority of the community (whether defined in local *or* in translocal terms). Do the non-migrants play any role in determining how to invest collective remittances? How are choices weighed between infrastructure projects that the migrants use on their annual visits home vs those that may have a greater impact on the daily lives of non-migrants (e.g., rodeo rings vs water systems)? It should be no surprise that relationships between migrants and mayors are not always easy, especially now that local elections are more democratic in many regions of Mexico.

THE THREE-FOR-ONE MIGRANT COMMUNITY DEVELOPMENT PROGRAMME

Mexico's Three-for-One community development matching fund programme is a rare example of a development programme that emerged in direct response to civil society pressures – in this case from migrant civil society, beginning with a state-level programme in Zacatecas. This programme allows organized migrants to propose community development project ideas, mainly for small towns and villages, to be funded by collective remittances. Federal, state and municipal governments then vet the proposals. If approved, each level of government contributes matching funds. In principle, local committees oversee project implementation. The programme therefore opens a window on the balance of power and negotiation between these different governmental and civil society actors.

It turns out to be difficult to assess the influence of migrants in the selection of Three-for-One projects. In the beginning, the Three-for-One programme's operating rules stated that any organized citizen group was eligible to submit a project, but that situation changed in 2004, when HTA federation lobbying limited access to organized migrant groups. As a result, the new system has generated some tension within rural communities that lack connections to organized migrant groups in the United States, since it excludes locally organized citizen initiatives from access to this potential source of project funding. As noted above, many low income rural communities do not experience high rates of out-migration – though they may have access to other, larger-scale anti-poverty programmes, such as regular municipal funds, or the federal Micro-Regions Programme.

Research on the relationships between HTAs and stay-at-home community members in the decision-making process remains incipient. However, in an

---

[21] See especially Burgess (2005, 2006) and Williams (2004).

official evaluation's 2004 survey of HTA members from six states, 62 percent of the club members interviewed declared that project selection was decided by unanimous consensus and 38 percent reported that project selection decisions were made by majority vote, indicating a degree of democratization within HTA structures (Servicios Profesionales para el Desarrollo Económico 2005).

Striking a balance between the participation of local government officials, project beneficiaries and migrant groups in selecting Three-for-One projects has proven to be quite challenging. For instance, municipal staff complained that local governments were obliged to choose projects that were not a priority and they expressed frustration at having to report to the migrants regarding project advances and spending. On the other hand, the beneficiaries' main complaint was that they could only submit projects with the approval of organized migrant groups (Servicios Profesionales para el Desarrollo Económico 2005). More generally, research and media reports on the role of HTAs tend to underestimate the active participation of stay-at-home community members in many community development projects. Indeed, Bada's (2008) fieldwork in Zacatecas and Michoacán revealed that more than half of the projects visited in both states involved funds contributed by both the stay-at home community and the organized migrants.

More recently, a study of 13 communities in three Zacatecas municipalities receiving Three-for-One funds found that their HTAs in the US have been offering their support to projects led by stay-at-home community members in order to get the approval for Three-for-One project funding. This strategy was devised to comply with the requirement that projects must be exclusively submitted by organized migrants (García de Alba Tinajero et al. 2006, 224). This pattern of collaboration between migrants and stay-at-home community members also emerges in a national survey, which found that 59 percent of villagers contributed some money to Three-for-One projects (Secretaría de la Función Pública 2006). Some HTAs respond that they finance these projects indirectly, since some local family contributions are made possible by remittances.

In Bada's research on dozens of projects financed through Three-for-One in the state of Michoacán, both mayors and community members reported that they contributed extensively to the infrastructure projects, either through local fund-raising efforts or with unpaid community labour (known as *faenas*, or *tequio*, in indigenous communities). Recognizing the participation of both the 'sending' and 'receiving' ends of the transnational community is an important step to sustain participation without provoking intra-community conflicts, especially in communities that have seen their social fabric weakened by the massive departure of so many working adults.

According to Social Development Ministry officials, the programme's requirement that organized migrant federations select the projects complicates efforts to maximize the funds' anti-poverty impacts. As the data in Table 1 indicate, many of the lowest-income municipalities do not produce large numbers of international migrants; others produce mainly domestic migrants, who send fewer remittances and are not subjects of the programme.[22] The federal representative for the matching fund programme in Morelia reinforced this concern about the difficulty for channelling resources to the most impoverished municipalities:

> One of the problems that we face in channelling resources to these 35 [poorest] municipalities is that, for instance, these communities don't have potable water but migrants say that they want to fix the village square or they want to fix the church. They have problems of sewage but the migrants want to build a rodeo ring. We try to encourage them to fund projects that focus on immediate and basic needs but we can't obligate them.... We let the [state-level] validation committee choose the projects with the highest merits to support with public funds.[23]

Yet local authorities also share responsibility for many 'community development' investment decisions that have little to do with poverty reduction. Mayors tend to be more interested in financing more easily visible public infrastructure projects. In contrast, sewage and drinking water projects in outlying areas are not easily visible. Moreover, the requirement that both migrant committees and local authorities must agree on project proposals may also encourage a 'lowest common denominator' approach.

Nevertheless, in spite of these obstacles, a substantial fraction of the Three-for-One projects do address basic infrastructure needs. Table 2 shows the sectoral distribution of projects, according to an official external evaluation. The data clearly show a strong preference for public goods such as roads, drinking water, welfare services, paving and electrification, with only a small fraction invested in productive projects.

The power relationship between the organized migrants and the municipal authorities is also influenced by the parallel project committees, which are citizen groups organized around approved infrastructure projects funded through the Three-for-One programme. These committees are often elected in a community assembly or are chosen by HTA leaders to represent them during their dealings with the three levels of government. Their main function is to supervise the construction process. In the year 2006, a national government survey found that 87 percent of 91 Three-for-One projects had a formally constituted parallel committee (Secretaría de la Función Pública 2006). Nevertheless, most of these committees are weak, due in part to poor training and low literacy levels. Currently, in the state of Michoacán, not more than 10 percent of these committees are working properly and very few have effective bargaining power with

---

[22] Some mayors of low-income municipalities also report that even when they find groups of expatriates in the US, they 'don't want to participate because they are afraid to give out their personal information. They believe that if they send a letter committing to donate funds, the Mexican government will report them to the Immigration office in the United States and they will be deported' (Bada interview, Morelia, Michoacán, July, 2004). All interview translations were done by the authors.
[23] Bada interview with Social Development Ministry official in Morelia, May 2005.

Table 2. Categories of Three-for-One community development investments, 2002–2005

Type of project	Average share of 2002–2005 projects (%)
Food marketing	0.32
Drinking water	7.12
Drainage	4.95
Support for primary production	1.02
Social welfare and community services	15.66
Rural roads	8.80
Feeder roads	10.68
Health centre	2.20
Regional development planning	1.31
Irrigation works	0.96
Electrification	7.55
Production and productivity support	1.88
Sports infrastructure	3.39
Educational infrastructure	4.32
Livestock infrastructure	0.34
Paving	14.60
Historic and cultural sites	1.58
Urbanization	14.11
Protection of federal areas and watersheds	0.15
Housing	0.20

Source: Servicios Profesionales para el Desarrollo Económico (2005, pp. 38–40).

municipal authorities and HTA leaders.[24] They are also hindered by the lack of a clearly defined division of labour with US-based HTAs. In addition, HTAs still have limited accountability mechanisms vis-à-vis their own constituents. Nevertheless, the existence of these new trans-locally based oversight structures represents first steps towards representing the voices of previously underrepresented communities in municipal governments – especially given the broader context of the subordination of villages to municipal authorities based in the town centres.

Indeed, when one examines whether Three-for-One community development projects are located in the municipal centres vs the outlying communities, the pattern clearly favours the smaller villages. This is consistent with the widely held view that out-migration rates for the outlying communities are higher than for town centres. As indicated in Table 3, the emphasis on outlying communities holds for all of the principal states involved in the Three-for-One programme – with the notable exception of the state of Jalisco, where field reports indicate that

---

[24] Personal communication with a staff member from the Migrant Affairs State Office, 8 January 2006.

Table 3. Percentage of migrant community development projects outside the municipal centre

State	2002–2005 (%)
Guanajuato	82
Guerrero	84
Hidalgo	66
Jalisco	48
Michoacán	73
Oaxaca	58
Veracruz	65
Zacatecas	65

Source: Burgess (2006, p. 113).

the mayors are often more influential than the migrant organizations in making project decisions.[25]

This pattern of favouring the lower-income outlying communities in the programme as a whole is consistent with Bada's field interviews in the state of Michoacán, where HTA capacity to mobilize and lobby increases the voice and standing of the outlying communities vis-à-vis the municipal authorities. The most important tool that HTAs have to improve the allocation of funds for underserved communities turned out to be their capacity to negotiate directly with the *state* government, and to a lesser degree with the federal Social Development Ministry, and thereby pressure unresponsive municipal authorities.[26] HTAs have been effective in informing the state government about the needs of their communities and the unfulfilled promises that many municipalities have long made on issues regarding deficient elementary schools, water, electricity, roads, etc. In response, the state government has tried to raise awareness among the municipal presidents on the pressing conditions that many communities are facing outside the municipal centres. However, the success of the HTAs sometimes ends here due to their lack of capacity in project supervision and a poor understanding of their role as public accountability actors. This was evident in the results of a survey applied to Three-for-One beneficiaries and conducted by the Public Management Ministry, which revealed that only 43 percent understood their rights to have access to information regarding projects financed with

---

[25] Note also that the programme resources remain highly concentrated geographically, with 72 percent of the 2006 funding focused on the four states considered historic 'sending states' in the central-western region. Zacatecas leads with 27.39 percent, followed by Jalisco (26.96 percent), Michoacán (10.24 percent) and Guanajuato (8.03 percent) (Sagarnaga Villegas et al. 2006, p. 32). Not coincidentally, these four states together account for 52 percent of the more than 600 officially registered HTAs, as of 2003, the most recent year for which data are available (Bada et al. 2006, p. 7).

[26] This dynamic could be understood as an example of a much broader process, the 'boomerang effect' in transnational civil society campaigning (Keck and Sikkink 1998).

federal funds and only 4 percent had ever submitted a formal complaint to the appropriate authorities (Secretaría de la Función Pública 2006).

One of the main sources of HTA influence on behalf of their communities of origin is their institutionally recognized voice in the Three-for-One committee for project evaluation and approval. However, their effectiveness as new power brokers is limited by problems of 'excessive representation' after long absences as community members and the imposed silence associated with those absences. When the opportunity to recover their voice becomes available, some absent members want to have direct representation at every decision-making opportunity. For instance, in 2005, Michoacano HTAs were allowed to have a seat on the Three-for-One committee for project evaluation and approval and they chose a representative from Chicago. Soon, complaints from representatives from Las Vegas, Los Angeles, Texas, Alaska and many other places with vast HTA representation from Michoacán also wanted to have a seat on this committee. As a result, in 2007, the committee in Michoacán had 12 people representing all HTA associations and federations in the United States, but they only had one vote.[27] Reaching a consensus is a challenge when not all representatives can afford the trip to the committee meetings in Morelia (those are not usually paid by the government) and they live several thousand miles apart from each other in the United States. So far, they have been able to offer a unified vote in the first meeting of the committee, but the long-term success of this model has yet to be seen. Despite their increased presence and participation in the evaluation committee of the Three-for-One programme, they have not always been successful in convincing municipal authorities to carry out all the Three-for-One projects that are needed in remote communities and rejection rates remain high. Between 2002 and 2004, 192 Three-for-One projects were rejected in the state. Of those, 140 (73 percent) were located outside the town centre (Bada 2008). In the long term, however, the most significant impact of the Three-for-One Programme's increased leverage for outlying villages may unfold in other local civic arenas – if they manage to exercise greater voice in the rest of the municipal decision-making process. In other words, the civic *spillover* effects may turn out to be the most significant.

THE PERSISTENT DISCONNECT BETWEEN MIGRATION AND DEVELOPMENT[28]

In light of the clear overlap between the challenges of migration and rural development, one might expect high levels of dialogue and convergence between the analysts and social actors involved. After all, the growth in migrant worker remittances, combined with the spread of organized hometown associations, has provoked widespread optimism about prospects for investing in cross-border

---

[27] Bada interview with Social Development Ministry official in Morelia at the First Latin American Migrant Community Summit, Morelia, Michoacán, 10–13 May 2007.
[28] This section draws on Fox (2006b, 2007).

community development. Yet analyses of Mexican migration and development continue to engage at most sporadically, for reasons that are not well understood. Each agenda tends to treat the other as a residual category, while fully integrated approaches have yet to be developed. One factor may be that specifying the nature of the linkages between migration and development turns out to be easier said than done. For example, does sustainable/fair trade coffee production and marketing provide an alternative to migration, does it serve as a source of funding for marketing, or do remittances end up subsidizing coffee production because demand at fair trade prices is insufficient? Available research finds little evidence that fair trade/organic coffee slows migration.[29]

So far, the huge volumes of economic remittances have attracted most of the public and policy attention. The framing of migration and development issues through the lens of remittances draws attention to questions of how financial institutions can capture the funds. While 'banking the unbanked' is certainly very important to those both sending and receiving remittances, the connection to broader development remains uncertain. For migrants and their families, the most tangible impact of the widespread public discussion has been the significant recent reduction in transaction costs, driven largely by increased private sector competition. The remittance focus also draws attention to collective remittance investments, primarily for social infrastructure rather than economic development (as indicated in Table 2).

The potential of remittances to generate economic development alternatives has been discussed for more than a decade, but in Mexico there is still little tangible evidence of sustainable jobs beyond a few micro-level cases. The challenge of finding and managing economically viable projects is compounded by the structure of the decision-making process. When migrants pool their hard-earned money for hometown projects, they place a premium on those investments that provide benefits to the community *as a whole*. Most job-creating investments, in contrast, directly affect only a small subset of the community. In addition, their benefits may be perceived as at risk of being captured by local elites or well-connected kinfolk – in a context in which 'long-distance accountability' is difficult. This dilemma suggests the importance of identifying those productive investments that can also have 'public goods' effects, such as improved coffee-processing infrastructure in those communities where most people depend on coffee and already have years of experience working together in a marketing coop whose leadership is publicly accountable. Yet this category of potential investment projects has yet to be linked to migrant collective action.

Efforts to bring migrant organizations into the broader development policy debate are still incipient, as their Mexican policy agenda continues to be dominated by the traditionally bounded 'migration policy' framework, limited to the

---

[29] For one of the few studies to directly address the relationship between migration and fair trade/organic coffee initiatives, see Lewis and Runsten (2005).

Three-for-One programme, the Institute for Mexicans Abroad and Mexico's approach to US immigration and border policies. Besides, locally based-NGOs are not always aware of the existence of HTAs and therefore there is no communication or common agenda to develop shared sustainable rural development goals. Even at the level of local and trans-local policy agendas, few cross-border membership organizations support grassroots development agendas *both* in communities of origin *and* in communities of settlement. Mexico's Association of Social Sector Credit Unions has worked with migrant organizations to launch a network of rural micro-banks, which could provide working capital for local economic development. The Bi-national Front of Indigenous Organizations (FIOB) is another exception, as it builds a participatory grassroots microcredit network back home, to make a locally accountable institutional base that could eventually receive and invest remittances (Domínguez Santos 2004).

In an effort to craft a new way of framing the relationship between migration and development, Mexican rural development strategist Armando Bartra (2003) bridges the migration, development and rights agendas with his call for respect for 'the right to not [have to] migrate'. After all, the Mexican Constitution's Article 123 still speaks of citizens' right to 'dignified and socially useful work'. The 'right to not migrate' can be a useful bridging concept for promoting reflection and discussion between diverse and sometimes disparate actors who see the process differently. This principle recognizes that while migration is an option, it is a choice made within a context imposed by public policies that enable some development strategies over others. Yet the apparently limited impact of the 'right to not migrate' concept suggests that translating an evocative frame into practical strategies for grassroots organizations turns out to be a serious challenge.

What might explain this persistent disconnect between migration and development? Migration is increasingly recognized as spreading throughout Mexico, remittances are widely seen as a development resource, and those practitioners and analysts working on migration increasingly acknowledge the need to take into account dynamics in communities of origin. Perhaps the roots go deeper and one needs to look at the basic frameworks used to define strategies for change. Most of Mexico's rural development practitioners and analysts implicitly treat migration as an external process happening 'outside' the grassroots development process, a de facto residual category – whereas for *campesino* families, migration is *inside* the box, a central component of a diversified survival strategy. For most practitioners and analysts who are working on migration, in contrast, the development dimension of the relationship between receiving and sending community is understood in terms of 'philanthropy from below', an approach that tends to prioritize high profile, 'something for everyone' projects over policy advocacy for job creation and sustainable development.

One indicator of the challenge of engaging the migration and development agendas involves the uneven landscapes of the relevant community-based organizations. Mexican migrants, for example, have generated a broad and diverse array of membership organizations, but they vary widely in their density and distribution. To contribute more directly to grassroots development strategies on the ground, a next stage of mapping is necessary. At the level of a state or a region, it would be very useful to take a map of those communities whose migrants have generated hometown associations and lay it over a map of those communities of origin that have also generated the social, civic and economic development organizations that could serve as counterparts with the organized migrants. Some 'sending' communities in the state of Oaxaca have very limited economic development prospects, but others have significant, scaled-up community-based enterprises, such as organic coffee and timber cooperatives. Imagining alternatives with those organized migrants who come from hometowns with community-based economic development track records could go a long way toward addressing the issues that make productive investments of remittances difficult. Those issues include the need for viable investment prospects, for entrepreneurial experience and reliable technical support, for public accountability to the communities of origin, and for positive social spillover effects beyond the local interested parties. Yet this social-geographical convergence between territorially-based migrant organizations and grassroots-led community economic development initiatives remains incipient.

CONCLUSIONS

The emergence of Mexican migrant civil society suggests that exit can be followed by voice. For many Mexican migrants, autonomous collective action begins as they look homeward. For those who were active before they left, civic life back home may be undermined, at least in the short term – though some later provide community service, directly or indirectly. Reflecting on those Mexicans active in migrant civil society who had track records of collective action before leaving suggests that many find new pathways for expressing their commitments, following Hirschman's (1984) principle of the 'transformation and mutation of social energy'. This idea refers to the ways in which activists often draw on their formative experiences with collective action, even after major changes in their political context and social terrain, and draw on these legacies to inform their new initiatives in different arenas.

The preceding analysis of the project decision-making dynamics within the Three-for-One community development investment programme indicates that migrant organizations have some capacity to bolster the representation of their often-subordinated home communities within municipal, state and federal politics. Yet this programme represents a tiny fraction of Mexico's overall social investment spending. More importantly, this programme has not managed to leverage substantial, sustainable productive investments. Without economically viable, broad-based, socially credible job alternatives, out-migration will continue to deepen in those communities not yet considered to have reached 'high migration intensity'. As a result, the migrant hometown associations that are tithing themselves to invest back home in public goods may face the dilemma of building basketball courts and baseball stadiums with very few players, except for the 3–4 weeks each year when expatriates return to visit.[30] The Three-for-One programme's investment in roads may facilitate migrants' return home over Christmas, but they also lead the next generation north.

In conclusion, Mexican experiences with organized migrant involvement in hometown community development initiatives show that voice sometimes can follow exit. Yet voice that is limited to addressing the symptoms rather than the causes of exclusion is unlikely to lead to sustainable community development.

REFERENCES

Bada, Xóchitl, 'Reconstrucción de identidades regionales a través de proyectos de remesas colectivos: la participación ciudadana extraterritorial de comunidades migrantes michoacanas en el área metropolitana de Chicago'. In *Organizaciones de Mexicanos en Estados Unidos: la política transnacional de la nueva sociedad civil migrante*, eds Guillaume Lanly and M. Basilia Valenzuela, 175–224. Mexico: Universidad de Guadalajara.

Bada, Xóchitl, 2004b. 'Clubes de michoacanos oriundos: Desarrollo y membresía social comunitarios'. *Migración y Desarrollo*, April, No. 2.

Bada, Xóchitl, 2004c. 'Las remesas colectivas de las organizaciones de migrantes mexicanos: Participación cívica transnacional y estrategias comunitarias de desarrollo'. Paper presented at the 4° Congreso sobre la Inmigración en España, Ciudadanía y Participación, Girona, 10–13 November.

Bada, Xóchitl, 2008. 'Transnational and Trans-local Sociopolitical Remittances of Mexican Hometown Associations in Michoacán and Illinois'. PhD Dissertation in Sociology, University of Notre Dame, forthcoming.

Bada, Xóchitl, Jonathan Fox and Andrew Selee, eds, 2006. *Invisible No More: Mexican Migrant Civic Participation in the United States*, Washington, DC: Woodrow Wilson Center, Mexico Institute/University of California, Santa Cruz, Latin American and Latino Studies, August. http://www.wilsoncenter.org/migrantparticipation, Accessed 1 February 2008.

Banco de México, 2007. 'Las remesas familiares en México. Inversión de los recursos de migrantes: resultados de las alternativas vigentes'. Mexico City: Banco de Mexico.

Bartra, Armando, 2003. *Cosechas de ira: Economía política de la contrareforma agraria*. México: Ed. Ithaca/Instituto Maya.

Brooks, David and Jonathan Fox, eds, 2002. *Cross-Border Dialogues: US–Mexico Social Movement Networking*, La Jolla, CA: University of California, San Diego, Center for US–Mexican Studies.

Burgess, Katrina, 2005. 'Migrant Philanthropy and Local Governance in Mexico'. In *New Patterns for Mexico: Remittances, Philanthropic Giving and Equitable Development*, ed. Barbara Merz, 99–124. Cambridge, MA: Harvard University Press.

Burgess, Katrina, 2006. 'El impacto del Three-for-One en la gobernanza local'. In *El Programa 3x1 para Migrantes, ¿Primera politica transnacional en Mexico?*, eds Rafael Fernández de Castro, Rodolfo García Zamora and Ana Vila Freyer, 99–138. Mexico: ITAM/UAZ/Miguel Angel Porrúa.

---

[30] The latter scenario was depicted in the 2003 documentary The Sixth Section (see http://www.sixthsection.com)

Carral Dávila, Alberto, 2006. 'Migración Rural'. In *Escenarios y actores en el medio rural*, ed. Héctor Robles Berlanga, 89–118. Mexico City: Centro de Estudios para el Desarrollo Rural Sustentable y la Soberanía Alimentaria, Dec.

CONAPO, 2000. *Indices de marginación*. Mexico City: Consejo Nacional de Población. http://www.conapo.gob.mx, Accessed 2 February 2008.

Cordero-Guzmán, Héctor R. and Victoria Quiróz Becerra, 2005. 'Mexican Hometown Associations (HTA) in New York'. New York City: Baruch College, Working paper. http://www.wilsoncenter.org/news/docs/Mexican-HTA-NYC-CQ-10-05.pdf, Accessed 1 February 2008.

Cornelius, Wayne and David Myhre, eds, 1998. *The Transformation of Rural Mexico*. La Jolla, CA: University of California, San Diego, Center for US–Mexican Studies.

Domínguez Santos, Rufino, 2004. 'The FIOB Experience: Internal Crisis and Future Challenges'. In *Indigenous Mexican Migrants in the United States*, eds Jonathan Fox and Gaspar Rivera-Salgado, 69–80. La Jolla, CA: University of California, San Diego, Center for Comparative Immigration Studies and Center for US–Mexican Studies.

Escala Rabadán, Luis and Carol Zabin, 2002. 'Mexican Hometown Associations and Mexican Immigrant Political Empowerment in Los Angeles'. *Frontera Norte*, No. 27.

Fitzgerald, David, 2000. *Negotiating Extra-Territorial Citizenship: Mexican Migration and the Transnational Politics of Community*. La Jolla, CA: University of California, San Diego, Center for Comparative Immigration Studies, Monograph 2.

Fitzgerald, David, 2004. 'Beyond Transnationalism: Mexican Hometown Politics and an American Labour Union'. *Ethnic and Racial Studies*, March, 27 (2): 228–47.

Fox, Jonathan, ed., 1990. *The Challenge of Rural Democratisation: Perspectives From Latin America and the Philippines*. London: Frank Cass.

Fox, Jonathan, 1994. 'The Politics of Mexico's New Peasant Economy'. In *The Politics of Economic Restructuring: State-Society Relations and Regime Change in Mexico*, eds Maria Lorena Cook, Kevin J. Middlebrook and Juan Molinar, 243–76. La Jolla, CA: UCSD, Center for US–Mexican Studies.

Fox, Jonathan, 1996. 'How Does Civil Society Thicken? The Political Construction of Social Capital in Rural México'. *World Development*, June, 24 (6): 1089–103.

Fox, Jonathan, 2005a. 'Unpacking "Transnational Citizenship"'. *Annual Review of Political Science*, 8: 171–201.

Fox, Jonathan, 2005b. 'Mapping Mexican Migrant Civil Society'. Presented at 'Mexican Migrant Social and Civic Participation'. Woodrow Wilson Center for Scholars, November. http://www.wilsoncenter.org/migrantparticipation, Accessed 1 February 2008.

Fox, Jonathan, 2006a. 'Rethinking Mexican Migration as a Multi-Ethnic Process'. *Latino Studies*, 4 (1): 39–61.

Fox, Jonathan, 2006b. 'Mexican Migration and Development: Encounters and Disconnects'. *Grassroots Development*, 27 (1): 2–5.

Fox, Jonathan, 2007. *Accountability Politics: Power and Voice in Rural Mexico*. Oxford: Oxford University Press.

Fox, Jonathan and Elizabeth Haight, forthcoming. 'El condicionamiento político del acceso a programas sociales en México'. In *Candados y derechos: El blindaje de la política social en México desde una perspectiva comparada*, eds Alejandro Grinspun and David Gómez Alvarez, United Nations Development Programme in preparation.

Fox, Jonathan and Gaspar Rivera-Salgado, eds, 2004. *Indigenous Mexican Migrants in the United States*. La Jolla, CA: University of California, San Diego, Center for Comparative Immigration Studies and Center for US–Mexican Studies.

García de Alba Tinajero, María, Leticia M. Jáuregui Casanueva and Claudia Núñez Sañudo, 2006. 'Liderazgos y nuevos espacios de negociación en el Programa 3x1 para Migrantes. El caso de Zacatecas'. In *El Programa 3x1 para Migrantes. ¿Primera Política Trasnanacional en México?*, eds Rafael Fernández de Castro, Rodolfo García Zamora and Ana Vila Freyer, 223–48. Mexico, DF: ITAM-Universidad Autónoma de Zacatecas-Miguel Ángel Porrúa.

García Zamora, Rodolfo, 2005a. 'Migración internacional y remesas colectivas en Zacatecas'. *Foreign Affairs en Español*, July–Sept. http://www.foreignaffairs-esp.org, Accessed 1 February 2008.

García Zamora, Rodolfo, 2005b. 'Collective Remittances and the 3 X 1 Program'. Washington, DC: Woodrow Wilson Center and Latin American and Latino Studies Department, UC Santa Cruz, November. http://www.wilsoncenter.org/migrantparticipation, Accessed 1 February 2008.

García Zamora, Rodolfo, 2006. 'El Programa 3 X 1 y los retos de los proyectos productivos en Zacatecas' In *El Programa 3x1 para Migrantes. ¿Primera Política Transacional en México?*, eds R. Fernández de Castro, R. García Zamora and A. Vila Freyer, 157–70. Mexico, DF: ITAM-Universidad Autónoma de Zacatecas-Miguel Ángel Porrúa.

Gledhill, John, 1995. *Neoliberalism, Transnationalization and Rural Poverty. A Case Study of Michoacán, Mexico*. Boulder, CO: Westview Press.

Goldring, Luin, 2002. 'The Mexican State and Transmigrant Organizations: Negotiating the Boundaries of Membership and Participation'. *Latin American Research Review*, 37 (3): 55–99.

Goldring, Luin, 2004. 'Family and Collective Remittances to Mexico: A Multidimensional Typology'. *Development and Change*, 35 (4): 799–840.

González Amador, Roberto, 2005. 'En lo que va del sexenio emigraron a Estados Unidos 400 mil personas al año'. *La Jornada*, 15 April.

González Gutiérrez, Carlos, 1997. 'Decentralized Diplomacy: the Role of Consular Offices in Mexico's Relations with its Diaspora'. In *Bridging the Border: Transforming US–Mexican Relations*, eds Rodolfo O. De la Garza and Jesús Velasco. Lanham, MD: Rowman and Littlefield.

Goodman, Gary and Jonathan Hiskey, 2008. 'Exit Without Leaving: Political Disengagement in High Migration Municipalities in Mexico'. *Comparative Politics*, January, 40 (2): 169–88.

Hernández Navarro, Luis, 2006. 'Optimismo y cambio en América Latina'. *La Jornada*, 31 January: 27.

Hirschman, Albert, 1970. *Exit, Voice and Loyalty: Responses to Decline in Firms, Organizations and States*. Cambridge, MA: Harvard.

Hirschman, Albert, 1981. *Essays in Trespassing: Economics to Politics and Beyond*. Cambridge: Cambridge University Press.

Hirschman, Albert, 1984. *Getting Ahead Collectively: Grassroots Experiences in Latin America*. Elmsford, NY: Pergamon Press.

Holt-Gimenéz, Eric, 2006. *Campesino a Campesino: Voices from Latin America's Farmer to Farmer Movement for Sustainable Agriculture*. Oakland, CA: Food First Books.

INEGI, n.d. (Instituto Nacional de Estadística y Geografía) 'Población ocupada según sector económica (nacional)'. Encuesta Nacional de Ocupación y Empleo. http://www.inegi.gob.mx, Accessed 2 February 2008.

INEGI, 2006. 'Resultados Definitivos del II Conteo de Población y Vivienda 2005 para el Estado de Michoacán de Ocampo'. Mexico City: Instituto Nacional de Estadística Geografía e Informática.

Kearney, Michael and Federico Besserer, 2004. 'Oaxacan Municipal Governance in Transnational Context'. In *Indigenous Mexican Migrants in the United States*, eds Jonathan Fox and Gaspar Rivera-Salgado, 449–68. La Jolla: University of California, San Diego, Center for Comparative Immigration Studies and Center for US–Mexican Studies.

Keck, Margaret and Kathryn Sikkink, 1998. *Activists Beyond Borders*. Ithaca, NY: Cornell.

Lanly, Guillaume and M. Basilia Valenzuela, eds, 2004. *Clubes de migrantes oriundos mexicanos en los Estados Unidos: la política transnacional de la nueva sociedad civil migrante*. Guadalajara, Mexico: Universidad de Guadalajara.

Laxer, Gordon and Sandra Halperin, eds, 2003. *Global Civil Society and its Limits*. London: Palgrave.

Levitt, Peggy, 2001. *Transnational Villagers*. Berkeley, CA: University of California.

Lewis, Jessa and David Runsten, 2005. 'Does Fair Trade Coffee Have a Future in Mexico? The Impact of Migration in a Oaxacan Community'. Presented at 'Trading Morsels' Conference, Princeton University, February 2005.

Leyva Reus, Jeanette, 2005. 'Motivan polémica modelos para clasificar remesas familiares. Reportan Sedesol y Colef diferencia de 41.9 porciento con el Banxico. Imposible que los hogares hayan captado 16 mil mdd'. 23 June. In *El Financiero* online. México, DF. http://www.elfinanciero.com.mx, Accessed 3 February 2008.

Lozano Ascencio, Fernando, 2003. 'Discurso Oficial, Remesas, y Desarrollo en México'. *Migración y Desarrollo*, 1: 23–31.

Lozano Ascencio, Fernando and Fidel Olivera Lozano, 2005. 'Impacto económico de las remesas en México: un balance necesario'. Paper presented at the Seminario Internacional: Problemas y Desafíos de la Migración y el Desarrollo en América. Cuernavaca, Morelos: Red Internacional Migración y Desarrollo. http://www.migracionydesarrollo.org, Accessed 3 February 2008.

Maldonado, Centolia and Patricia Artía Rodríguez, 2004. '"Now We are Awake". Women's Participation in the Oaxacan Indigenous Binacional Front'. In *Indigenous Mexican Migrants in the United States*, eds Jonathan Fox and Gaspar Rivera-Salgado, 495–510. La Jolla, CA: University of California, San Diego, Center for Comparative Immigration Studies and Center for US–Mexican Studies.

Martin, Philip, 2005. 'NAFTA and Mexico–US Migration'. Presented at Consejo Nacional de Población, Mexico City, 16 December 2005.

Moctezuma Longoria, Miguel, 2003a. 'The Migrant Club El Remolino: A Binational Community Experience'. In *Confronting Globalization: Economic Integration and Popular Resistance in Mexico*, eds Timothy Wise, Hilda Salazar and Laura Carlsen. West Hartford, CT: Kumarian Press.

Moctezuma Longoria, Miguel, 2003b. 'Territorialidad de los clubes zacatecanos en Estados Unidos'. *Migración y Desarrollo*, October, No. 1.

Moctezuma Longoria, Miguel, 2005. 'Transnacionalismo, agentes y sujetos migrantes. Estructura y niveles de las asociaciones de mexicanos en Estados Unidos'. Presented at 'Problemas y Desafíos de la Migración y Desarrollo en América', Red Internacional de Migración y Desarrollo. Centro Regional de Investigaciones Multidisciplinarias de la UNAM y CERLAC, Cuernavaca, 7–9 April.

Muñoz, Alma, 2004. 'Niega Sedeso que remesas de migrantes ayuden a paliar la pobreza del país'. *La Jornada*, 5 December.

Mutersbaugh, Tad, 2002. 'Migration, Common Property and Communal Labor: Cultural Politics and Agency in a Mexican Village'. *Political Geography*, 21 June.

Mutersbaugh, Tad, 2008. 'Serve and Certify: Paradoxes of Service Work in Organic Coffee Certification'. In *Confronting the Coffee Crisis: Fair Trade, Sustainable Livelihoods and Ecosystems in Mexico and Central America*, eds Christopher Bacon, V. Ernesto Méndez, Stephen R. Gleissman, David Goodman and Jonathan A. Fox, 261–88. Cambridge, MA: MIT Press.

Orozco, Manuel with Michelle LaPointe, 2004. 'Mexican Hometown Associations and Development Opportunities'. *Journal of International Affairs*, 57 (2): 31–49.

Passell, Jeffrey, 2005. 'Unauthorized Migrants: Numbers and Characteristics'. Washington, DC: Pew Hispanic Center, 14 June 2005. http://www.pewhispanic.org/reports, Accessed 1 February 2008.

Passell, Jeffrey and Roberto Suro, 2005. 'Rise, Peak and Decline: Trends in US Immigration, 1990–2004'. Washington, DC: Pew Hispanic Center, 21 March 2005. http://www.pewhispanic.org/reports, Accessed 2 February 2008.

Polaski, Sandra, 2003. 'Jobs, Wages and Household Income'. In *NAFTA's Promise and Reality: Lessons from Mexico for the Hemisphere*, eds Demetrious Papademetriou, John Audley, Sandra Polaski and Scott Vaughn, Carnegie International Endowment for Peace, November 2003. http://www.cargenieendowment.org, Accessed 2 February 2008.

Ramos, Jorge, 2007. 'Migró más gente de la que murió en el país en 2006'. *El Universal*, 4 May, Mexico City. http://www.eluniversal.com.mx, Accessed 1 February 2008.

Rivera-Salgado, Gaspar and Luis Escala Rabadán, 2004. 'Collective Identity and Organizational Strategies among Indigenous and Mestizo Mexican Migrants'. In *Indigenous Mexican Migrants in the United States*, eds Jonathan Fox and Gaspar Rivera-Salgado, 145–78. La Jolla, CA: University of California, San Diego, Center for Comparative Immigration Studies and Center for US–Mexican Studies.

Rivera-Salgado, Gaspar, Xóchitl Bada and Luis Escala Rabadán, 2005. 'Mexican Migrant Civic and Political Participation in the US: the Case of Hometown Associations in Los Angeles and Chicago'. Presented at 'Mexican Migrant Civic and Political Participation', Washington, DC: Woodrow Wilson Center and Latin American and Latino Studies Department, UC Santa Cruz. http://www.wilsoncenter.org/migrantparticipation, Accessed 2 February 2008.

Sagarnaga Villegas, Leticia Myriam, et al., 2006. 'Evaluación Externa del Programa 3 x 1 para Migrantes 2006'. Mexico: Universidad Autónoma Chapingo/Sedesol, December. http://www.coneval.gob.mx/evaluaciones, Accessed 1 February 2008.

Schwentesius, Rita, Miguel Angel Gómez, José Luis Calva Téllez and Luis Hernández Navarro, eds, 2004. *¿El campo no aguanta más?*, 2nd edn. Chapingo: Universidad Autónoma de Chapingo, CIESTAAM.

Secretaría de la Función Pública, 2006. 'Encuesta de Opinión al Programa 3x1 para Migrantes. Beneficiarios'. Mexico City: Secretaría de la Función Pública.

SEDESO, Michoacán, 2004. 'Crecimiento poblacional estatal 1990–2000'. In Excel database of migration trends between 1980 and 2000. Morelia: Secretaría de Desarrollo Social.

SEDESOL, 2006. 'Programa 3x1 para Migrantes'. Mexico City: Secretaría de Desarrollo Social, Dissemination Brochure.

Servicios Profesionales para el Desarrollo Económico, 2005. 'Evaluación Externa del Programa Iniciativa Ciudadana 3x1, 2004'. Mexico City: Secretaría de Desarrollo Social (SEDESOL). http://www.coneval.gob.mx/evaluaciones, Accessed 1 February 2008.

Smith, Michael Peter, 2003. 'Transnationalism, the State and the Extraterritorial Citizen'. *Politics and Society*, December, 31 (4): 467–502.

Smith, Michael Peter and Matt Bakker, 2008. *Citizenship across Borders: The Political Transnationalism of El Migrante*. Ithaca, NY: Cornell University Press.

Smith, Robert, 2003. 'Migrant Membership as an Instituted Process: Transnationalization, the State and the Extra-Territorial Conduct of Mexican Politics'. *International Migration Review*, Summer, 37 (2): 297–343.

Smith, Robert, 2006. *Mexican New York. Transnational Lives of New Immigrants*. Berkeley, CA: University of California Press.

Stephen, Lynn, 2007. *Transborder Lives: Indigenous Oaxacan Migrants in the US and Mexico*. Durham, NC: Duke University Press.

Suro, Roberto, 2005. 'What Do Surveys Tell Us about Mexican Migrant Social and Civic Participation'. Presented at 'Mexican Migrant Civic and Political Participation'. Washington, DC: Woodrow Wilson Center and Latin American and Latino Studies Department, UC Santa Cruz, November. http://www.wilsoncenter.org/migrationparticipation, Accessed 1 February 2008.

Waldinger, Roger, 2007. 'Between Here and There: How Attached Are Latino Immigrants To Their Native Country?' Washington, DC: Pew Hispanic Center. http://www.pewhispanic.org, Accessed 1 Februay 2008.

Williams, Heather, 2004. 'Both Sides Now: Migrants and Emerging Democratic Politics in Mexico'. Presented at the Latin American Studies Association, Las Vegas.

World Bank, 2004. *Mexico: Public Expenditure Review* (Two Volumes), Report No. 27894-MX, 10 August. http://www.worldbank.org, Accessed 1 February 2008.

11 From Covert to Overt: Everyday Peasant Politics in China and the Implications for Transnational Agrarian Movements

KATHY LE MONS WALKER

INTRODUCTION

In the mid-1980s, with the publication of his path-breaking and enormously influential *Weapons of the Weak: Everyday Forms of Peasant Resistance* (1985), James Scott shifted the analytic lens from peasant revolution to resistance. Discarding the focus on collective action, rebellion and revolutionary mobilization that had preoccupied much of the prior research on peasant politics, Scott chose to spotlight the covert, informal and often individual acts through which, reinforced by a popular culture of resistance, peasants attempted to maintain or better their position in agrarian class struggles. In emphasizing that when 'multiplied many thousand-fold' individual acts of everyday resistance – foot-dragging, arson, sabotage, desertion, pilfering, feigned ignorance, and so on – might have significant political or socioeconomic consequences, he thus opened up new ground for conceptualizing protest and analyzing the contours and consequences of agrarian struggles. On the other hand, conceived at a time when radical social movements were on the wane and ideological pessimism was spreading among academics and organizers alike, Scott's theory of everyday resistance came to have a broader, more negative impact. It contributed to a recasting of peasants as not those who were capable of transforming new historical conditions, but rather as relatively disempowered agents whose struggles were mostly defensive adaptations to change.

The global resurgence over the last two decades of grassroots rural movements, as well as the phenomenon of organized transnational agrarian movements, problematize Scott's formulation. In particular, they prompt us to consider the question of why everyday resistance must be covert and deferential, and to consider historically the circumstances in which covert resistance prevails or in which peasants move to more overt collective action. In this contribution I do not intend to enter into an extended discussion or critique of Scott's sociological model. Rather my aim is to draw attention to the above questions by focusing empirically on China, where since the mid-1980s rural collective protests and instances of outright insurgency have become commonplace. Involving tens of millions of participants, contentious political activism, I argue, has become the everyday form of Chinese peasant politics.

In analyzing the development of this sustained overt resistance among Chinese peasants, the chapter takes as its starting point the conjunctural coincidence of two separate developmental trends: the turn to 'market socialism' in China in 1978 and the shift to neoliberalism in the advanced capitalist world. It first considers the features and negative consequences for the countryside resulting from the intertwining of these two trends, arguing that in the post-socialist period China's rise as a growth engine for the world economy has been achieved to a considerable degree both through the labour and at the expense of rural people. It then examines the collective protest of the last two decades, highlighting first the contradictory interplay of corrupt local and central state power in the initial development of the rural movement; and focusing, at greater length, on land seizures, which over the last decade have become the principal focus of contention. The final portion of the chapter considers the implications of these findings on the recent struggles of Chinese peasants – whose 210 million households represent one out of every three farming families in the world – for transnational agrarian movements.

CONTOURS: CHINA'S POST-SOCIALIST PATH

Since China's leaders abandoned the Maoist road to socialism and adopted a programme of marketization in 1978, China's spectacular emergence as one of the main growth engines for the world economy has been chronicled widely. More recently scholars and the press have begun to focus as well on the darker aspects of that global ascendance, including the growing social inequalities, the layoff and displacement of 60 million urban and industrial workers, and, especially, the fact that the countryside – where the majority of Chinese still live – seems to have been left behind. In this section so as to provide historical contextualization for the growth of widespread rural collective action, I wish to consider several of the interrelated aspects of China's post-socialist path that have influenced negatively rural people.

First, in making the post-socialist turn China's leaders, headed by Deng Xiaoping, expected to maintain a substantial degree of control over the marketization process, conceived initially as one that would combine state-owned enterprises (SOEs) – the 'stable centrepieces' of the state-run economy – with gradual marketization and privatization. But each stage in the process generated new contradictions that were resolved only through a further expansion of market power, which in turn encouraged the privileging of private over state enterprises, urban over rural development, and, increasingly, foreign enterprises and markets over domestic ones (Hart-Landsberg and Burkett 2004, 9, 31).

Second, China's post-socialist turn coincided with a parallel movement in the advanced capitalist world – the shift to neoliberalism. Born from a crisis of overaccumulation among the advanced capitalist regimes (Brenner 2002; Harvey 2003), the reorganization of the accumulation process under the neoliberal strategy depended, in the last analysis, on substantial transfers of property and property rights on a world scale and on a dramatic increase in the international competition of workers through the global expansion of cheap labour. Precisely because the post-socialist developments in China generated enormous possibilities for the release of 'hitherto unavailable assets into the mainstream of capitalist accumulation', the neoliberal players used all of their organizational capacity to facilitate and create favourable conditions for China's re-entry into the global capitalist system (Harvey 2003, 149; 2005, 121). The restoration of class power and exclusion of massive numbers of the world's people, especially the rural poor, from the accumulation process underpinning the neoliberal project have thus also shaped the contours and outcomes of China's increasingly 'foreign-driven, export-led' development (Amin 2003; Harvey 2003, 137–82; 2005, 120–1).

Third, if the global neoliberal project set itself up as a modern form of 'primitive accumulation' or enclosure targeting globally both customary/established property relations and state-created 'social commons' (De Angelis 2001), in China it found willing accomplices and allies: an alliance of officials and businessmen who were anxious to carry out their own variant of primitive accumulation and thus benefit directly from the interlinked accumulation processes. Officials and cadre initially assumed the leading role. In the breakneck race to get rich, they resorted illicitly to a 'wanton plunder' and 'wholesale pillaging' of public property and state assets (Hu 2005). In the 1980s, this occurred mainly through the illicit reselling on the market of state sector raw materials and commodities. In the late 1980s and 1990s, the trend deepened as SOEs were transformed into private shareholding companies and were stripped of their assets, as state funds were embezzled for the setting up of private businesses, and as officials sold off state-owned (in urban areas) or collectively-owned (in the countryside) land use rights (He 2000, 38, 45, 50, 59; 2001a, 27; 2001b, 68). This pattern of primitive accumulation, albeit to a lesser degree, was replicated in cities, towns and villages across the country in 'wicked coalitions' (*hei tongmeng*), as the Chinese press has termed them, of corrupt party and government officials and eager entrepreneurs (Cody 2005, 1A; Kahn and Yardley 2007; Li C. 2006, 6). In the process, especially after 1992 when capitalists and businessmen were first allowed to become party members, the Communist Party was transformed into a 'ruling syndicate' of the representatives of the wealthy and privileged social layers. By 2007, one-third of China's richest people were party members (Chan 2007, 1).

Flanked by and intertwining with massive amounts of foreign direct investment, which from 1997 to 2005 totalled over US$465 billion, the post-socialist path has generated not only nearly double-digit growth figures for more than a decade, but also unprecedented levels of wealth. By the end of 2004, there were 236,000 millionaires (in US dollars) in China – more than in India and Russia combined (Barboza 2005, 1). By 2007 the number had risen to 345,000. At the same time, the number of billionaires moved from 15 in 2006 to 106 in 2007 (Cheng and Lawrence 2007). The children of China's senior leaders had also joined the ranks of the super rich. Of the 3,220 Chinese with a personal wealth of 100 million yuan[1] or more, 2,932 were children of high-level cadres (Holz 2007, 38).

---

[1] The rate of exchange is 1 Yuan=US$0.1354, or US$1=7.385 Yuan.

Against the story of economic growth and the accumulation of great wealth by the bureaucratic-capitalist elite stands that of rapidly increasing inequalities, human degradation and other social costs. In less than three decades, China moved from being the most egalitarian society in the world to one in which the gap between the rich and poor is among the highest worldwide (Fewsmith 2007, 9). Spatial cleavages and regional inequalities also deepened as the coastal areas became the principal sites of direct foreign investment and both urban and rural industrialization, while the vast agricultural hinterland remained mostly peripheral to these processes. Between 1978 and 2003, for example, 86 per cent of total foreign direct investment went to the coastal areas of the south and east, while only 9 per cent went to the central region and 5 per cent to the west (Li C. 2006, 21). Equally glaring has been the growing disparity in rural–urban income – urban incomes by 2005 averaging 3.2 times those in the rural areas (Kahn 2005, 3).

Indeed, it was only in the first few years of China's turn to marketization that the countryside as a whole saw real gains. From 1978 to 1984, rural incomes rose at a rate of 15 per cent per year. After 1984, however, rural producers encountered multiple difficulties. The initial productive increases gained from family farming following the dismantling of the people's communes could not be sustained, the government lowered quota prices, the cost of inputs rose, and the tax burden of many peasants increased substantially. Consequently, increases in rural incomes began to slow, contract and in some areas even reverse (Hart-Landsberg and Burkett 2004, 34, 40). The over-taxation, embezzlement and de facto privatization by corrupt local cadres of township and village enterprises (TVEs), which were created from communes' assets and helped to subsidize agriculture, exacerbated this trend (Greenfield and Leong 1997, 107–8). Consequently, from the mid-1980s there was a growing popular perception in rural areas that in China's rush to get rich the countryside was being left behind. In typically sardonic fashion, rural dwellers quipped that while salaries were 'surging in the cities', in the rural villages there were only 'rising prices, taxes, and fees' (Goodspeed 1993, 2F; Poole 1993, 10).

As the rural situation deteriorated in the decade from 1985 to 1995, a 'floating population' of more than 100 million poverty-stricken peasants appeared. It revealed in bold relief the actual impact and consequences of the urban-centred developmental path for the countryside. By 2003, its ranks had swelled to 150 million and included many of the tens of millions of peasants who had lost their land through illegal or under-compensated seizures (Greenfield and Leong 1997, 100; 'An Overview of Unemployment 2003' 2004; Yardley 2004). Many members of this population made their way to the cities, development zones and suburban districts of the coastal areas in search of work. Despite millions of laid-off industrial workers by the mid-1990s, the poor and dispossessed rural migrants became the preferred labour force for the burgeoning export and construction industries. In the eyes of employers they were at once more controllable and exploitable, since in the cities they lacked formal residential status and were treated with contempt as second-class citizens, while in the factories they worked without contracts or legal protection for extremely low wages (Li

1993; Solinger 1999; Yu 1994). The remittances they sent back to the countryside helped to keep family members there afloat, but as rural conditions continued to deteriorate rather than single men (or women) entire families increasingly made their way to urban enclaves. Employers also increasingly used the tactic of withholding migrants' wages as a means of exerting control over them (a practice reminiscent of 'bondage labour' (*baosheng*) in pre-revolutionary China), making life more precarious both for families who had migrated and family members in rural areas who depended on remittances. According to official estimates, by early 2005 up to US$12 billion in unpaid wages were owed to rural migrant workers (Chan 2005b, 1). Thus rooted deeply in processes that impoverished and forced rural dwellers to leave the countryside, China's emergence as the so-called workshop of the world was connected centrally to the creation of a massive cheap labour platform for global capitalism (Chan 2006, 3).

Finally, in recent years China's entry into the WTO has compounded rural problems. After entering the organization in 2001 the government slashed its average agricultural tariff from 54 per cent to 15.3 per cent, compared to the world average of 62 per cent. According to Commerce Minister Bo Xilai: 'Not a single member in the WTO history has made such a huge cut [in tariffs] in such a short period of time' (Chan 2006, 2). To increase access for its manufacturers to developed countries and, thereby, maintain its rate of growth, China's leaders were apparently willing to once again sacrifice the countryside. By 2004, China had an agricultural trade deficit of US$5.5 billion resulting from a jump in imports. Cotton imports alone increased 175-fold, from 11,300 tons in 2001 to 1.98 million tons in 2004. For China's cotton farmers and others (especially those growing soybeans and sugarcane) the results have been devastating. Oxfam Hong Kong reported in 2006, for example, that the import of cheap US cotton into China in 2005 resulted in the loss of US$208 million in income for peasants and 720,000 jobs (Chan 2006, 2–3). Under the impact of the WTO, China's peasant farmers have thus experienced the rising imports, drop in domestic prices, difficulty in selling their crops on domestic markets, and, increasingly, devastation and ruin that have both affected small peasants on a global scale and formed key issues in the development of transnational agrarian movements.

RURAL COLLECTIVE ACTION, CORRUPT LOCAL POWER AND THE CENTRAL STATE

The broad outline presented above provides the context for analyzing the metastasizing over the last 20 years of overt rural resistance. According to the Ministry of Public Security, the number of 'mass incidents', including protests, demonstrations, road blockages and 'disturbances' that produced severe damage to persons and/or property, stood at 8,700 in 1993. That number rose to 13,000 in 1996; to 58,000 in 2003; to 74,000 in 2004 (involving 3.7 million people); and to 87,000 in 2005, with perhaps as many as 5 million participants (Li L. 2006, 250). Forty to fifty per cent of the contention has occurred in the countryside, urban labour protests making up most of the remainder (Tanner 2005, 196–97; 'Highlights' 2004).

Growing social inequality has been a persistent underlying theme of the protests. In the mid-1980s when resistance began to surface with some regularity it was comprised of mostly covert, individual acts of 'revenge' (*baohu*) or violence, which targeted tax collectors, corrupt village and township cadres, and the newly wealthy in the villages (often also cadres) by destroying their property, poisoning their livestock and poultry, and cutting down their trees (Bernstein 1994, 63; Bernstein and Lu 2003, 120; Li and O'Brien 1996, 29). It corresponded in almost textbook fashion to Scott's conceptualization of everyday resistance. In the latter half of the 1980s overt collective action also appeared. Through the 1990s, the issues of tax abuse and various other aspects of corrupt local power dominated the protest agenda. Thereafter, land seizures by local officials and cadres became the major focus of protest. As I have written in some detail elsewhere on the struggles of the earlier period (Walker 2006), I will present only a brief overview of them here, and then will turn, in greater detail, to the issues of land seizures and the resistance to them.

From the 1980s down to the present, most of the rural contention has been localized and small-scaled, often involving only the residents of several or even a single village. Formal organizational structure has also been notably absent. This pattern of development can be explained by the fact that because the problems peasants face – whether originating in global or domestic accumulation processes or the often associated policies of the state – play out in the context of local power, resistance mostly assumes local form. At the same time, both localized 'cellular' contention, which has also characterized collective action among urban workers (Lee 2007), and the lack of formal organizational development reflect the restraints on dissent imposed by the central Chinese state.

In the first instance, the Chinese government bans the formation of autonomous organizations in politically sensitive issue areas, such as human rights or labour. China has, for example, only one legal trade union, the All China Federation of Trade Unions, which is controlled by the communist party, and workers are prohibited from establishing independent unions. Those who have attempted to do so have been charged with subversion or treason (Lee 2007, 57–8). Similarly, peasants who have organized independent political organizations have been charged with subverting the state (Holland 1999, 10). The Chinese constitution grants all citizens freedom of association, and thus technically peasants should be able to form organizations to protect their rights and interests. In practice this seldom occurs because to do so, as imposed by State Council regulations, they must first register the organization at a local civil affairs bureau and to be able to register they must have already secured a government department to act as their 'professional supervisory unit' (*yewu zhuguan danwei*) (Li L. 2006, 252). The likelihood of finding such a government sponsor is practically non-existent if the organization is intended to protect peasants' rights and interests against state infringement. Under these circumstances although contention has become overt, organizational linkages generally have not. They have developed only clandestinely or informally, which has contributed to the localized character of much of the protest. On the other hand, police reports and other data indicate that as the collective protests increased in number and frequency they also increased in size, accompanied by an escalation in their levels of organization and violence (a point to which we will return below) (Tanner 2004, 140).

The central state also played a pivotal, if contradictory, role in both the deepening of corrupt local power and the resistance to it. Its implementation of fiscal decentralization after decollectivization, which allowed the newly created townships to manage their own finances by retaining locally a portion of the taxes they collected for the central government, effectively 'opened the door to [the] proliferation of fines, payments, and extortion under various pretexts' by corrupt local administrations (Chen and Wu 2006, 171). In 1994 when the central state transferred responsibility for health care, education and other public services to counties and townships, local officials began to squeeze peasants even more in order to run their under-funded and increasingly bloated bureaucracies and for personal gain (Chen and Wu 2006, 151–2). The peasants' tax 'burden' began to spiral out of control, especially in the grain-producing central provinces and mountainous western regions where alternate sources of income were restricted.

On the other hand, the central state balanced its fiscal and tax policies with a series of laws and regulations that gave peasants the right to reject unlawful taxes and fees and to bring collective lawsuits against local governments, and which set a 5 per cent limit on local taxes and fees (Bernstein 1994, 71; 'Decision on Alleviating' 1997; Li L. 2006, 253; Tang 2005, 28). These initiatives enabled the central government to promote itself as a guarantor of peasants' interests and to direct 'blame' downward to the local level. They also appear to have been intended to channel peasant activism into the legal arenas of petitioning and the courts, and thus away from protest and collective action. Indeed, coinciding with a proliferation of NGOs in China in the 1990s, a host of 'rights protection' (*weiquan*) advocates, lawyers and groups appeared to spearhead that effort (Chen 2005, 2, 4, 7; Li C. 2006, 4).

The state's intentions notwithstanding, by legitimating resistance to illegal taxes and fees, its initiatives also fuelled the broadening and deepening of rural protest. Employing the central state's rhetoric and policies as a tactical weapon, in the growing wave of collective action peasants used the 5 per cent limit and other laws to contest tax abuse and the use of force in tax collection, as well as other aspects of corrupt local power (Li and O'Brien 1996; O'Brien and Li 2006). Much of this collective action occurred through the official petitioning process, as peasants appealed to Beijing and other administrative levels to intercede against the local government. But mediated contention such as petitioning represented only one tactical weapon in the arsenal and everyday development of sustained collective action. It was flanked from the outset by more transgressive protests, demonstrations and risings that often turned violent, resulting in injuries to and the deaths of officials, as well as the occupation or destruction of government buildings and property. Some of these risings attained significant size, involving up to 20,000 participants, and were only quelled by the use of paramilitary troops (Bernstein 1994, 70–5; Divjak 2000, 1; Goodspeed 1993, F2).

From early on, as well, more conflictive forms of resistance appeared, especially in the grain belt and west, where tax resistance was most pronounced and conflict and fighting became almost commonplace. A top-level government report revealed that in 1993 a total of 8,200 township and county officials were injured or killed. It also noted, with alarm, that 'in some villages peasants have spontaneously founded organizations of various types, including religious or armed organizations, to replace the party and government. They have established taxation systems on their own' (as quoted in Perry 2003, 264). In other 'runaway' villages, by unanimous agreement villagers stopped paying taxes and everyone in the village, including women, took up arms to ward off tax collectors or officials who tried to enter. In Wugao in Anhui Province, where this type of resistance prevailed, for example, by the late 1990s 22 of the 29 villages in the township had flatly refused to continue paying taxes or to make grain contributions (Chen and Wu 2006, 163–4).

New collectivities and forms of organization also appeared in the 1990s, particularly during the latter half of the decade. Some operated at the village and transvillage level under such names as 'Committee of Peasant Autonomy' and 'Peasant Unity Committee'. Some, such as the 'Peasant Revolution Committee' and 'Peasant Rebellion Command Committee', assumed a more radical political thrust. Larger clandestine rebel organizations also formed, along with large apparently coordinated actions that spread through several counties and/or occurred in several provinces simultaneously (Holland 1999, 10; Thornton 1999, 25, 2004, 93, 98). The major actions during the summer of 1997 illustrate the point. In those risings 70,000, 200,000, 120,000 and 200,000 participants in the central provinces of Anhui, Henan, Hubei and Jiangxi respectively attacked government buildings, took party secretaries hostage, burned government vehicles, wrecked roads, commandeered government cement and fertilizer, and in at least two instances seized guns and ammunition (Becker 1997; Li 1997; Yue 1997).

There is also strong evidence that over the course of the 1990s both the party and central state declined in credibility and legitimacy in the eyes of many peasants. It reflected growing dissatisfaction with the government's inability and/or unwillingness to rein in and control corrupt local cadres and officials. It also intertwined with a resurgent popularity of Mao Zedong and a counter-remembrance among both peasants and workers of the socialist era as a time when there were social guarantees, society was free of corruption and a vast polarization of wealth, and they exercised the greatest degree of collective social power and democratic control over cadres who were accountable to the people. The revival of Maoist ethics and values in juxtaposition to the 'immorality' and 'exploitation' of post-socialist present thus became a means of denouncing that present, keeping socialist ideals alive, and further legitimating the rural contention (Hurst and O'Brien 2002; Jacka 1998; Lee 2007, 140–7; Weil 2006). Viewed in light of the above, in the short span of a decade in both thought and action Chinese peasant politics had not only traversed the divide between covert and overt as the everyday form of resistance, but it had done so in ways that challenge and contradict standard scholarly assumptions about the state's ability to determine and control the foci and contours of peasant protest.

By the end of the 1990s the mounting frequency and militancy of the protests and risings prompted the central leadership to acknowledge its growing lack of credibility and to state publicly that the social unrest in the countryside was threatening it rule. In response, the central government has adopted, especially in recent years, a number of populist measures designed to 'ease the peasants' burden' and, thereby, dissipate dissent. These have included tax reductions culminating in the elimination of the agricultural tax in 2006, substantial subsidies for the rural sector (mostly for under-funded local administrations), direct subsidies to peasant farmers for grain production and labour legislation for migrant workers (Li C. 2006, 2; 'Rural Tax' 2003; 'Threat of Rural' 2002; 'Wen Jiabao' 2003).

On one level, these policies stand as an important outcome of the sustained tax resistance and rural contention of the 1980s and 1990s. Although individual protests often met with only partial, if any, success, their cumulative effect was to force the state to offer concessions. On the other hand, it did so precisely at the moment when agrarian conflict was intensifying over the issue of land seizures. Moreover, it balanced those concessions with a new programme of increased repression and control, and thus to a considerable extent jettisoned the tolerance it had shown in the 1980s and 1990s for rural protest that remained small-scaled, targeted only local leaders, and did not assume explicitly political form. The crackdown has included greater use of armed police, paramilitary troops, tear gas and other weapons, more frequent arrests, the use of sophisticated new techniques, such as 'snatch squads' to apprehend protest leaders, the formation of specialized, heavily armed riot police units stationed in 36 cities, and the creation of 30,000 new police stations in rural areas for both control and surveillance (Armitage 2004; Buckley 2006, 1; 'China Reports' 2006; 'Five Thousand PRC Farmers' 1999; Manthorpe 2004, A7; 'Public Security Organs' 1999).

Even so, during 2005 (the last year for which the government has released statistics) not only did the number of protests continue to rise, but of the 87,000 incidents that year, those involving violent confrontations or attacks on government property still surged forward at the fastest rate (Kahn 2006a, 10). Those statistics reflected the intensification of agrarian conflict as land struggles moved to centre stage and the use of private, criminally-linked coercive force by local governments developed alongside and in concert with the central state's programme of repression and control.

LAND SEIZURES AND THE ROUTINIZATION OF OFFICIALLY SANCTIONED CRIMINAL VIOLENCE IN THE COUNTRYSIDE

The 'land enclosure movement' (*quandi yundong*) of the 1990s was prompted in 1987 after the government issued new land management regulations. Subsequently the National People's Congress amended the constitution so as to allow for the transference of land ownership, thereby endorsing the existence of a land market

(Miao 2003). The ensuing 'frenzy over land enclosure' initially fuelled a real estate boom centered in urban areas, but included suburban and rural land. It powered both domestic and foreign accumulation. In the early 1990s, 90 per cent of all direct foreign investment flowed into the newly opened land market (He 2000, 59).

Scope and Scale of the 'Enclosure Movement'

The growing possibilities for foreign investment underlay the craze, especially in 1992 and 1993, for making tracts of land available for lease in 'economic development zones' (*jijngji kaifa qu*). To create these zones corrupt officials, often in collaboration with foreign real estate partners, abused their authority and carried out forced evictions of urban residents as well as seizures of rural farmland. Guangdong Province paved the way, with land enclosures eventually spreading through the entire province as Hong Kong real-estate companies poured investment into development zones. Other coastal provinces and cities soon followed suit, such as Zhejiang Province, where over 600 development zones and industrial parks were established (Miao 2003).

From the latter half of the 1990s, rapid urbanization and the rush for real estate development, declining local government revenues resulting from the changes in tax policies, and the personal greed of cadres combined to accelerate the pace of land seizures. Figures for the loss of arable land suggest this trend. According to the State Statistics Bureau, 27.5 million square *mu* of land were lost between 1986 and 1995 (Gilboy and Heginbotham 2004, 256); and, as reported by the Ministry of Agriculture, total arable land declined by almost 120 million *mu* in the eight years from 1996 to 2004, or an average of 14.25 million *mu* annually (He 2007, 51; Cody 2004, 1A).

In the brief span of 20 years, then, China experienced an 'enclosure movement' of unprecedented proportion worldwide, resulting in the dispossession – and in many cases impoverishment – of tens of millions of peasant households. Calculations of the total number of dispossessed peasants have been contradictory. Official government statistics indicate that by 2005 more than 40 million households had lost their land, and that this number was continuing to increase at approximately 2 million per year. But many other estimates place the number much higher. Using the Ministry of Agriculture's figures of 2 *mu* of farmland per household and 120 million *mu* for the amount of arable land seized from 1996 to 2004, economist He Qinglian calculates that during those years 60 million peasant households lost their land (2007, 52). Based on the same procedure and using the figure of 147.5 million *mu* for the amount of arable land lost from 1987 to 2004, the number of dispossessed households totals 73.75 million. This figure resonates closely with other expert estimates that place the total number of dispossessed at 70 million by 2004 (Yardley 2004). The estimate of 70 million households becomes even more staggering when considered in terms of the actual number of people involved. Calculated at 4.5 persons per household, the standard figure used for average rural household size in China, then by 2004 as many as 315 million people may have been displaced through land seizures.

The Legal Basis and Contours of Land Seizures

The relative ease with which the seizures have been carried out is a direct consequence of both ambiguities in ownership rights in the countryside and the state's right to appropriate land. The Land Administration Law (2004) places ownership in the peasants' 'collective of the village', but it does not define the organization or structure of that collective. Actually, in most localities since the dissolution of the communes, there has been no functioning village collective. Instead, village committees or the 'administrative village' manage and administer the land contracted to peasants under the Household Responsibility System (Ho 2001, 401–2; Pils 2005, 240–1). The village committees and administrative villages, however, have no right to transfer land for compensatory use. That right resides with the state, which may, in accordance with the law, expropriate land that is under collective ownership, if 'it is in the public interest' (Cai 2003, 664–5; 'Economists' 2004; Guo 2001). Thus although technically the peasants own the land and have guaranteed 30-year contracts to its use, under the guise of 'public interest' officials simply seize it. As one official put it, 'the ownership rights to land have been silently stolen from the natural village and vested in a higher level' (Ho 2001, 404).

The chief players in that 'higher level' have been the township administrations in the countryside and municipal governments in towns and cities. By law the land of the 'city's urban area' is state-owned, while that of the 'city's suburbs' belongs to the rural collective. But because of continuous urbanization, in many localities much of the suburban land has been subsumed within the limits of the city or town, including, in some instances, the expropriation of all of the land belonging to a collective (Ho 2001, 405). On the other hand, since township and municipal government administrations do not have direct access to the collective land of the villages, they have had to include village committees in land deals – but only as 'extension[s] of their political power' (Ho 2001, 405; see also Guo 2001, 425).

In the countryside as well as the cities, these 'state agents' have used rapacious and violent means both to garner personal wealth for themselves and to gain new sources of revenue for their under-funded and bloated local administrations. Consequently, the illegal transferring, leasing or requisitioning of land contracted to peasants without giving them prior knowledge of the transaction, the coercive expropriation and occupation of villagers' farmland, the retention and withholding of all or part of funds designated for peasants' compensation and/or resettlement, the lack of job placement for villagers who have lost their land and the use of force or threats to gain compliance from villagers who resist their demands have all become commonplace among the officials and cadres who have seized peasants' land (Tang et al. 1993; Wen and Zhu 1996; Yao 2003, 1).

Officially Sanctioned Criminal Violence

Furthermore, local governments, or their business allies with government approval, have increasingly resorted to hiring privately organized forces – often members of criminal organizations – to carry out both the forcible expropriation and the occupation of farmland, as well as violent attacks on villagers to force them into submission. Organized crime (*heishehui zuzhi*, or literally, organized black societies) appeared in China in the late 1980s, and mafia-style criminal organizations (*heishehui xingzhi de fanzhuituanhuo*), including the internationally prominent Triad organization whose origins lie in China's imperial history, grew in strength and number in the 1990s (He 2006, 93; Wen 2004). By 1992, as a Ministry of Public Security circular states, 'there are more than 1,830 underworld organizations, gangs and associations bearing the character of secret societies . . . A small secret society may have dozens of members, while a large secret society may have 5,000 to 30,000 members with transregional branch organizations existing throughout the country' (cited in Perry 2003, 265).

Local governments first began to resort to underworld or thug violence when collecting taxes and fees and for ransacking peasants' grain stores if they refused to pay their taxes (Chen and Wu 2006, 159). But as land seizures and peasants' resistance to them intensified, and as the influence of gangsters, bandits and the criminal underworld became more pronounced in various arenas, including the economy and public life, the routinization of officially sanctioned criminal violence became more widespread. As He Qinglian states:

> An alliance between gangsters and local officials has led to the increase in officially sanctioned crime, especially at the grass-roots administrative level. In many localities, criminal organizations protected by local officials have taken over control of certain government functions and key sectors of the economy. They are so powerful that local people refer to them as a 'second government'. Since the late 1990s in particular, local governments throughout China have used criminal organizations as goon squads to force urban residents from their homes and seize farmers' land. (2006, 93)

The result in many localities, as another author puts it, is 'an emerging "elite alliance" that includes township government officials, business companies, village leaders, government officials in the county or higher level, and hired thugs – [who] are called by the farmers . . . the "black force." This alliance controls the rural areas by violence' (Li F. 2006, 2).

Land Struggles

It has been, then, within a web of growing state repression and the new levels of violence, brutality and human degradation generated by corruption and officially sanctioned crime, that since the late 1990s land struggles have surged ahead as the central grievance of the rural protests. Because the seizures destroy peasants' livelihoods and basis for survival, they have had a more profound impact than tax and fee abuses, and thus have lain at the root of an escalating spiral of violence and resistance. But as the figures for the massive number of families who have lost their land suggest, these struggles have usually ended in failure.

The three-year struggle of Sanchawang villagers in Shaanxi Province – one example among thousands – is typical. In late 2002 after the local government seized a portion of the villagers' land and they learned that the officials had leased it for 50 times what they had been paid, nearly 800 of them blocked the construction of a development zone on the land. They organized 16 teams that alternated the sit-ins. But the following spring the police and more than 300 construction workers moved in to break up the occupation. After that the villagers drew up a petition and took it to Beijing, but subsequently learned that officials in Beijing had simply sent their complaints back to the officials who had seized the land. In the spring of 2004 when the local government seized another strip of land, the villagers decided to protest by leaving their remaining fields untilled and survive by eating from the village grain reserves. The refusal to plant was provocative, since officials are held responsible for local production. Villagers also began a new wave of sit-ins to block construction on the strip of land, and once again the police came in and started to arrest people. By that point the villagers were so frustrated, angry and desperate that hundreds of them seized the communist party's village headquarters and held the walled compound for five months. But their effort produced no positive result and in the end the government sent 2,000 paramilitary troops to forcibly remove the protestors and arrest the leaders. The assault broke the protests. As one woman said: 'Nobody will dare protest now. Everybody is afraid' (Yardley 2005, 4A; 2004, 2).

In Sanchawang hundreds of villagers were wounded but no one was killed. In the last several years, however, as thug violence has increased and the government has armed paramilitary troops with real rather than rubber bullets, villagers have been killed in a number of confrontations. In Shengyou village in Hebei Province in 2005, with the approval of local authorities, a construction contractor sent in 300 helmeted thugs armed with hunting rifles, metal pipes and shovels to remove villagers who were occupying land that had been seized by the local government. In the confrontation that followed, the thugs shot and killed six villagers and wounded over 100 (Saiget 2007; 'Turning Ploughshares into Staves' 2005). In Shanwei in the southern province of Guangdong, paramilitary troops, who were sent in late 2006 to uproot and disperse participants in another land occupation and struggle, killed as many as 20 villagers in what became known as the 'Shanwei Massacre' (Gu 2006).

Because of mounting resistance and violence, as well as its concern that the loss of farmland from land seizures could affect the country's food security, in 2003 the central government began to make land seizures a centrepiece of its policy. It limited the number of development zones and announced a crackdown on illegal seizures. In 2004 it suspended all non-urgent conversion of agricultural land for six months, and then issued new regulations requiring any such change to be approved at a high level (He 2007, 48–9; Rural Development Institute 2005; 'Turning Ploughshares' 2005). But as in its campaigns against corruption and tax abuse, and the examples above suggest, these policies had only minimal effect. According to figures released by the Ministry of Land and Resources, there were, for example, 168,000 cases of illegal land deals in 2004 (Cody 2004, 1A).

By 2006, in an address in which he stated 'we absolutely cannot commit a historic error over land problems', Premier Wen Jiabao bluntly admitted that illegal seizures without adequate compensation were still a key source both of instability and of uprisings in the countryside (Kahn 2006b, 3A). Indeed in that year, according to the Ministry of Land and Resources, 131,000 illegal seizures involving 100,000 hectares of land still took place ('Illegal Land Grabs Up' 2007). In early 2007, the National People's Congress passed a law limiting the conditions for the transfer of rights of contracted rural land and prohibiting transfers of land with existing homes ('NPC' 2007). Yet it appears that these regulations will do little to deter 'wicked coalitions' of officials and developers in illegal expropriations, or their use of thug violence to carry them out. Many reports continue to surface of such seizures, as in the case of Yutian village near Harbin in Heilongjiang Province, where in August 2007 under-compensated residents clashed with armed police and thugs sent by developers over the seizure of their farmland and the demolition of their homes ('Farmers in Revolt' 2007; see also 'Farmers Protesting' 2007).

In the meantime, many peasants and dispossessed migrants are well aware that the government's refusal to establish secure land rights in the countryside continues to provide the framework for and ease the takeover of rural land for industrial use, urban expansion or the construction of transport infrastructure. This contradiction underpins the cynicism and anger among them about the party and the central as well as the local state (Pils 2005, 288; Cody 2006, 14A; Thornton 1999, 27–8). The lack of redress in the courts and petition halls only reinforces the government's declining legitimacy in their eyes. Comments such as 'People can see how corrupt the government is while they barely have enough to eat' and 'The officials despise ordinary people and are not afraid to bully them' indicate both derision and disgust. The language of protest has also changed. In tax resistance, as one Chinese scholar puts it, 'because there were clear central policies to support these causes, slogans would often be "implement the centre's policies . . . reduce the peasant burden". Since the land issues are a matter of survival for the peasants, their language has become more direct. For instance, [they say:] "How shall we live without land!"' (Pils 2005, 288). Slogans such as 'Return land and property to the peasants' and 'Establish peasants' own political power' are equally direct (Thornton 2004, 98), but they do not capture the fundamental contradiction of neoliberal capitalism perceptively embedded within 'how shall we live?' Peasants who have lost their land through seizures or who otherwise have been made landless now say they belong to a new 'class' of 'three nothings' – no land, no work, no social security. Implicit in this epithet is a critique of the state and society that are not only permitting, but also propelling that process of class formation.

The indignation and anger of many rural residents regarding the growing inequalities of the post-socialist path have been echoed in recent years by the 'eruption' of numerous large-scale societal outbursts. Not unsurprisingly, the rural, suburban and urban areas in which these risings occurred often had large rural migrant populations. Most of these 'spontaneous' outbursts involved the amassing of tens of thousands of people within a matter of hours. They generated considerable violence, including attacks on police headquarters, police vehicles, state property and the property of the wealthy or officials whom the police protected, and were contained only by the state's deployment of large numbers of paramilitary forces. Most were triggered by incidents in which officials, the newly wealthy or even minor state employees acted with contempt and brutality toward migrants, peasants and the urban poor, or in which the protestors apparently viewed state policies as representing yet another assault on poor people. In 2004, for example, in the Wanzhou district of Chongqing in Sichuan Province – where many rural residents had been resettled due to the Three Gorges Dam project – perhaps as many as 80,000 people 'rioted' after a tax official brutally beat a migrant worker who had accidentally bumped into the official's wife (Chan 2004a, 2; 2004b, 1). Similarly in 2005, 50,000 migrant workers in Guangdong also 'rioted' after a security guard killed a migrant youth accused of stealing a bicycle (Chan 2005a, 4). The rapidity with which such large numbers of people have acted in unison in these risings suggests that among the urban as well as the rural poor, a shared class perspective and widespread sense of injustice have deepened to the point that they have overridden both fear and deference (cf. Thompson 1971, 78).

Indeed, born from the class contradictions of and their marginalization in post-socialist society, the outbursts of indignation and anger among members of the rural poor, the 'have nothings', migrant workers, disenfranchised workers and urban poor suggest that a 'socio-cultural unity' based on a common conception of the world has developed among them. Social movement theorists posit that 'mobs, riots, and spontaneous assemblies' indicate that a social movement is in the process of formation (Tarrow 1994, 5). If this formulation is correct, can we expect to see further development of that 'movement' in the near future, either through rural–urban linkage or in the countryside alone? Most China scholars would, I think, answer 'no' on both counts, given the lack of an organizational nexus, the greater capacity for repression that the state has assembled in recent years and the possibility that the populist policies of the central government may defuse contention.

On the other hand, the historical record for China indicates that, as in the present, in pivotal moments of the past when sustained rural collective action – rooted in local struggles – developed on a transregional or even national scale, it did so with a unified discourse of dissent and, as such, although unorganized, assumed the character of a movement. In those pivotal moments new organizational forms appeared through which peasants changed and made history. In the Yangzi delta, arising from the sustained collective action in the decade before the 1911 revolution, for example, a host of new organizations enabled peasants to carry out rent and tax strikes over an entire county or multi-county area and to develop a political agenda in which, as in the Thousand Person Society (*Qianrenhui*) Uprising in Changshu, Wuxi and Jiangyin counties in 1911, they declared their independence from the new Republican state and established an autonomous domain (Walker 1999, 168–9). Similarly, near the end of the

Ming dynasty (1368–1644) peasants moved from relative passivity to sustained resistance, and then on to a militant organized offensive in which their armed legions struggled against and effectively ended bondservantry/serfdom in the delta (Walker 1999, 47–8). Given this historical legacy, it is probably best not to conclude too quickly that the protests in China for the last 20 years have *only* been local struggles and thus will not assume a broader social or political form.

PEASANTS' LAND STRUGGLES IN CHINA AND TRANSNATIONAL AGRARIAN MOVEMENTS

As analyzed by Karl Marx (1976, 873–940), the separation of producers from their means of production during the (so-called) primitive accumulation of capital, which underlay and culminated in capitalist development in England, was established principally through the extra-economic force of the state and fractions of social classes rather than through the 'silent compulsion' of the market. As such it was achieved to a considerable degree through predation, robbery, fraud, brutality and violence. The fusion of global and local capitalist interests and the emergence of a bureaucratic-capitalist elite in China over the last 25 years have generated processes of accumulation that not only resemble the primitive accumulation described by Marx, but which in their attack on and disadvantaging of rural people have been equally rapacious. The resulting dispossession – and in many cases pauperization – of tens of millions of China's peasants links their plight and politics to those of the rural poor in other parts of the world who have borne the brunt of global neoliberal policies, as actualized in nationally and locally specific forms, and who have been contesting these policies in a resurgent agrarian politics of overt collective action, social movements and alternative visions of the future. Many have interlinked with expanding transnational agrarian networks and movements.

Due to the lack of organizational linkage, the tidal wave of overt everyday resistance in China has rarely been analyzed as forming part of the social base for the movement against capitalist globalization or for transnational agrarian movements. But, clearly, along with the struggles of the 'unorganized' poor and dispossessed elsewhere, the recent agrarian contention in China may be understood as a component of the global groundswell that has formed in reaction and opposition to neoliberal capitalism.

Viewed in this light, a key question is the extent to which the structural conditions and struggles facing the unorganized rural poor in China and other parts of the world are addressed in transnational agrarian movements. Stated differently, does the issue of the link between capitalist accumulation, on the one hand, and land seizures, dispossession and social pauperization on the other, inform their struggles strategically? Under the banner of neoliberalism, the collective imperialist assault on rural people has been multifaceted, but from Asia and Africa to Latin America a massive transfer of property and land use rights from the rural poor to those with capital has been a central axis. Although that process has been carried out on an unprecedented scale in China, dispossessed Chinese peasants now occupy the same structural space as their counterparts in India, where large-scaled land seizures and evictions are also taking place – often, as in China, in the context of officially sanctioned violence (Communist Party of India n.d.); as those in Africa, who have been the victims both of corrupt government officials' transference of public land to politically connected individuals and private corporations and of the World Bank-IMF offensive to destroy communal land arrangements and effect a widespread reorganization of class and property relations (Federici 2001); and as their counterparts in Latin America who, especially in coastal areas, have lost their land to property developers (Vía Campesina 2006). Land seizures, forced evictions, dispossession, pauperization and peasants' land rights have thus intertwined as crucial global issues.

How are transnational agrarian movements addressing these issues, and are there gaps between their programmes and campaigns and the actual everyday land struggles of peasants on the ground? These questions immediately raise the issues of accountability and 'representativity' of movements to their organized constituencies, which have been a concern of this volume. They also raise the matter of *responsibility* of transnational movements to the unorganized poor who form the majority of the world's peasants, another concern of this issue. Vía Campesina, one of the most dynamic of the transnational agrarian organizations (and perhaps the only truly transnational agrarian movement – as opposed to network – approaching global reach), provides an excellent focal point for considering these questions. Since its founding in 1993, Vía Campesina has been a strong voice in opposition to neoliberalism and in support of the poor and landless. Although it now has member organizations in 56 countries, it has developed unevenly with social network bases in the Americas and Europe and expanding organizational ties in Asia, but until recently virtually no affiliates in Africa (it admitted seven African organizations as members in 2004) and no presence in China at all (Borras 2004, 25; Edelman 2003, 213). Given the political situation in China and the fact that Vía Campesina's principal political strategy has been collective action and mobilization (while also pursuing negotiation and collaboration 'on select issues with select agencies and institutions'; Borras 2004, 21–2), it is perhaps unsurprising that it has not attempted to develop an organizational presence in China. But China likewise remains a 'blank' area in its reports and analyses of global conditions. Even its annual reports of 'Violations of Peasants' Human Rights' virtually ignore this area of the world where one in three peasants reside. In the space that remains, then, I wish to consider the Vía Campesina position on land issues and then return to the question of its silence on China.

Following an initial period of internal organizational consolidation, in 1999 Vía Campesina in conjunction with the Foodfirst Information and Action Network (FIAN), an established international human rights organization, launched its first major initiative on land issues – the 'Global Campaign for Agrarian Reform'. The campaign initially involved mobilizations, land occupations and public events that emphasized 'the right to land and security of land tenure' as prerequisites to realizing the human right to food stipulated in Article 11 of the International Covenant on Economic, Social and Cultural Rights (ICESCR) (Desmarais 2002, 108).

Over the next several years, although Vía Campesina continued to link the issues of land and food to the ICESCR, the central thrust of its agrarian campaign became political opposition to 'market-assisted agrarian reform' (Borras 2004), which was developed and introduced by the World Bank in 1998. Against the principle of 'land . . . is for those who own the capital to buy it', which in its view underpinned the World Bank's land policies and placed them in the service of transnational capital, Vía Campesina proposed that rural land ownership should be submitted to the criterion 'only those who work the land have the right to land', and on that basis articulated an alternative principle of the 'maximum size of social ownership of land' (Borras 2004, 11). In 2002, with other organizations it rejected explicitly the World Bank's land policies, arguing that they were designed to at once 'evade the true redistribution of landed property' and 'distract, undermine, divide, and curb the movements of landless peasants' (as quoted in Borras 2004, 13–14). By focusing on the World Bank, the Vía Campesina-FIAN agrarian campaign thus exposed the intent and aims of neoliberal land policies at the international institutional level. But it did little to spotlight land seizures, forced evictions and dispossession as central to the neoliberal accumulation strategy as well.

Indeed, the link between capitalist accumulation, dispossession and rural pauperization emerged more prominently in Vía Campesina's other major campaign – that for 'food sovereignty',[2] in which the WTO became the major institutional target. In that campaign it argued that as a result of their implementation by the WTO, neoliberal policies had 'forced hundreds of millions of farmers to give up their traditional agricultural practices, to rural exodus or to emigration' (as quoted in McMichael 2007, 7). In linking the movement of food to the dispossession, displacement and circulation of people, Vía Campesina brought into fuller view the extent to which the current regime of capital accumulation both depends upon and is forcing the creation of an expendable global wage-labour force (McMichael 2007, 9). What is still missing, however, is an emphasis on the massive, and in many instances forced, transfer of property rights – and thus on peasants' land rights – that likewise has at once underpinned the making of a marginalized global labour force and been a key arena of neoliberal accumulation.

In recent years the Vía Campesina has begun to devote greater attention to the issues of land rights, land seizures and dispossession. But as in the case of the food sovereignty campaign, this has occurred mostly in venues other than the agrarian reform campaign. In its 2006 report on 'Violations of Peasants' Human Rights', for example, which in its words was written so as 'to shed more light

---

[2] The concept of food sovereignty, which the VC first articulated at the 1996 World Food Summit, involves the right of each nation to maintain and develop its capacity to produce its basic foods 'in its own territory' (Desmarais 2002, 104).

on the everyday situation of peasants worldwide', land seizures and dispossession emerge as important themes. As the report states:

> Land use conflicts are on the rise in many parts of the world ... Impunity for action taken against landless labourers or small holder farmers in rural areas is, in many countries, one of the biggest and most profound human rights violations. Symptomatic of these are forced evictions and land grabbing, without adequate government response. (Vía Campesina 2006, 1–2)

In 2007 at the World Social Forum (WSF) in Nairobi, Rafael Alegria, Vía Campesina's international coordinator of the Global Campaign for Agrarian Reform, announced the launching of the 'African part' of the campaign. In outlining this new initiative, he stated that large amounts of land were being seized in Africa by the World Bank and that the African campaign would include missions of Vía Campesina leaders to support and accompany Africans in their land struggles and to seek meetings with government representatives (Duraes 2007).

The brief outline presented above highlights that as in its construction of an overall banner of 'the struggle against neoliberalism', Vía Campesina shaped and aggregated specific grievances and issues pertaining to the land question into a broad campaign on agrarian reform. That campaign enabled it to target neoliberal land policies at the international institutional level (and therewith at the level of the national state) and to develop an alternative conceptualization of land reform based on human rights. As such it appears to have created an oppositional pole to the neoliberal policy model on land resources in localities where land reform is an issue (Borras 2004, 26). On the other hand, to return to our original query, the way in which neoliberal capitalist accumulation has actually been taking place on the ground in much of the countryside of the global South, that is, both through land seizures, forced evictions and the dispossession of tens of millions of peasant families and through the resistance of the organized and unorganized poor, remained largely a silent issue in the agrarian campaign.

Consequently, a challenge for Vía Campesina appears to be to broaden and deepen its land agenda so as to at once strengthen its opposition to neoliberalism and speak directly to the widespread struggles of those who are resisting land seizures and forced evictions. In Nairobi during his announcement of the African campaign, Rafael Alegria also indicated that after Africa Asia would become the geographic focus of the global campaign. Given the growing prominence of land seizures, forced evictions and dispossession in India and Indonesia, where the social networks affiliated with Vía Campesina have been expanding, it seems likely that if it does develop an Asian campaign those issues may (finally) emerge as chief areas of concern.

If Vía Campesina should make land seizures, forced evictions and dispossession a focus of its Asian campaign, it could begin to reverse its silence on land struggles in China. Because of the Chinese state's restrictions on politically sensitive issue areas, Vía Campesina could not of course develop a militant campaign within China. But it could include China in its analyses, discussions and initiatives and, importantly, also extend solidarity to Chinese peasants as it has done for others in the past. Chinese rural activists have desperately sought to transmit news of their struggles outside of China in the hope of gaining support and allies who might exert pressure on the Chinese state. Since 2006, when the central authorities blocked hundreds of international websites, shut down thousands of Chinese websites, and placed new restrictions on both Chinese and foreign journalists (Amnesty International 2007), communications and information transmittance have been more difficult than previously, but they are far from impossible.

Since it has based its agrarian reform and food sovereignty campaigns on the issue of human rights and has established a durable alliance with the human rights organization FIAN, Vía Campesina likewise might be able to form alliances with groups that are based outside of, but whose focus is human rights in China. The Asian division of the well-established Human Rights Watch, for example, has displayed considerable interest in land seizures and forced evictions in China and has already published a 40-page report on forced evictions and (what it terms) the 'tenants' rights movement' in urban China. Human Rights in China, founded by Chinese students and scholars in 1989, publishes daily and monthly news briefs that summarize news about human rights covered in the local and regional Chinese and English press. The monthly brief, in which the topical areas include protests and petitions, labour, human rights defenders and media censorship, is one of the best available resources providing systematic coverage of peasants' land struggles.

Scholars of social movements pinpoint the importance of political opportunity structure in the development of social movements. They emphasize, as well, that opportunity structure applies not only to the formation of movements. 'Movements *create* opportunities for themselves or others', writes Sidney Tarrow. 'They do this', in his words, 'by diffusing collective action through social networks and by forming coalitions of social actors; by creating political space for kindred movements . . . and by creating incentives for elites to respond' (1994, 82). Viewed in this light, an Asian campaign in which Vía Campesina broadened its land agenda so as to include the issues of land seizures, forced evictions and dispossession, brought the rural struggles in China (and among the 'unorganized' elsewhere) on these issues into full view, and developed a coalition of like-minded allies could invigorate both Vía Campesina and 'kindred movements' among the unorganized poor.

On the other hand, although human rights broadly speaking is considered to be a sensitive issue area by the Chinese state, focused issues such as migrant workers rights, sustainable agriculture, the environment, poverty, rural gender relations and most recently peasants' rights all appear to be 'acceptable' issues, and thus are being addressed by the proliferation of domestic, foreign and international NGOs that are now operating in China. They likewise form foci around which Vía Campesina could perhaps begin to develop connections with Chinese organizations. Although transnational projects of Chinese NGOs cannot involve any 'political' or 'militant' international networks or groups, their linkages with transnational civil society have been steadily expanding. In 1986, for example, only 484 Chinese NGOs were members of international NGOs. By 2002 that number had risen to 2297 (Chen 2005, 7).

Other transnational agrarian networks, such as the Reseau Mondial Agricultures, Paysannes, Alimentation et Mondialisation (World Network of Agriculture, Peasants, Food and Globalization) (APM Mondial), have already begun to establish organizational ties in China. Growing out of agriculturalists' organizations established in the 1980s, the APM-Mondial first developed as a regional network in Africa (Reseau APM-Afrique) and then in the 1990s internationalized. Although it favours educational and activist development programmes over the direct action and 'pressure tactics' employed by Vía Campesina, its analysis of the WTO is similar to that of Vía Campesina and the two organizations have collaborated on projects in the past (Edelman 2003, 213). Its China network (Reseau APM-Chine) was established in the late 1990s and operated initially only in association with individuals and institutions in China designated by the Chinese government (Edelman 2003, 213). Although the China network remains informal, more recently it has collaborated with other international organizations to produce workshops on a sustainable food system (Citizens' Earth – APM Network 2006). Similarly, the International Forum on Globalization, a forum of activist-scholars that opposes the globalization of industrial agriculture because of its destruction of small farms and rural communities and its links to rising food insecurity, is planning a series of seminars in China in 2008 that, according to its announcement, will be 'co-sponsored by leading Chinese environmental and agricultural groups'. The above examples suggest that there are multiple avenues through which Vía Campesina could perhaps begin to establish connections with a peasantry that has been engaged in intense agrarian struggle over the last 20 years. The challenge here would seem to lie in the realms of flexibility and coalition building.

Developing ties with rural Chinese seems particularly important in view of the fact that the problems of land seizures and dispossession described in the preceding pages will in all likelihood continue for some time. Experts project that at least 30 million more rural households will lose their land through seizures in the coming years, and they expect that between 2004 and 2020 300 million of China's rural dwellers will have moved to cities and towns (Gilboy and Heginbotham 2004, 258; Yardley 2004). If urban-centred neoliberal planners within and close to the government have their way, that figure could be even higher. Some, such as Pan Wei, an influential government adviser and Beijing University professor, think that the state should 'encourage' even greater and a more rapid acceleration of peasant migration to urban centres so that China can develop an additional 100 cities of 5 million people or more over the next 30 years, either by building entirely new cities or expanding existing ones (Gilboy and Heginbotham 2004, 257).

Finally, Chinese peasants and the rural dispossessed have forged collective identities that resonate with transnational agrarian movements such as Vía

Campesina and provide a basis for conceptualizing alternatives to neoliberal capitalism. In their assertions of 'peasantness', they have articulated a sustained critique of urban-centred development and linked the class conflicts of post-socialism to a position of autonomous authority and worth for themselves – both of which correspond to a striking degree to Vía Campensina's defence of a 'peasant way' of rural peoples (Desmarais 2002, 96). At the same time, Chinese peasants have experienced the most thorough social revolution and truly redistributive agrarian reform undertaken anywhere thus far. Although their numbers are rapidly dwindling, there are still a sizeable number who experienced the first stage of that revolution/reform, and many more who were present during its collectivization process. As noted, many now look back to that earlier period as a better time when the moral discourse of social justice and the actually existing forms of production ensured greater dignity and security of livelihood. Along with dispossessed migrants, members of the (largely disenfranchised) urban working class and the urban poor, as a result of their lived experience with so-called globalization – as well as their struggles in the countryside – and with the earlier period as a point of reference with which to assess the present, many peasants are keeping the egalitarian ideals of socialism alive (cf. Chen 2006). This is a crucial alternative ideological space that capital cannot fully capture or control, and thus it forms a vital terrain for the further development of transnational agrarian movements.

REFERENCES

Amin, Samir, 2003. 'World Poverty, Pauperization, and Capital Accumulation'. *Monthly Review*, 55 (5): 1–9.

Amnesty International, 2007. '2007 Human Rights Report, Annual Report for China'. http://www.amnestyusa.org/annualreport, Accessed 29 December 2007.

'An Overview of Unemployment 2003', 2004. *China Labor Bulletin*, 14 July. http://www.china-labor.org.hk, Accessed 24 July 2004.

Armitage, Catherine, 2004. 'Winter of China's Discontents'. *The Australian* (Sydney), 20 December.

Barboza, David, 2005. 'In China, a New Capitalist Beachhead'. *International Business*, 18 January: 1.

Becker, Jasper, 1997. 'Rural Revolt Forces Action on Excessive Tax'. *South China Morning Post*, 20 November: 9.

Bernstein, Thomas P., 1994. 'In Quest of Voice: China's Farmers and Prospects for Political Liberalization'. Paper presented to the University Seminar on Modern China, Columbia University, 10 February.

Bernstein, Thomas and Xiaobo Lu, 2003. *Taxation Without Representation in Contemporary Rural China*. New York: Cambridge University Press.

Borras, Saturnino M. Jr, 2004. 'La Vía Campesina: An Evolving Transnational Agrarian Movement'. *TNI Briefing Series* 2004/6: 1–30. Amsterdam: Transnational Institute.

Brenner, Robert, 2002. *The Boom and the Bubble: The U.S. in the World Economy*. London: Verso.

Buckley, Chris, 2006. 'China to "Strike Hard" Against Rising Unrest'. *Boston Globe*, 26 January.

Cai, Yongshun, 2003. 'Collective Ownership or Cadres' Ownership? The Non-Agricultural Use of Farmland in China'. *China Quarterly*, 175: 663–80.

Chan, John, 2004a. 'Martial Law Declared as Unrest Deepens in Rural China'. *World Socialist Web Site* (http://www.wsws.org), 15 November: 1–4. Accessed 16 February 2006.

Chan, John, 2004b. 'China: Riot in Guangdong Province Points to Broad Social Unrest'. *World Socialist Web Site* (http://www.wsws.org), 30 November: 1–2. Accessed 16 February 2006.

Chan, John, 2005a. 'Beijing Tightens Political Control Over Dissent'. *World Socialist Web Site* (http://www.wsws.org), 6 January: 1–5. Accessed 1 March 2007.

Chan, John, 2005b. 'New Year for China's Rural Migrant Workers'. *World Socialist Web Site* (http://www.wsws.org), 22 February: 1–4. Accessed 1 March 2007.

Chan, John, 2006. 'Beijing Abolishes Centuries-Old Agricultural Tax'. *World Socialist Web Site* (http://www.wsws.org), 17 January: 1–3. Accessed 1 March 2007.

Chan, John, 2007. 'An Explosion of Billionaires in China'. *World Socialist Web Site* (http://www.wsws.org), 4 November: 1–2. Accessed 10 November 2007.

Chen, Feng, 2006. 'Privatization and Its Discontents in Chinese Factories'. *The China Quarterly*, March (185): 42–60.

Chen, Guidi and Chuntao Wu, 2006. *Will the Boat Sink the Water? The Life of China's Peasants*. New York: Public Affairs.

Chen, Jie, 2005. 'Community in China: Expanding Linkages with Transnational Civil Society'. *Asia Research Center*, Murdoch University, Working Paper No. 128, December: 1–18.

Cheng, Allen T. and Dune Lawrence, 2007. 'China Has 106 Billionaires, Up From 15 Last Year'. 14 October. http://www.Bloomberg.com/apps/news, Accessed 14 October 2007.

'China Reports Drop in Rural Riots, Protests', 2006. *China Confidential*, 7 November.

Citizens' Earth – APM Network, 2006. 'The Development of the Citizens' Earth – APM Network in the Wake of the APM Program and the Current Situation'. http://www.terre-citoyenne-apm.org/img/html/bip-3063.html, Accessed 27 December 2007.

Cody, Edward, 2004. 'China's Land Grabs Raise Specter of Popular Unrest'. *The Washington Post* (Foreign Service), 5 October: 1A.

Cody, Edward, 2005. 'For Chinese, Peasant Revolt is Rare Victory; Farmers Beat Back Police in Battle Over Pollution'. *The Washington Post* (Foreign Service), 13 June: 1A.

Cody, Edward, 2006. 'China Leader Makes Appeal on Corruption; President Sounds Alarm Loudly on Party Anniversary'. *The Washington Post* (Foreign Service), 1 July: 14A.

Communist Party of India (Marxist-Leninist), n.d. 'Agrarian Crisis and Agrarian Struggles'. http://www.cpiml.org/index.html, Accessed 5 June 2007.

De Angelis, Massimo, 2001. 'Marx and Primitive Accumulation: The Continuous Character of Capitalist "Enclosures"'. *The Commoner*, September (2). http://www.thecommoner.org, Accessed 3 March 2005.

'Decision on Alleviating Peasants' Burdens', 1997. *Xinhua* (Beijing), FBIS Translated Text (AFS Doc. No. 0W0404115697).

Desmarais, Annette-Aurelie, 2002. 'The Vía Campesina: Consolidating an International Peasant and Farm Movement'. *The Journal of Peasant Studies*, 29 (2): 91–124.

Divjak, Caarol, 2000. 'Rural Protests in China Put Down By Riot Police'. *World Socialist Web Site* (http://www.wsws.org), 7 September: 1–3. Accessed 16 February 2006.

Duraes, Suzanne, 2007. 'La Vía Campesina Launches a Global Campaign for Agrarian Reform in Africa'. 24 January. http://www.viacampesina.org/main_en/index.php?option=com_content&task=viewoid=271&itemid=1, Accessed 27 December 2007.

'Economists Examine Rural Land Problems, Offer Diverse Solutions', 2004. FBIS Report, FBIS Document No. FBIS-CHI-2004-0722, 8 July.

Edelman, Marc, 2003. 'Transnational Peasant and Farmer Movements and Networks'. In *Global Civil Society*, eds Mary Kaldor, Helmut Anheier and Marlies Glasius, 185–220. London: Oxford University Press.

'Farmers in Revolt Near Harbin Against Land Seizures', 2007. *Asia News*, 27 August: 1.

'Farmers Protesting the Theft of Their Lands Are Jailed for Extortion', 2007. *Asia News*, 13 April: 1.

Federici, Silvia, 2001. 'The Debt Crisis, Africa, and the New Enclosures'. *The Commoner*, September (2): 1–13. http://www.thecommoner.org, Accessed 3 March 2005.

Fewsmith, Joseph, 2007. 'Assessing Social Stability on the Eve of the 17th Party Congress'. *China Leadership Monitor*, 20: 1–24.

'Five Thousand PRC Farmers Protest Investment Firms' Collapse', 1999. Hong Kong AFP, FBIS Transcribed Text (AFS Doc. No. 0W2809121399), 28 September.

Gilboy, George J. and Eric Heginbotham, 2004. 'The Latin Americanization of China?'. *Current History*, September (103): 256–61.

Goodspeed, Peter, 1993. 'China's Peasants Get Restless: Rural Protests Send Shock Waves Through Beijing's Urbanized Elite'. *The Toronto Star* (Toronto), 11 July: 2F.

Greenfield, Gerard and Apo Leong, 1997. 'China's Communist Capitalism: The Real World of Market Socialism'. In *The Socialist Register 1997*, ed. Leo Panitch, 96–122. New York: Monthly Review Press.

Gu, Qing-er, 2006. 'Communist Authorities Sentence Survivors of Shanwei Massacre'. *The Epoch Times*, 28 May. http://en.epochtimes.com/news/6-5-28/42034.html, Accessed 4 October 2007.

Guo, Xiaolin, 2001. 'Land Expropriation and Rural Conflicts in China'. *The China Quarterly*, June (166): 422–39.

Hart-Landsberg, Martin and Paul Burkett, 2004. 'China and Socialism: Market Reforms and Class Struggle'. *Monthly Review*, 56 (3): 7–123.

Harvey, David, 2003. *The New Imperialism*. New York: Oxford University Press.

Harvey, David, 2005. *A Brief History of Neoliberalism*. New York: Oxford University Press.

He, Qinglian, 2000. 'China's Descent into a Quagmire: Part I'. *The Chinese Economy*, 33 (3): 32–88.

He, Qinglian, 2001a. 'China's Descent into a Quagmire: Part II'. *The Chinese Economy*, 34 (2): 6–96.

He, Qinglian, 2001b. 'China's Descent into a Quagmire: Part III'. *The Chinese Economy*, 34 (4): 6–94.

He, Qinglian, 2006. 'Officially Sanctioned Crime in China: A Catalogue of Lawlessness'. *China Rights Forum*, (3): 92–3.

He, Qinglian, 2007. 'Officially Sanctioned Crime and Property Rights'. *China Forum*, (2): 44–57.

'Highlights: PRC Civil Disturbances, 1 Jan–15 Dec 04', 2004. FBIS Report, 23 December.

Ho, Peter, 2001. 'Who Owns China's Land? Property Rights and Deliberate Institutional Ambiguity'. *The China Quarterly*, June (166): 394–42.

Holland, Lorien, 1999. 'China Cracks Down on Secretive Army of Peasants'. *The Independent* (London), 27 August.

Holz, Carsten A., 2007. 'Have China Scholars All Been Bought?'. *Far Eastern Economic Review*, 170 (3): 36–40.

Hu, Ping, 2005. 'Taishi Village: A Sign of the Times'. *China Rights Forum*, (4).

Hurst, William and Kevin J. O'Brien, 2002. 'China's Contentious Pensioners'. *The China Quarterly*, (170): 345–60.
'Illegal Land Grabs Up', 2007. *Asia News*, 22 March: 1.
Jacka, Tamara, 1998. 'Working Sisters Answer Back: The Representation and Self-Representation of Women in China's Floating Population'. *China Information*, 13 (1): 43–75.
Kahn, Joseph, 2005. 'China to Cut Taxes on Farmers and Raise Their Subsidies'. *The New York Times* (New York), 3 February: 3A.
Kahn, Joseph, 2006a. 'Pace and Scope of Protest in China Accelerated in '05'. *The New York Times* (New York), 20 January: 10A.
Kahn, Joseph, 2006b. 'Chinese Premier Says Seizing Peasants' Land Provokes Unrest'. *The New York Times* (New York), 21 January: 3A.
Kahn, Joseph and Jim Yardley, 2007. 'As China Roars, Pollution Reaches Deadly Extremes'. *The New York Times* (New York), 26 August: 1A, 10A.
Lee, Ching Kwan, 2007. *Against the Law: Labor Protests in China's Rustbelt and Sunbelt*. Berkeley, CA: University of California Press.
Li, Cheng, 2006. 'Think National, Blame Local: Central-Provincial Dynamics in the Hu Era'. *China Leadership Monitor*, Winter (17): 1–24.
Li, Debin, 1993. 'Liudong renkou yu shehui wending' (The floating population and social stability). *Shehui*, 8: 42–3.
Li, Fan, 2006. 'Unrest in China's Countryside'. *China Brief: The Jamestown Foundation*, 6 (2): 6–8. http://www.jamestown.org/china_brief/article.php?articleid=2373163, Accessed 9 October 2007.
Li, Lianjiang, 2006. 'Driven to Protest: China's Rural Unrest'. *Current History*, September: 250–4.
Li, Lianjiang and Kevin J. O'Brien, 1996. 'Villagers and Popular Resistance in Contemporary China'. *Modern China*, 22 (1): 28–61.
Li, Zijing, 1997. 'Si sheng wushi wan nongmin kangzheng' (500,000 peasants resist in four provinces). *Zheng Ming*, August (238): 19–21.
Manthorpe, Jonathan, 2004. 'Unrest Sharply Increasing Through Much of China'. *Vancouver Sun* (Vancouver), 22 August.
McMichael, Philip, 2007. 'Revisiting the Peasant Question: A Lens on the Making of Another World'. http://devsoc.cats.cornell.edu/cats/devsoc/upload/mcmichal.pdf, Accessed 28 October 2007.
Marx, Karl, 1976. *Capital, A Critique of Political Economy*, Vol. 1. London: Pelican Books.
Miao, Ye, 2003. 'China's Land Market Creates Pressures'. *Asia Times*, 20 August.
'NPC: In the End China Opts for Private Property', 2007. *Asia News*, 9 March: 1.
O'Brien Kevin J. and Lianjiang Li, 2006. *Rightful Resistance in Rural China*. New York: Cambridge University Press.
Perry, Elizabeth J., 2003. 'To Rebel Is Justified: Cultural Revolution Influences on Contemporary Chinese Protest'. In *The Chinese Cultural Revolution Reconsidered: Beyond Purge and Holocaust*, ed. Kam-yee Law, 262–81. New York: Palgrave Macmillan.
Pils, Eva, 2005. 'Land Disputes, Rights Assertion, and Social Unrest in China: A Case From Sichuan'. *Columbia Journal of Asian Law*, Spring–Fall (19): 235–92.
Poole, Teresa, 1993. 'Chinese Leaders Told to Heed Disgruntled Farmers'. *The Independent* (London), 20 October.
'Public Security Organs Strike Hard in 98', 1999. *Xinhua* (Beijing), FBIS Translated Text (AFS Doc. No. 0W0901013099), 8 January.

Rural Development Institute, 2005. 'China to Deepen Land Management Reform', Monitoring Report, 1 November 2004–31 January 2005. http://www.rdiland.org/RESEARCH/Monitoring_Archives_01_05.html, Accessed 2 April 2005.

'Rural Tax Reform to Reduce Financial Burden on Farmers', 2003. *Xinhua* (Beijing), FBIS Translated Text (AFS Doc. No. CPP20030528000024), 28 May.

Saiget, Robert J., 2007. 'AFP Interviews Residents of Northern Chinese Village 2 Years After Deadly Protest'. Hong Kong Agence France-Presse (English), 10 October.

Scott, James C., 1985. *Weapons of the Weak: Everyday Forms of Peasant Resistance*. New Haven, CT: Yale University Press.

Solinger, Dorothy J., 1999. *Contesting Citizenship in Urban China: Peasant Migrants, the State, and the Logic of the Market*. Berkeley, CA: University of California Press.

Tang, Hongqian, Guo Xiaoming and Shen Maoying, 1993. 'Dangqian nongyong tudi feinonghua wenti de diaocha yu fenxi' (An investigation and analysis of the current non-agricultural uses of farmland). *Nongye jingji wenti*, 3: 45–51.

Tang, Yuen Yuen, 2005. 'When Peasants Sue En Masse: Large Scale Collective ALL Suits in Rural China'. *China: An International Journal*, 3 (1): 24–49.

Tanner, Murray Scott, 2004. 'China Rethinks Unrest'. *Washington Quarterly*, 27 (3): 137–56.

Tanner, Murray Scott, 2005. 'Rethinking Law Enforcement and Society: Changing Police Analysis of Social Unrest'. In *Engaging the Law in China: State, Society, and Possibilities for Justice*, eds Neil J. Diamant, Stanley B. Lubman and Kevin J. O'Brien, 193–212. Stanford, CA: Stanford University Press.

Tarrow, Sidney, 1994. *Power in Movement: Social Movements, Collective Action and Politics*. Cambridge: Cambridge University Press.

Thompson, E.P., 1971. 'The Moral Economy of the English Crowd in the Eighteenth Century'. *Past and Present*, 50: 83–103.

Thornton, Patricia, 1999. 'Beneath the Banyan Tree: Popular Views of Taxation and the State during the Republican and Reform Eras'. *Twentieth-Century China*, 25 (1): 1–42.

Thornton, Patricia, 2004. 'Comrades and Collectives in Arms: Tax Resistance, Evasion, and Avoidance Strategies in Post-Mao China'. In *State and Society in 21st Century China: Crisis, Contention and Legitimation*, eds Peter Hays Gries and Stanley Rosen, 87–104. New York: Routledge Curzon.

'Threat of Rural Unrest Forces China to Push Ahead with Tax Reform', 2002. Hong Kong AFP, FBIS Transcribed Text (AFS Doc. No. CPP20020623000008), 23 June.

'Turning Ploughshares into Staves: China's Land Disputes', 2005. *The Economist*, 25 June.

Vía Campesina, 2006. 'Violations of Peasants' Human Rights: A Report on Cases and Patterns of Violence'. http://www.viacampesia.org, Accessed 28 December 2007.

Walker, Kathy Le Mons, 1999. *Chinese Modernity and the Peasant Path: Semi-Colonialism in the Northern Yangzi Delta*. Stanford, CA: Stanford University Press.

Walker, Kathy Le Mons, 2006. '"Gangster Capitalism" and Peasant Protest in China: the Last Twenty Years'. *The Journal of Peasant Studies*, 33 (1): 1–33.

Weil, Robert, 2006. 'We Have Been Here Before: The Cultural Revolution in Historic Perspective in the Global Struggle for Socialism'. *China Study Group*, 31 December: 1–13. http://www.chinastudygroup.org, Accessed 4 May 2007.

Wen, Dale, 2004. 'China Copes with Globalization: A Mixed Review, Part IV'. *International Forum on Globalization*, 31 December. http://www.chinastudygroup.org, Accessed 4 May 2007.

'Wen Jiabao Leading Move to Reduce Rural Taxes', 2003. *Wen Wei Po* (Hong Kong), FBIS Translated Text (AFS Doc. No. CPP20030804000045), 3 August.

Wen Tiejun and Zhu Shouyin, 1996. 'Xian yixia difang zhengfu ziben yuanshi jilei yu nongcun xiaochengzhen jianshe zhong de tudi wenti' (The primitive accumulation of capital by sub-county local governments and land problems under rural urbanization). *Jingji yanjiu ziliao*, 1: 20–5.

Yao, Runfeng, 2003. 'Encroachment on the Rights and Interests of Peasants Regarding Land Has Become the Most Glaring Current Problem in the Infringement of Peasants Interests'. *Xinhua* (Beijing). FBIS Translated Text (AFS Doc. No. CPP20030820000137), 20 August.

Yardley, Jim, 2004. 'In Rural China, Unrest Over Land Seizures'. *New York Times* (New York), 9 December: 2.

Yardley, Jim, 2005. 'A Chinese Court Sentences Farmers Who Protested a Land Seizure'. *The New York Times*, 21 January: 4A.

Yu, Depeng, 1994. 'Chengxiang guanxi zhong de "cheng che xiaoying"' (The 'riding the train' effect in relations between city and countryside). *Shehui*, 3: 37–9.

Yue, Shan, 1997. 'Gan-E wushi wan nongmin baodong' (Five hundred thousand peasants violently rebel in Jiangxi and Hubei). *Zheng ming*, September (239): 21–23.

12 Where There Is No Movement: Local Resistance and the Potential for Solidarity

KEVIN MALSEED

INTRODUCTION

At an international conference on 'Land, Poverty, Social Justice and Development' in 2006,[1] I asked what I thought was a simple question. The panel consisted of representatives from transnational agrarian movements and academia, in front of their colleagues, funders and supporters, development agency officials, and others. 'I work among Karen villagers in Burma who are actively struggling for land and human rights at the village level, but because of the repressive nature of the military regime they cannot form an overt movement. Do your movements have any policies of solidarity or engagement with people in situations like this?' As this was translated into Spanish I thought I noticed perplexed looks from the panel, and when the time for answers came my question was nowhere in sight, dropped in favour of firing more broadsides at the WTO, World Bank and others. In fact, for the remainder of the conference no one seemed able to give me an answer. While peasant leadership in these movements had become expert in the workings of international institutions, I began to wonder whether they had given as much thought to how to engage people at the grassroots in places where they have no members. This chapter aims to raise my question again along with some others, in the hope of stimulating discussion and attention toward people around the world whose struggles in many respects resemble movements, but who lack the political freedom to form a 'Movement'.

Writing on agrarian movements generally ignores states like Burma,[2] where any overt movement must either cooperate with the state or face annihilation. In spite of this, local and everyday resistance against state and other actors is much in evidence throughout rural Burma. How does this resistance compare to the objectives, tactics and achievements of more structured agrarian movements elsewhere? Can it be considered a form of movement itself? And what, if anything, do transnational movements and Burma's rural people have to offer

---

[1] Held 9–14 January 2006 at the Institute of Social Studies, The Hague, Netherlands.
[2] Burma was renamed 'Myanmar Naing-Ngan' by the ruling military junta in 1989. This name change has been rejected as illegitimate by the leadership democratically elected in 1990 (but never allowed to form a government), and as assimilationist by most ethnicity-based opposition groups; this study therefore retains 'Burma'.

each other? This chapter examines these questions as they apply to Karen hill villagers, who practise widespread non-compliance and resistance despite living in a context that denies them any space to organize an overt movement.

'WE WILL STAND HERE':[3] THE CONTEXT

Just two weeks before the October 2005 rice harvest, a column of over 300 Burmese Army soldiers came up the Shwegyin River to force Karen villagers to move to state-controlled areas to the west. About 1,000 villagers fled their villages as the troops approached. On 19 September, the column shot dead a villager they spotted in the ricefields, then began shelling Ler Wah and other villages in the area. The 35 Karen resistance soldiers based nearby fought briefly and then withdrew as the Army moved through the villages, tearing down and burning houses, slashing the villagers' winnowing trays and puncturing their water tins to prevent them living there. For the villagers this was nothing new: the Army first burned their villages in 1975, causing them to disperse into smaller settlements hidden in the forest where they can keep working their land but disappear whenever the columns come, which usually happens once or twice a year. They prefer this to the forced labour and repression they say they would face under state control.

This time they headed uphill to the east, the men to a nearby hillside where they could monitor the Army's movements, the families higher into the hills. While adults quickly built shelters, teenaged students were dispatched to retrieve rice from hidden storage barns. Schoolteachers leaned blackboards against trees and resumed school for younger children. Village elders went to contact Karen resistance forces to obtain information and a few homemade landmines for use in defending their hiding places should the Army attempt pursuit. The Army never came up the hill, probably afraid of ambush, and withdrew on 3 November without having captured a single villager. People immediately returned to begin their overdue harvest, while Karen resistance forces swept the villages and fields for any landmines left by the departing column.[4]

Meanwhile, villagers living in 20 state-controlled villages just to the west – where the Ler Wah villagers had been ordered to go – were being forced to maintain a military access road without pay. Through their organized flight and evasion, the Ler Wah villagers had not only retained their harvest and access to their land, but also evaded unpaid forced labour for the state. This insubordination came at a price: it was the second time their land had been attacked in 2005, and in early 2006 state troops came and burned their villages yet again. The *Tatmadaw* (Burmese military) then established a permanent post in the area, creating a food crisis and causing many villagers to head for refugee camps in Thailand; but large numbers remain there, monitoring Tatmadaw movements while encamped in

---

[3] Interview, Karen villager, Nyaunglebin District, November 2005.
[4] Interviews, with villagers in the affected area, November 2005.

the forests to continue farming and to exert their continuing claim on their lands (KHRG 2006b).

Ethnicity and Oppression: The Conflict Over Sovereignty

At least 80 per cent of Burma's estimated population of 50 million is rural and agrarian.[5] Karen people probably number between four and seven million, concentrated in Karen State and Pegu, Irrawaddy and Tenasserim Divisions.[6] Their main unit of social organization is the village (Marshall 1997, 127), which can range from five to several hundred families and is guided by a village head and a council of elders chosen through various forms of village consensus. Ethnicity and religion are probably the two strongest components of most people's self-identification in Burma. Other aspects of identity include livelihood: in Karen regions most villagers are swidden hill rice or irrigated paddy farmers, but some are landless labourers, merchants or traders, or mix these with rice farming to various extents. In plains areas family-scale commercial agriculture and mixed livelihoods are common, whereas in smaller villages (generally under 50 households) in the hills families focus more on subsistence and barter, each having swidden rice fields, small gardens and livestock, and possibly a small cash crop plantation.[7]

Karen identity has largely been formed as a defence against 'Burmanness', a reference to the dominant ethnicity of central Burma. Writers have attributed this to historical migration of many Karen to remoter hill areas to escape appropriation or enslavement by Burman kingdoms, engendering a self-identification as hill people living beyond the reach of the state, accustomed to outside threats and oppression (Keyes 1977; Hinton 1979). With no higher political structure than the village (Marshall 1997, 143), Karen regions have always been vulnerable to domination by more hierarchically structured valley societies like the Burman kingdoms. In the 1800s, the Sgaw, Pwo, Bwe, Pa'O and various 'Karenni' groups, despite linguistic, religious and geographic differences, forged a more unified 'Karen' identity for survival (Hinton 1979, 92–3), a process encouraged by foreign missionaries (Cheesman 2002, 203). Central to this identity are a 'sense of oppression at the hands of their neighbors' (Keyes 1977, 51) and self-characterization as 'oppressed, uneducated and virtuous' (Cheesman 2002, 204). They have defined themselves largely based on 'structural opposition to other similar groups . . . The Karen also have distinct myths and folklore that have a common theme – their inferior structural position vis-à-vis the lowland peoples, compensated for somewhat by belief in their moral superiority' (Fernando 1982, 130).

---

[5] No reliable census data exist due to manipulation of data by military governments since 1962 (see Smith 1991, 30).
[6] State censuses have claimed 'between 2 and 5 million, whereas Karen nationalists claim between 7 and 12 million' (Cheesman 2002, 203). On the regions associated with Burma's major ethnic groups, see Smith (2002).
[7] Participant observation, Karen villages, 1991–2005.

This identity could exist alongside central authority, first dominated by Burman kingdoms and then by British colonialists, as long as traditional Asian forms of sovereignty prevailed, i.e. the central authorities control only limited territory within reach of the centre of power, while content to extract tribute from or neutralize the 'non-state' spaces further afield (Scott 1998, 185–7). Since Burmese independence in 1948, however, the centralized state has progressively attempted to enforce territorial sovereignty over the entire geography of 'Burma', leading it into direct conflict with Karen and other villagers accustomed to exercising local control over their own affairs (Malseed 2006, 10). The result is a state–society conflict pitting state military encroachment against a spectrum of resistance ranging from widespread civilian non-cooperation to active support for and participation in Karen armed resistance.

Armed resistance began immediately after Burma's independence in 1948, including a Communist insurgency and several armed resistance movements built around ethnicity and seeking either independence or autonomy. The military seized state power in 1962 and has held it ever since, while fighting as many as 20–30 armed resistance groups at a time in Burma's ethnically diverse remoter regions. In 1988 the regime violently crushed urban pro-democracy demonstrations, then annulled an election held two years later. The Communist insurgency imploded and became an ethnic Wa army in 1989, and since that time this and most of the other armed groups have been pressed into informal 'ceasefires' with the regime, now calling itself the State Peace and Development Council (SPDC). Less than ten armed groups are still fighting the SPDC, none of which poses an immediate threat to state power; most operate as small guerrilla resistance forces in low intensity conflict, trying to minimize state power within their respective regions.

In Karen-populated regions of Southeastern Burma, the main armed group is the Karen National Union/Karen National Liberation Army (KNU/KNLA), still in armed conflict with the SPDC. The Democratic Karen Buddhist Army (DKBA) and several other smaller Karen splinter groups have entered ceasefires with the SPDC and sometimes work as proxy armies against the KNU/KNLA. The SPDC conscripts most of its troops in central Burma, but the Karen groups rely on local civilian support and recruits, sometimes voluntary, sometimes coerced (HRW 2007). Though grossly outnumbered and outgunned,[8] the KNU/KNLA survives by local civilian support, but more importantly by civilian non-cooperation with state forces. Most families have immediate or close relatives in the KNU/KNLA ranks, which have downsized and become predominantly voluntary since 1995. Civilian support is also predicated on the KNU's health and education programmes, its fight against state encroachment and its ongoing actions to protect displaced civilians, though it is also criticized for taxation and occasional forced recruitment. The DKBA has a smaller support base, largely because it has alienated many civilians through its heavy demands

---

[8] The Tatmadaw Army numbers approximately 350,000, the KNLA 3,000–5,000 (HRW 2002, 19,121).

for forced labour, extortion and forced recruitment; it seldom protects civilians against state predation. Other armed groups in Karen regions are small and localized, and focus much of their energy on extortion and business.[9]

The Tatmadaw continues to expand and to establish bases throughout rural areas nationwide, including areas where there is no armed resistance (Selth 2002, 35–6,165–6). These bases radiate power over the surrounding villages, imposing restrictions on the activities and movements of civilians and extorting resources, crops and labour. Where populations do not submit easily to control, the Tatmadaw orders civilians to move to army-controlled sites while destroying their villages, crops and food supplies. Civilians in these areas are also forced to do labour as military porters, guides and as servants at army bases (TBBC 2007, 2–4; HRW 2002, 20).

These operations have been backed up by laws established in 1974 that cede control over all land to the state (Hudson-Rodd et al. 2003). At present,

> The State controls all land. Farmers have rights only to cultivation, which household members can inherit if permitted by the authorities . . . The State can revoke landuse ownership rights if the farmers do not grow the crops specified by the authorities or use the land as specified. Land sales and transfers are illegal but tenancy and land sales and transfer of land to non-household family members do exist at the informal level. (Hudson-Rodd and Nyunt 2001, 6)

In most areas the state has decreed that villagers must grow paddy as the dominant crop and must maximize output; those failing to do so have been stripped of their land rights and/or jailed (Hudson-Rodd et al. 2003). Production increases have been attempted through forced double- and triple-cropping schemes, which often fail when corrupt officials steal the required fertilizers and money for irrigation infrastructure, leaving villagers to pay quota penalties at harvest time (Thawnghmung 2004, 1, 156–7; KHRG 2007, 43–5). Though the regime claims to have abolished in 2003 its paddy procurement system, which forced villagers to hand over roughly 20 per cent of each crop at well below market prices (Fujita and Okamoto 2006, 9–10), Karen villagers in most areas say that this has only resulted in increased *ad hoc* demands on their harvests by local military officials.[10] Military and civil authorities routinely confiscate land and demand uncompensated labour whenever required for roads, Tatmadaw bases or Tatmadaw supply farms (KHRG 2007, 20, 57–8). Trade in rice and other commodities, though no longer tightly controlled by state monopolies, is still restricted by controls on moving goods in 'sensitive' areas like the Karen hills, and by the high costs of bribing officials and checkpoints to move produce to market.[11]

---

[9] Field Research, 1991–2007; supporting interviews can be found in KHRG reports since 1995 (see http://www.khrg.org).
[10] Interviews, Karen health workers, September 2005.
[11] See World Food Programme web site, http://www.wfp.org/country_brief/indexcountry.asp?country=104.

These policies have placed the state in direct conflict with villagers' traditional systems of land management. In Karen villages, some land around the village is communal while cropping land is held under traditional tenure within families; land allocation and disputes are handled by village elders, and if they occur between villages they are dealt with at meetings of elders within the 'village tract', a unit of several villages in an area. A 2005 study among Karen hill villagers found that only 23 per cent held any government-issued documents granting them some form of tenure over their land, while over 70 per cent held land rights through customary ownership or the permission of village elders (TBBC 2005).

Subsistence farming forms an important component of identity in rural villages. Ties to the land are both material, as an essential component of survival, and spiritual. Traditional Karen animism is based around forces residing in all things, which must be appeased (Marshall 1997, 210–11), beliefs that have been partially assimilated into local Buddhism and Christianity. State policies on land and crop production are therefore commonly seen as an attack on the very core of village life, survival and spirituality. Combined with demands for various kinds of forced labour (including work on confiscated village land), regular extortion of food and money, and violent punishments for non-compliance, this pushes villagers into a corner where they feel little choice but to resist.

Village Responses

Responses have evolved over the decades in different directions, most involving forms of resistance and evasion. For example, when villagers are ordered to plant a dry season rice crop they often buy the seed as decreed but leave it unsown, knowing that the crop would fail because the required irrigation infrastructure is not in place. Even the amount of seed they buy is reduced because they have underreported the amount of land they till. When quotas are demanded at the projected harvest time, they negotiate reductions by pleading illness, lack of water or other disasters, and if this fails they bribe the officials. These tactics are presently undermining the state's nationwide agrofuel project, which requires each village to plant thousands of castor and jatropha bushes and harvest the bean to provide agrofuel for the Tatmadaw (KHRG 2007, 45–6). One village headwoman expressed a commonly held opinion: 'I think this castor bean will not do us any good. No one wants to plant it and we don't know how to plant it. . . . We think we won't plant it, we'll just pay the cost they demanded and store the seed in a safe place' (KHRG 2007, 50).

When village leaders receive written demands for forced labourers or materials, it is common for them not to respond, then to have a relative plead their absence or illness, then to object that the villagers are too busy with crucial farm work; eventually, many demands either evaporate or end up greatly delayed and reduced. In one example in 2004, a junior officer demanded forced labour of a village day after day without anyone ever showing up, until the village received an exasperated letter in which he wrote, 'I was punched out [by higher officers] because of you' and threatened to raid the village if no one came for labour. The village finally 'complied' only when physical violence appeared imminent, by sending half the requested number of labourers an entire week late (KHRG 2005, 93).

Many resistance tactics are not entirely spontaneous, and here the role of village heads and village elders is key because they act as liaison with all armed groups and many villagers follow their cues. This gives resistance strategies a degree of coordination and unity of direction, and also transmits strategies between villages via 'village tract' meetings where village heads can exchange experiences. Village leaders are regularly summoned by both Tatmadaw and non-state forces to receive orders for their villages, so they have to be canny negotiators. The reports and lists they are required to provide are usually heavily edited to suit their village's interests. In effect they act as tactical coordinators of villagers' non-compliance and resistance, though when they decide a particular demand must be obeyed they sometimes find themselves resisted by their own villagers. Regardless, they are generally held responsible by the armed groups for any non-compliance and are usually the first to be detained or tortured for any failure to obey.

Negotiating power is augmented by appointing elderly women as village leaders. Exploiting the reverence for mother-figures in Burmese cultures, these women routinely scold or challenge the young military officers who give the orders, knowing that their sense of power and authority becomes confused when confronted with a mother-figure. In written orders to villages, it is very common for Tatmadaw officers to address village headwomen as 'Mother' and refer to themselves as 'Son'.[12] In one such case, a village headwoman in Papun district received an order in late 2005 to send several villagers as 'guides' for a military patrol. Knowing that this meant forced labour as human minesweepers, she went to the base to confront the officer. As she related afterward, she told him, 'You know I cannot ask my villagers to do this. I cannot ask them to walk in front of your troops to step on mines.' He apologized and said there was nothing he could do; 'It is my duty and these are orders from above, you will have to do it or your village will be punished.' She said, 'Then take me instead. I will go. But on one condition. I'm afraid of mines, and I'm sure you're afraid of mines too. So let's walk in front together, hand in hand. If I step on a mine or you step on a mine, we'll both die together. I can be content with that.' The officer eventually responded, 'I'll think about it, go home Mother and I'll tell you my decision later.' The demand didn't come again.[13] She knew that the officer had impunity to kill her on the spot, or detain her indefinitely without charge, and that if he did so the villagers could do little but plead or pay for her release, because any other form of protest would be violently punished. Yet she knew him sufficiently and was confident enough in her maternal authority to gamble that he would not do so. Similar instances occur almost weekly in many areas.

[12] Many examples of this appear in KHRG (2002) and KHRG (2003).
[13] Interview, Karen human rights researcher, November 2005.

Perhaps most significant are villagers' strategies for retaining control over land they have been ordered away from. When villagers are ordered to move off their land and into Army-controlled sites to bring them under control, they usually respond with flight and displacement – but not beyond reach of their land. Instead, they adopt mobile livelihoods allowing them to evade authority, monitoring military movements so that they can continue working their lands. This usually involves living in the village whenever soldiers are not around and in the forests at other times; caching food in hidden locations or with relatives living in more stable villages; shifting to more durable root crops or concealable cash crops (such as cardamom) if necessary; and trading with sedentary villages for dry goods in pre-arranged covert 'jungle markets' which spring up for a day and then as quickly disappear (Malseed 2006, 18). All of these are arranged informally by the villagers themselves, with village heads as facilitators when necessary. Family links to armed groups are exploited for intelligence or protection; but if a local KNLA or DKBA commander makes excessive demands, he usually finds the same villagers' tactics turned against him.

Monitoring military movements, sharing out food when conditions are hard, setting up covert markets, planting and harvesting in groups and by night if necessary with armed guards, are operations which require levels of preparation, coordination and ongoing cooperation that debunk outside perceptions of these villagers as starving, powerless 'frightened chickens' huddled in the forest. Many hill villagers have been practising these fluid forms of displacement since the mid-1970s, and this has been a key factor in preventing military control of the hills (Malseed 2006). At a typical meeting between the author and displaced villagers in late 2005, they began by characterizing themselves as helpless and vulnerable; yet when asked what specific forms of control the military exercised over them, they began recounting how their displacement strategies had allowed them to evade forced labour, extortion and torture, and were soon discussing how this must have undermined the Tatmadaw's plans and angered the officers in their area. This is displacement practised not as the abandonment of land in terror, but as a strategy to retain control of land; a strategic displacement, which falls within what Adas (1986, 64) appropriately characterizes as 'avoidance protest'.

Village heads frequently exploit contacts with resistance forces to strengthen their villagers' ability to evade abuses, whether by obtaining intelligence on Tatmadaw movements, landmines or guns to arm village patrols, or playing sides off against each other (for example, refusing a demand by claiming that another armed group has already placed a demand on that resource, be it food, oxcarts or people's work time). As the Tatmadaw landmines fields and destroys more crops, more displaced villagers tend toward using landmines and asking for weapons;[14] thus, more violent oppression and a more desperate material situation force villagers' resistance toward the direct and overt.

---

[14] Interviews, displaced villagers in eastern Pegu Division, November 2005.

Despite their key role, I have yet to hear of a village head formalizing this or attempting to structure a movement with themselves at its head, possibly for the reasons discussed below. On the contrary, many villages have decided to share the burden of being village head by regularly rotating the position. As explained by a woman villager in southern Karen State,

> Village heads . . . are elected by the villagers themselves. They are usually women, because men cannot survive the repeated beatings and punishments by the soldiers. Therefore, nobody wants to be a village head throughout the whole region. Some villages operate a rotation system for the position, and change the village head as often as every two weeks or every month. As a result, even 17- or 18-year-old girls sometimes act as village heads, but they can control the villagers and will be obeyed because everyone knows that they are being instructed and guided by the village elders, usually monastic leaders, and so they never misuse their powers. (KHRG 2006a, 68)

Resistance strategies have acquired strong resemblances across time and space, with examples like those above repeated in regions separated by 1,000 kilometres or more.[15] These may have evolved independently as needed, but communication between village heads and across regions via military and governance structures set up by the KNU/KNLA has almost certainly played a role. All of this is achieved without any formalized structure, but with a degree of consensus that usually centres on the person of the village head. As noted by Scott (1985, 35), the involvement of a large proportion of the community in such acts combined with a 'supportive subculture' make it 'plausible to speak of a social movement. Curiously, however, this is a social movement with no formal organization, no formal leaders, no manifestoes, no dues, no name, and no banner.' Moreover, these tactics are deployed for objectives that would sound familiar to any activist in a transnational agrarian movement: control over village lands, retention of traditional livelihoods, and maintenance of identity and community.

An unusually overt expression of group resistance occurred in September 2007 in Northeastern Karen State, in a region where villagers have been repeatedly displaced into nearby forests by Tatmadaw encroachment since 2005. Though displaced, these villagers can usually produce a rice crop when the Tatmadaw withdraws during the monsoon growing season, but in 2007 the Tatmadaw established several small but permanent posts, leading to a food crisis. At a meeting with displaced people from four villages in September, KNU officials pressed them to move to a safer area, but village spokespersons responded 'If we move to another area the KNLA cannot secure it for us, and the SPDC will make bases around our villages so we won't be able to come back'. They said they would rather stay on their land and flee when necessary, and that they are calling themselves 'Gher Der' ('Defend [our] homes'). No

---

[15] This is reflected in oral histories gathered by the author from, southern Tenasserim Division to Northern Kayah State, 1991 to 2007.

leaders were named.[16] Naming their strategies may not change them, but it reflects clear recognition of common interests and collective action. As these villagers live by evading state contact, this initiative has occurred beyond state control. Whether it will spread, and whether it will be targeted for destruction by the Tatmadaw or cooptation by the KNU, remains to be seen.

Problems of Forming Overt Movements

This begs the question whether villagers' resistance could be more effective if it became more structured and systematic, or coalesced into a named, identifiable 'movement'. In most local contexts there are possibilities for greater coordination between families and villages, and since 2005 field researchers with the Karen Human Rights Group have been encouraging discussions in villages as to where these potentials might exist.[17] Possibilities vary between villages depending on cropping systems, the character of the local military officers and other factors, but creating a more formalized 'movement' with recognizable structures or hierarchies is generally not seen as an option, for the following reasons.

Firstly, villagers' non-compliance strategies have flourished for so long largely because there are no leaders or structures for the state to kill, arrest or buy off. Contrast this with Burma's several dozen armed groups, most of which have been coerced, surrounded or bribed into 'peace' agreements with the regime, many of them mutating into exploitative business entities or proxy armies of the state. The SPDC is not prepared to allow any form of civil society that it cannot control or effectively restrict (Steinberg 1999, 13). Among the only organizations allowed to operate without direct state intervention are religious organizations, and the regime's violent repression of the Buddhist Sangha in September 2007 after monks dared to protest against economic policies provides a clear example of the regime's response when even revered religious institutions call for reform. The present state, dedicated to controlling all productive land, crops and trade through the mechanisms outlined earlier, would certainly not allow any form of independent movement representing rural villagers or advocating agrarian reforms; as partial evidence of this, there are no legal trade unions or agrarian organizations in Burma today.[18] The Tatmadaw has been known to specifically target even local health and relief operations in Karen

---

[16] Information and quotes translated from field report of KHRG researcher who was present. Village names are omitted to protect the villages concerned.
[17] Author's participant observation, November 2005, 2006; April 2007.
[18] In 2006 the International Labour Organization's Committee of Experts on the Application of Conventions and Recommendations wrote, 'the information provided by the Government [of Burma] continues to demonstrate a total lack of progress towards establishing a legislative framework under which free and independent workers' organizations can be established'. 'CEACR: Individual Observation concerning Freedom of Association and Protection of the Right to Organise Convention, 1948 (No. 87) Myanmar (ratification: 1955)', (see http://www.ilo.org/ilolex/cgi-lex/pdconv.pl?host=status01&textbase=iloeng&document=8454&chapter=6&query=%28C087%29+%40ref+%2B+%28Myanmar%2CBurma%29+%40ref&highlight=&querytype=bool&context=0).

State.[19] Though armed organisations have managed to survive, they have done so either by making deals with the regime, or by militarily defending their existence against the overwhelming military offensives that inevitably ensue otherwise.[20]

The KNU/KNLA and DKBA leadership themselves grew up under rigidly hierarchical state–society relations, and as a result are suspicious of any civilian movement developing in their areas of operation without submitting to their leadership. The KNU has occasionally put pressure on Karen civilian groups to submit to oversight,[21] while the DKBA is reported to have made violent threats against local human rights groups documenting DKBA demands on civilians. Karen-based relief organisations such as the Karen Office of Relief and Development (KORD) and activist groups such as Karen Rivers Watch work in Karen State but are run by refugees based in Thailand, partly to facilitate access to resources and advocacy but also for greater independence from armed groups. Political space is far more constricted for villagers within Burma itself, even in resistance-controlled areas.

Second, the villagers' strategies often take the form of non-compliance concealed by a fig-leaf of compliance. The village headwoman refusing to provide human minesweepers mentioned earlier did not challenge the officer's authority openly or even refuse his demand; she simply suggested an alternate method, giving him no grounds to punish her for insubordination but knowing he would balk. Resistance generates a less violent reaction the less it is seen as challenging structures of authority (Scott 1985, 33). If the objectives or structure of resistance become more visible, however, state resources would probably be directed toward undermining or crushing it.

Finally, a corollary to the state's control over civil society is that civilians throughout Burma tend to be intensely distrustful of organizations. State-run organizations operate largely as instruments of repression and predation, while armed resistance groups have themselves frequently exploited villagers and forcibly conscripted them. Moreover, villagers know that membership in any group or initiative often results in punitive retaliation from some opposing group, so they are hesitant to join anything structured. The 'Gher Der' example mentioned earlier could develop into an exception, growing out of a context where villagers see no other viable option; but even 'Gher Der' is notable for its lack of any formalized structure.

WHAT MAKES A 'MOVEMENT'?

Without structure and overall coordination, is it possible to speak of a 'movement'? Just what constitutes and drives a 'movement' has long been debated. Traditional

---

[19] Backpack Health Worker Teams, a medical relief organization, reports that its teams are routinely tracked and targeted by the military, and that from 1998 to 2005 seven of its medics were executed (BPHWT 2006, 22).
[20] The DKBA is an example of the former, the KNU/KNLA of the latter.
[21] Observed by the author in Karen civil society meetings, and in KNU statements, for example: 'We, the KNU, earnestly urge the entire Karen revolution and Karen people to struggle on, in unity, with the KNU, and resolutely oppose all activities destroying the revolutionary and national unity' (KNU 2007).

analysis focused on collective behaviour, and often characterized movement activity as irrational explosions of mob anger (Cohen 1985, 672). Raising empirical evidence that contradicted this, 'resource mobilization' theorists went to the other extreme, applying a 'neoutilitarian logic imputed to collective actors' (ibid., 674), treating movement participants as utilitarian rational actors and emphasizing interests, strategies and contention between organizations for 'recognition of the group as a political actor or ... increased material benefits' (ibid., 675). This is contrasted by 'identity-centred' or 'new social movements' theorists, emphasizing collective identity formation and studying loosely networked movements seeking 'to create democratic spaces for more autonomous action' (Escobar and Alvarez 1992, 5). They characterize 'new' social movements by 'an open, fluid organization, an inclusive and non-ideological participation, and greater attention to social than to economic transformations' (Della Porta and Diani 1999, 12).

Both resource mobilization and identity-centred paradigms assume a 'contestation between organized groups' (Cohen 1985, 673). Tarrow (1998, 4) defines social movements as 'collective challenges, based on common purposes and social solidarities, in sustained interaction with elites, opponents, and authorities'; yet his 'collective challenge' is to be mounted first (presumably by a vanguard), *after* which others are mobilized to the cause through 'social networks' and 'building solidarity' (ibid.). Others declare a need for leaders more bluntly: 'A social movement is at least minimally organized. If we cannot identify leaders or spokespersons, members or followers, and organizations or coalitions, the phenomenon under investigation is a trend, fad, or unorganized protest, not a social movement' (Stewart et al. 2001, 2). Early on, Piven and Cloward challenged such definitions, arguing that, 'the stress on conscious intentions in these usages reflects a confusion in the literature between the mass movement on the one hand, and the formalized organizations which tend to emerge on the crest of the movement on the other hand – two intertwined but distinct phenomena' (1977, 5). They consider 'collective defiance' to be the 'key and distinguishing feature of a protest movement'. Defiance here refers to people's acts of non-compliance or resistance when they no longer perceive rules and demands as legitimate; 'apparently atomized acts of defiance can be considered movement events when those involved perceive themselves to be acting as members of a group, and when they share a common set of protest beliefs' (ibid., 4).

Though focused on 'organized groups', most 'new social movements' theorists allow for great fluidity in organizational structure and objectives. Escobar and Alvarez (1992, 7), for example, criticize the 'empirical simplification and political reductionism that lead researchers to focus their attention on the measurable aspects of protest, such as confrontation with the political system and the impact on state policies; consequently, they disregard the less visible effects at the levels of culture and everyday life' that are the actual objectives of much social protest. Access to or replacement of the state is only a means, not an end. Some writing on 'identity-centred' or 'new social movements' goes further, arguing that the tendency toward openness and fluidity over structure makes these movements resistant to any definition whatsoever.[22] A more moderate position would be that if social movements themselves are evolving, then the way they are defined must evolve as well.

Della Porta and Diani (1999, 16) have synthesized several perspectives to put forward an inclusive definition of social movements that seems apt for the networked movements of today, and that allows room for comparisons with Karen village-level resistance: 'We will consider social movements – and, in particular, their political component – as (1) informal networks, based (2) on shared beliefs and solidarity, which mobilize about (3) conflictual issues, through (4) the frequent use of various forms of protest.' They then elaborate on this definition, adding that movements are 'not organizations' but 'networks of interaction between different actors which may either include formal organizations or not, depending on shifting circumstances' (1999, 16). This emphasizes the fluidity of social movements; 'one of their characteristics is, indeed, the sense of being involved in a collective endeavour – without having automatically to belong to a specific organization. Strictly speaking, *social movements do not have members, but participants*. . . . the membership of movements can never be reduced to a single act of adhesion. It consists, rather, of a series of *differentiated acts, which, taken together, reinforce the feeling of belonging and of identity*' (1999, 17; emphasis added). By shifting the focus away from organizational structure to shared belief and action, this definition allows for comparison between different forms of like-minded and collective resistance, while still accepting that there is a definable boundary between movements and non-movements.

Structure and Spontaneity: Comparing Karen Villagers' Resistance to an Internationally Known Movement

Brazil's Movimento dos Trabalhadores Rurais Sem Terra (MST), or Landless Workers' Movement, is an internationally known agrarian social movement. Formed in 1985, MST now claims 1.5 million members throughout Brazil, making it the largest social movement in Latin America. Operating on principles such as 'land to the tiller' and an article in Brazil's constitution stating that unproductive land should be put to social purposes, it seeks sustainable livelihoods for landless families through peaceful occupations of unused parcels of land, where it establishes collective or family farms supported by schools, clinics and other facilities. It claims to have secured land titles for over 350,000 families in 2,000 settlements, with 180,000 more families currently settled on land awaiting titles.[23]

The MST falls easily within Della Porta and Diani's (1999) definition of social movements. It has become far more than an 'informal network', with provincial and national offices, hierarchies and boards (MST 2001), yet it is still propelled

---

[22] Escobar and Alvarez (1992, 6–7) cite writings of Alain Touraine, Ruth Cardoso and Elizabeth Jelin to this effect.
[23] MST web site (http://www.mstbrazil.org/?q=about), accessed October 10 2007.

by the commitment and work of its landless members. Shared beliefs and solidarity exist in the philosophy of land rights and the establishment of collectively-run farms and social services. It mobilizes around the very conflictual issue of land rights, sometimes leading to violent attacks on its members by state and other forces (MST 2001; Filho and Mendonça 2007, 83–4). Finally, it is involved in various forms of protest, such as land occupations and demonstrations. The MST also satisfies Piven and Cloward's definition, though they would differentiate the 'collective defiance' of landless workers rising against a land tenure system that they no longer view as legitimate from the formalized movement structures that have emerged on the 'crest' of this defiance.

The widespread but unstructured resistance of Karen villagers is more difficult to place within Della Porta and Diani's definition, unless its terms are interpreted broadly. Villagers do operate as part of informal interaction networks, whether through kinship, communication or cooperation networks between families and villages. Most of these networks were not established for purposes of resistance, though they serve that purpose as resistance develops. Similarly, shared beliefs and a sense of belonging do exist, though not perceived as membership of a movement. Instead, belonging exists through the sense of shared identities, as Karen, as hill or valley villagers, as co-residents of a particular village or area, as swidden rice cultivators, etc., while shared beliefs include ideas of local sovereignty and human rights, traditions, sometimes religion, opposition to the military dictatorship and sympathy for the armed resistance. Though not codified as a movement philosophy, these elements create what Scott (1985, 35) called the 'supportive subculture' essential for acts of everyday resistance to succeed, and qualify villagers' strategies as what Della Porta and Diani (1999, 17) call 'differentiated acts, which, taken together, reinforce the feeling of belonging and of identity'. Regarding the third requirement, for collective action focusing on conflictual issues, Karen villagers often act collectively to evade or mitigate demands and abuses and to preserve what they see as their Karen identities and livelihoods against state efforts to assimilate them. In a broad sense, this action aims to produce social change: an end to state assimilationist practices, and the preservation of local sovereignty over land and livelihoods. Finally, the requirement that movements use protest as a method does not apply if this means overt, public protest. But if protest is interpreted more broadly to incorporate widely practised forms of 'everyday resistance' (Scott 1985), then flight can be considered a form of protest ('voting with one's feet', or 'avoidance protest'), as can refusing *en masse* to show up for mandated forced labour, or sending village representatives to the local authorities to oppose and negotiate a particular demand.

From Piven and Cloward's perspective, whether Karen villagers' acts of non-compliance could be considered a 'movement' depends on whether they are driven by the illegitimacy of the exploitation, and by a common set of beliefs. If asked, most Karen villagers would be unlikely to describe themselves as part of an agrarian social movement. However, when asked how they have responded to specific recent abuses, awareness of their own acts of resistance promptly emerges, usually accompanied by expressions of community pride and solidarity. Though it could be argued that many acts are performed more for defence than resistance, the line between the two is seldom clear (Adas 1986). Villagers may withhold their forced labour primarily because they need to work in the fields, but they also resent and see no legitimacy in the demand and frequently say so when interviewed. Similarly, villagers flee into the forest to avoid being shot on sight by Tatmadaw troops, but this only occurs because they have refused to move to state-controlled sites as ordered. Even when non-compliance bears little intent to resist, the military authorities still view it as resistance, as evidenced in the threats they make when villagers disobey.[24] For such acts to be considered part of a 'movement' of resistance, villagers need not be conscious of each and every action as a step toward larger goals, but there should be an atmosphere of communal acceptance and support, a like-mindedness of action and consistency with commonly held ideas of social justice. As noted earlier, Karen village resistance strategies and their objectives – centred around retention of control over land and livelihoods, community and identity – show remarkable similarity among people within a single village and across diverse and non-contiguous territories.[25] The *collective* nature of these actions comes not by coordination or communication, but by similarity in causes, methods and objectives. The 'Gher Der' phenomenon mentioned earlier is merely an overt recognition of common cause that is usually left implicit.

Moreover, the very lack of structure, hierarchy or codified objectives can help keep resistance more spontaneous and adaptable to changing needs and conditions. Hierarchy and/or representative democracy within movements makes them subject to the personalities in leadership positions and to power relations, whereas the combined effect of uncoordinated peasant action by its very lack of leadership renders itself resistant to diverted or corrupted goals. Each act occurs with a specific goal of its own, and the overall effects on power structures are only manifest when all of these acts combine. While this can prevent resistance from becoming overt or coordinating itself, it seems an efficient way of keeping a movement 'on course', for all its appearance of a body without a head. 'Where everyday resistance most strikingly departs from other forms of resistance is in its implicit disavowal of public and symbolic goals' (Scott 1985, 33); which serves it well, because public and symbolic goals are always subject to interpretation (in the case of the symbolic) and to politics and conflicting interests (in the case of the public). Public and symbolic goals are also easy targets for attack or co-optation.

When villagers' responses to oppression follow similar patterns and are played out consistently on a large scale, it becomes possible to think of a movement without name or self-consciousness as such. In the Karen case there is a large

---

[24] Village leaders regularly testify that officers threaten to treat their village as 'enemy' should they fail to comply with orders.
[25] This can be seen in, *inter alia*, food caching and sharing strategies, monitoring military movements, evading forced labour demands and misleading state forces while informing human rights groups and resistance forces, observed by the author between 1991 and 2005 throughout Tenasserim Division, Mon State, Karen State and Eastern Pegu Division.

population spread over wide regions, yet identifying themselves as a 'people' and practising comparable strategies to evade or mitigate exploitation and abuses by commonly perceived 'oppressors', be these people, institutions or programmes. These apparently isolated acts combine with profound effect to undermine Tatmadaw control over villages and sabotage state 'development' programmes. This is grassroots action with no vanguard; existing as a multitude of individual and collective acts, it is much harder to steer off course and can achieve similar ends to those of more structured movements. Before we draw a rigid line between structured and unstructured movements, however, it is more useful to think of movement activity having a multitude of possible structural expressions, ranging from well-organized network structures with identifiable representatives, to leaderless and nebulous movements like Karen-style village resistance.[26]

Commonalities become clearer if we compare the objectives, tactics and achievements of Karen village resistance to those of a recognized agrarian movement like the MST. The specifics of course differ greatly because they exist in very different contexts for different purposes and with different participants, but the objective here is to compare the *types* of objectives, tactics and achievements.

- **Objectives**. The MST organizes around issues of rights, specifically land rights and equitable land redistribution. The struggle plays out primarily between the landless poor and large landowners. Karen villagers' acts of resistance also occur on a battlefield of rights, and many villagers explicitly state this. Specifically, they want control of the land they have always worked without undue interference or exploitation by the state or other powerful actors. The struggle plays out primarily between villagers and the largest landowner, the Burmese state. Individual acts may be motivated by a combination of self-interest and the desire to undermine oppression, but the same could be said of the MST; most people who participate in the MST are probably more interested in gaining access to land than in changing Brazil's system of land tenure.
- **Tactics**. The MST uses organized overt action, with a vanguard that negotiates with government or identifies land for seizure, plans the action and organizes people into large communities to seize land and establish rights over it (MST 2001). Karen villagers operate in a far less organized manner, their acts of resistance usually occurring within the village. When ordered off their land by the Tatmadaw, however, their response is often guided by village elders, who lead the move into the forest or to the state-decreed forced relocation site; if the former, KNLA protection is sometimes negotiated beforehand and survival activities are divided among the population as described earlier.
- **Achievements.** Through the MST, several hundred thousand people have gained access to land and formed new communities with support services; however, some have been killed by violent state or landowner retaliation. Though not seeking formal land titles, Karen villagers' resistance has enabled

---

[26] I am indebted to an anonymous reviewer for this insight.

hundreds of thousands of them to stay on or near their land despite state predation and forced relocation campaigns. In this process, many have been killed by violent military reprisals, illnesses and malnutrition. The covert nature of their resistance means that we may never know the full extent of its achievements or failures.

In summary, though very different in context and specifics, the overall nature of these two struggles makes them comparable, and suggests that if one qualifies as a 'movement' then the other should as well.

Whether Karen village resistance is considered a 'movement' is not important in itself, but it does have a strong effect on how other actors respond to it. Currently many external actors think of Karen villagers as helpless and passive actors, starving and in need of outside assistance;[27] whereas recognizing their resistance as effective, conscious and quasi-coordinated demands responses that acknowledge their agency, support instead of undermining their existing practices, and let them control any process of change.

LINKS TO OUTSIDE MOVEMENTS: THE POTENTIAL FOR SOLIDARITY

Domestic Networks

In August and September 2007 the outside world took a rare interest in Burma when processions of thousands of Buddhist monks took to the streets to protest state policies. As urban civilian support for the monks grew more overt, some commentators questioned why rural villagers and ethnicity-based armed resistance groups did not appear to be actively involved.[28] Before considering the potential for links between village-level resistance and transnational actors, it is worth discussing the accusation that rural villagers have been insufficiently supportive of urban 'movements' within Burma.

Here it is worth noting that the monks' uprising did not take place in isolation. It followed an escalation of more subtle methods of urban protest which had developed over the past three years. This included the 'white shirt' campaign, which involved people showing their opposition to the regime by wearing a white shirt on designated days; a prayer campaign, when people wore yellow to privately pray for political prisoners at Shwedagon Pagoda in Rangoon on Tuesdays; and petition campaigns, with covert petitions circulating throughout the cities.[29] Each time the regime attempted to shut down these campaigns (for example, by deploying paid thugs to throw water on those praying at the Shwedagon or by arresting those suspected of initiating petitions), a new form

---

[27] Such portrayals still dominate outside reports. Horton (2005, 14), for example, describes displaced Karen villagers as 'traumatised, weak, psychologically numb people'.
[28] See, for example, 'Lack of unity kept ethnic groups out of the showdown', The Irrawaddy, 11/10/2007 (http://www.irrawaddy.org/article.php?art_id=8984).
[29] See 'Junta warns of action against student group', The Irrawaddy, 3/11/2006 (http://www.irrawaddy.org/article.php?art_id=6307).

of protest arose to supplement or replace the old. This pattern has continued since the September 2007 violent crackdown on the monks' protests, with signs bearing junta leaders' names appearing around the necks of stray dogs, and boycotts of mandatory-attendance SPDC rallies.[30] Meanwhile, urban people continue their own daily strategies for evading and minimizing taxes and demands. Just like rural resistance, urban resistance is ever-present in myriad forms.

On the surface there appear to be few overt links between everyday urban and rural resistance, or between rural resistance in different regions. Karen villages face regular military curfews and travel restrictions limiting movement outside their villages, and travelling even short distances requires bribes to many 'checkpoints'. In addition, the lack of structured leadership inherent to village-level resistance precludes the possibility of formal connections. Both state and non-state armed groups are suspicious of civilian political activity that is outside their sphere of control, making it difficult to establish civilian-to-civilian links. When urban dissidents fleeing arrest arrive in rural villages, the villagers sometimes face violent punishments by state forces for harbouring them; the presence of overt resisters thus puts their daily covert resistance at risk. Finally, Karen villagers are fiercely protective of local sovereignty and suspicious of Burman assimilationist tendencies, leading to a great deal of distrust between Karen villagers and urban Burmans. The historical baggage of distrust, both inter-ethnic and rural–urban, can make rural villagers more open to solidarity with people in similar positions halfway around the world than with urban dissidents within their own state. As Fox (2005, 187) has pointed out, many transnational links are actually 'translocal', between local groups or polities scattered around the world. Sometimes establishing links with those in distant lands can be less fraught with political landmines than cooperating with those in a neighbouring town or village.

Despite these obstacles to overt links, connections with other domestic movements operate in more subtle ways. Karen villagers are aware of struggles occurring in other parts of the country via shortwave radio broadcasts,[31] human rights reports circulated in Karen and Burmese languages, information from armed resistance groups, and increasingly through participation in meetings and relief efforts coordinated by resistance groups or local human rights organizations. The information thus gained keeps people informed on how villagers elsewhere are resisting the state agrofuel project, or evading military forces through strategic displacement. Information from radio broadcasts resulted in a peaceful rally by villagers in Southern Karen State in September 2007 to express support for the nationwide monks' protests.[32] Traders and trade routes also

---

[30] See 'Protesting dogs are now on the regime's wanted list', The Irrawaddy, 12/10/2007 (http://www.irrawaddy.org/article.php?art_id=8998, accessed 31/10/2007), and 'Unlikely resistance in Burma's Mandalay', BBC News, 25/10/2007 (http://news.bbc.co.uk/2/hi/asia-pacific/7060424.stm).
[31] BBC and Voice of America broadcasts in Burmese are popular, as are Democratic Voice of Burma broadcasts in Burmese, Karen and other languages.
[32] See 'Protests spread in rural Karen State', KHRG Bulletin, 25 September 2007. (http://www.khrg.org/khrg2007/khrg07b3.html).

function as conduits of information and material support; for example, 'jungle markets' covertly organized between villagers in hiding and those in sedentary villages function as a survival strategy, a resistance tactic and a communication link. Thus networks of village resistance exist as part of the 'movement', but like the resistance itself they are unstructured and often transitory.

Transnational Links and Global Coalitions

Whether transnational links or solidarity would be useful to the local-level covert resistance of Karen villagers is a complex question. On this subject, Keck and Sikkink (1998, 12) have noted the 'boomerang effect', whereby domestic struggles that cannot get action from the state engage transnational networks to bring pressure upon that state from outside. They write that 'transnational advocacy networks' can 'make international resources available to new actors in domestic political and social struggles. By thus blurring the boundaries between a state's relations with its own nationals and the recourse both citizens and states have to the international system, advocacy networks are helping to transform the practice of national sovereignty' (1998, 1–2). This is particularly relevant to villagers who are essentially non-state actors in a struggle with the state over local sovereignty. Fox and Starn (1997, 4) claim of peasant resistance that 'it is impossible to broadcast a muffled resistance at any great distance, and certainly not around the world'. And yet Karen villagers have already transmitted their resistance strategies throughout widely separated areas within Burma, local human rights groups have dispersed knowledge of these strategies to institutions, governments and other actors worldwide, and even this chapter is an attempt to further transnational discussion of this case. Over the past 15 years I have witnessed issues of immediate concern to rural villagers, including forced labour, displacement and forced relocation, move from being ignored in international fora to being central to international discussions on human rights in Burma,[33] all because of such action; similarly, reports by local and international organizations now frequently acknowledge the survival strategies of displaced Karen villagers,[34] which was not the case five years ago. Contrary to Fox and Starn's claim, a 'boomerang effect' appears a very viable option, but if it is to occur it will require external actors to initiate engagement with the villagers rather than the other way round.

One of the 'transnational agrarian movements' represented on the conference panel that failed to answer my question on movement solidarity was Vía Campesina. Founded in 1993 by a group of peasant and farmers' movements and organizations in the Americas and Europe (Borras 2004, 3), Vía Campesina now declares itself to be

---

[33] As an example, pressure by the International Labour Organization since 1997 eventually forced the SPDC to formally outlaw forced labour (though still unenforced) (see http://www.ilo.org/public/english/region/asro/yangon/).
[34] See TBBC (2007, 4), South (2006, 87), BI/PWF (2006, 23–4).

the international movement of peasants, small- and medium-sized producers, landless, rural women, indigenous people, rural youth and agricultural workers. We defend the values and the basic interests of our members. We are an autonomous, pluralist and multicultural movement, independent of any political, economic, or other type of affiliation. Our members are from 56 countries from Asia, Africa, [E]urope, and the Americas.[35]

As one of the best known 'transnational agrarian movements', it will be used here as an example of a transnational actor.

Vía Campesina is an important step toward peasants representing themselves at international levels, rather than being represented. Its delegates are usually local peasant leaders in their own movements. Rather than a 'movement' itself, however, Vía Campesina is more accurately a coalition of movements, because it is made up of member movements rather than individuals. Membership in Vía Campesina is limited to movements with names and clear structures and acceptance of new members is decided among existing members, a process sometimes affected by the politics between rival movements (Borras 2004, 17). My repeated inquiries to Vía Campesina representatives regarding the details of the acceptance process and their policies on peasant struggles outside the context of membership received no reply whatsoever.[36] The position of Vía Campesina on solidarity with non-members therefore remains unclear. Though its primary function is to 'defend the values and basic interests of our members' (see above), the coalition's international advocacy statements frequently suggest a broader claim to represent 'peasants' worldwide, often citing its peasant membership in 56 countries. Even without this claim, advocacy aimed at changing the policies of global institutions and trade practices can potentially affect peasants and the rural poor in every region of the world. This creates a tension between the coalition's commitment only to its members, based in less than a third of the world's countries and with no representation across large regions such as Russia, the CIS, Central Asia, China and much of mainland Southeast Asia, and its advocacy on behalf of a much larger polity. As noted by Borras (2004, 24),

> The great majority of marginalised rural people, of course, remain outside formal organisations. While it is important for the cause of poor peasants and small farmers that Vía Campesina advocates positions that favour the marginalised social classes and groups more generally, it is important to be critically aware of the gap between the groups of peasants and farmers within the transnational reach of the Vía Campesina movement, and the greater number of rural people that are not.

The International Conference on Agrarian Reform and Rural Development (ICARRD) in March 2006, attended by representatives of Vía Campesina, other

---

[35] See Vía Campesina web site (http://www.viacampesina.org/main_en/index.php?option=com_content&task=blogcategory&id=27&Itemid=44).
[36] These and related questions were sent by email to two separate Vía Campesina representatives on 1 October and again on 17 October 2007. No response was received.

transnational coalitions, activist groups and United Nations and government officials, adopted a final declaration calling for 'international solidarity and support to *organizations* of small farmers, landless people and rural workers' (ICARRD, 2006; emphasis added). This statement indicates again the bias in favour of formally 'organized' peasantry, to the exclusion of those who do not have the political space to organize. Most transnational networks and coalitions other than Vía Campesina are similarly structured around the concept of membership, whereby structured movements must apply to join.[37] Exclusion from these structures has meant that the voices of those in less organized struggles have gone largely unheard in processes such as the drafting of the United Nations Declaration on Indigenous Peoples, passed in September 2007 after over 20 years of work.[38]

Vía Campesina has noted that 'there are systematic patterns of violations of the human rights of smallholder peasants' worldwide which require 'systematic policy responses in order to find adequate solutions' (Vía Campesina 2006, 1). To decide on and advocate 'policy responses' on behalf of peasants worldwide places a certain onus on the coalition to poll the interests of the peasantry where membership is weak or nonexistent. Otherwise, it could be accused of formulating and articulating peasant interests without consultation, which would run counter to its fundamental philosophy of peasants representing their own interests.

Expanding Vía Campesina's membership is an ongoing and gradual process complicated by repressive governments, local tensions and differing ideologies between rival or parallel movements, among other factors (Borras 2004, 17), and Karen village resistance and similar movements may never take a form in which they could be considered for membership. This is not an argument for the Karen village resistance 'movement' to join Vía Campesina, which even if it were desirable would be impractical because of the difficult questions of representation it would raise. Instead, a more practical question is whether there are forms of solidarity and consultation that could be helpful, what forms these could take and what dangers could exist.

Potentials and Risks of Solidarity

Forms of engagement based on solidarity and networking rather than formalized membership can provide a middle ground, which would further legitimize both Vía Campesina's representation of global peasantry and the local struggles of unrecognized movements, while potentially creating a 'boomerang effect', which could benefit Karen and other villagers. Admittedly, engaging less organized struggles may do little to bolster the international influence of transnational coalitions while burdening them by extending the range of interests they are

---

[37] The International NGO/CSO Planning Committee on Food Sovereignty, for example, advocates globally on land rights and related issues but limits its membership to non-governmental organizations and civil society organizations (see http://www.foodsovereignty.org/new/whoweare.php).

[38] The UN Permanent Forum on Indigenous Issues that drafted the Declaration consists of 16 'experts' who regularly consult with accredited organizations (see http://www.un.org/esa/socdev/unpfii/en/structure.html).

expected to serve, but if they are to claim legitimacy among peasants beyond their limited memberships then it is a necessary step.

What form such solidarity engagement could take is best determined by the participants, but some general possibilities can be identified here. Firstly, representatives of coalitions like Vía Campesina could visit Karen villagers to discuss what is happening globally, with other movements and institutions, and to hear about local struggles and interests. This local information could feed into the coalition's positions and advocacy both regionally and globally, while the coalition's international mobility and institutional access could promote greater recognition of and outside support for Karen village resistance – the 'boomerang effect'. Secondly, transnational coalitions could facilitate 'translocal' links between Karen villagers and people elsewhere currently or previously engaged in similar struggles, useful for exchange of experiences and more importantly as morale-boosters – a component often undervalued in outside analyses but extremely important in protracted and difficult struggles. Thirdly, transnational coalitions can act as watchdogs to help protect villagers from being ambushed by foreign development agendas implemented by international agencies in cooperation with states. Villagers in Burma have had little or no contact with such agencies and little opportunity to access critical views on international development, making them vulnerable to top-down programmes and destructive land or resource exploitation should their homelands open up to greater outside intervention and investment. To pre-empt this possibility, transnational coalitions can educate people about international forces they have not yet had to face, and monitor plans hatched in foreign capitals that could affect Karen villagers. Vía Campesina is well positioned to be part of such an initiative because activities could take the form of peasant-to-peasant information sharing.

Points and methods of contact already exist. Village heads are usually able to arrange meetings with their villagers. To contact village heads, local relief and human rights groups can facilitate access. Though structured for other purposes, several local groups already mentioned are staffed by villagers and refugees and can act as links.

Many factors still need consideration to prevent engagement causing more harm than good. To avoid military reprisals such initiatives would need to be low key and covert; local human rights and relief groups are already adept at this, and often facilitate similar visits by journalists and advocates. Another is the risk of endangering villagers' resistance strategies by drawing attention to them. Thus far, village strategies largely succeed by remaining covert and appearing non-threatening to the state. The Karen Human Rights Group has already faced this concern since 2005 in its reporting of resistance strategies, and has tried to focus on those already known to local military authorities and omit more sensitive strategies or details that could be used to target villages.[39]

---

[39] Observed through the author's participation in, KHRG management meetings and editing processes, October 2005 through April 2007.

Specific forms of engagement and solidarity would have to be designed by the participants, if consensus exists on their usefulness. With their greater mobility and access to communications, the transnational coalitions are best placed to initiate this contact, if necessary with logistics facilitated by local organizations working among the villagers. It can be argued that this goes beyond the mandate of these coalitions, which exist to serve their member organizations. Yet this has not prevented them from building networks 'upward' outside their membership, with institutions like the Food and Agriculture Organization of the United Nations, international non-governmental organizations, and even the World Trade Organization, and pressing these institutions to alter their *global* policies, not just their local policies in countries where the coalition has members. In doing so, they acquire a responsibility to respond to the expressed needs and goals of peasants globally, who would also be affected by any policy changes. Building additional networks 'downward' should therefore be a priority.

CONCLUSIONS

This chapter raises more questions than it can answer, in the hope of stimulating discussion rather than reaching conclusions. As to whether Karen village resistance and non-compliance can be considered a 'movement', I believe it can. There is sufficient evidence of widespread like-minded and directed action seeking identifiable goals on conflictual issues, primarily retention of land and livelihoods; that this action is based on commonly held beliefs and perceived identities; and that it is enacted through forms of protest including evasion ('avoidance protest') and systematic non-compliance which undermines a system seen as illegitimate and oppressive; to meet the requirements of a movement broadly defined. The lack of structure or identifiable leaders makes it viable and durable in Burma's repressive context, where a more structured movement would be vulnerable to violent attack or cooptation. This also keeps it spontaneous, and prevents it being redirected through misrepresentation or personal interests.

Considering this resistance as a form of movement calls for a reassessment of the agency of people in repressive situations where the space for overt action is restricted, and opens possibilities for engagement beyond traditional charity or sympathy. There are possibilities here for transnational agrarian coalitions to increase their legitimacy and coverage by engaging villagers where they currently have no membership, and for the villagers themselves to benefit from transnational actors' access and influence on global policies and their ability to generate pressure from outside, which Keck and Sikkink labelled the 'boomerang effect'. While accepting such struggles as formal 'members' of transnational coalitions is in most cases impractical and could even be harmful, there are forms of solidarity, engagement and advocacy that could be a constructive middle ground, while falling well short of formal 'membership'.

The suggestions presented above for initiating this solidarity are deliberately rudimentary, because the details can best be determined by the actors themselves. It must also be their decision whether such engagement would even be desirable.

While there are always risks, these can be managed, and in my opinion an initial contact between transnational coalitions like Vía Campesina and Karen villagers could and should be attempted, as an exchange of information and views and to evaluate the potential and form of further engagement. In recent years many Karen villagers have become accustomed to contact with outside agencies seeking to provide them with relief, but not with their peers from elsewhere offering them solidarity on an equal footing. As they cannot easily travel internationally and have little access to international communications, the onus falls on the transnational actors to initiate this first contact, possibly with the logistical assistance of local organizations.

The first step, however, is simply for transnational actors to increase their awareness of these less organized 'movements', and to begin a discussion on whether and how to engage them. Beginning with solidarity between transnational coalitions and Karen villagers, perhaps a future could exist wherein villagers in unstructured resistance struggles could connect with their peers elsewhere, just as transnational movement activists are able to do today.

REFERENCES

Adas, M., 1986. 'From Footdragging to Flight: The Evasive History of Peasant Avoidance Protest in South and South-East Asia'. In *Everyday Forms of Peasant Resistance in South-East Asia*, eds J.C. Scott and B.J. Tria Kerkvliet, 64–86. London: Frank Cass.

BI/PWF (Burma Issues/Peace Way Foundation), 2006. *Shoot on Sight: The Ongoing SPDC Offensive Against Villagers in Northern Karen State*. Bangkok: Peace Way Foundation.

Borras, S. Jr, 2004. 'Vía Campesina: An Evolving Transnational Social Movement'. TNI Briefing Notes No. 2004/6. Amsterdam: Transnational Institute.

BPHWT (Backpack Health Worker Teams), 2006. *Chronic Emergency: Health and Human Rights in Eastern Burma*. Chiang Mai: Wanida Press.

Cheesman, N., 2002. 'Seeing "Karen" in the Union of Myanmar'. *Asian Ethnicity*, 3 (2): 199–220.

Cohen, J., 1985. 'Strategy or Identity: New Theoretical Paradigms and Contemporary Social Movements'. *Social Research*, 52 (4): 663–716.

Della Porta, D. and M. Diani, 1999. *Social Movements: An Introduction*. Oxford: Blackwell.

Escobar, A. and S.E. Alvarez, 1992. 'Introduction: Theory and Protest in Latin America Today'. In *The Making of Social Movements in Latin America: Identity, Strategy, and Democracy*, eds A. Escobar and S.E. Alvarez, 1–15. Boulder, CO: Westview Press.

Fernando, T., 1982. 'Ethnic Adaptation and Identity: The Karen on the Thai Frontier with Burma (book review)'. *The Journal of Developing Areas*, 17 (1): 130–1.

Filho, J. and M.L. Mendonça, 2007. 'Agrarian Policies and Rural Violence in Brazil'. *Peace Review*, 19 (1): 77–85.

Fox, J., 2005. 'Unpacking "Transnational Citizenship"'. *Annual Review of Political Science*, 8: 171–201.

Fox, R.G. and O. Starn, 1997. 'Introduction'. In *Between Resistance and Revolution: Cultural Politics and Social Protest*, eds R.G. Fox and O. Starn, 1–16. New Brunswick, NJ: Rutgers University Press.

Fujita, K. and I. Okamoto, 2006. 'Agricultural Policies and Development of Myanmar's Agricultural Sector: An Overview'. Discussion Paper 63. Tokyo: Institute of Developing

Economies. http://www.ide.go.jp/English/Publish/Dp/pdf/063_okamoto.pdf, Accessed 25 October 2007.

Hinton, P., 1979. 'The Karen, Millenialism, and the Politics of Accommodation to Lowland States'. In *Ethnic Adaptation and Identity: The Karen on the Thai Frontier with Burma*, ed. C.F. Keyes, 81–94. Philadelphia, PA: Institute for the Study of Human Issues.

Horton, G., 2005. 'Burma's Generals Must Be Brought to Account'. *Burma Campaign News*, 10: 14–15.

Hudson-Rodd, N. and M. Nyunt, 2001. 'Control of Land and Life in Burma'. *Tenure Brief*, 3: 1–8. Madison, WI: Land Tenure Center, University of Wisconsin-Madison.

Hudson-Rodd, N., M. Nyunt, S.T. Tun and S. Htay, 2003. 'The Impact of the Confiscation of Land, Labor, Capital Assets and Forced Relocation in Burma by the Military Regime'. NCUB/FTUB discussion paper. http://www.ibiblio.org/obl/docs/land_confiscation-contents.htm, Accessed 16 December 2005.

HRW (Human Rights Watch), 2002. *'My Gun was as Tall as Me': Child Soldiers in Burma*. New York: Human Rights Watch.

HRW (Human Rights Watch), 2007. *Sold to Be Soldiers: The Recruitment and Use of Child Soldiers in Burma*. New York: Human Rights Watch.

ICARRD, 2006. 'International Conference on Agrarian Reform and Rural Development: Final Declaration'. Porto Alegre: ICARRD. http://www.icarrd.org/en/news_down/C2006_Decl_en.doc, Accessed 24 October 2007.

Keck, M.E. and K. Sikkink, 1998. *Activists Beyond Borders: Advocacy Networks in International Politics*. Ithaca, NY: Cornell University Press.

Keyes, C.F., 1977. *The Golden Peninsula: Culture and Adaptation in Mainland Southeast Asia*. New York: Macmillan.

KHRG (Karen Human Rights Group), 2002. *Forced Labour Orders Since the Ban*. Chiang Mai: KHRG.

KHRG (Karen Human Rights Group), 2003. *SPDC & DKBA Orders to Villages: Set 2003-A*. Chiang Mai: KHRG.

KHRG (Karen Human Rights Group), 2005. 'Forced Labour Orders to Villages 2004–2005: Papun, Toungoo, Thaton, Pa'an and Dooplaya Districts' (unpublished). Submitted to the International Confederation of Free Trade Unions by the Karen Human Rights Group for consideration by the ILO Committee of Experts. KHRG internal document.

KHRG (Karen Human Rights Group), 2006a. *Dignity in the Shadow of Oppression: The Abuse and Agency of Karen Women Under Militarisation*. Chiang Mai: KHRG.

KHRG (Karen Human Rights Group), 2006b. *SPDC Attacks on Villages in Nyaunglebin and Papun Districts and the Civilian Response*. Chiang Mai: KHRG.

KHRG (Karen Human Rights Group), 2007. *Development by Decree: The Politics of Poverty and Control in Karen State*. Chiang Mai: KHRG.

KNU (Karen National Union), 2007. 'The KNU Statement Regarding Brigadier General Htain Maung'. KNU Public Statement, 30 January 2007. E-mail copy received by the author from KNU representatives on 30 January 2007.

Malseed, K., 2006. '"We Have Hands the Same as Them": Struggles for Local Sovereignty and Livelihoods by Internally Displaced Karen Villagers in Burma'. Paper presented at Land Poverty, Social Justice and Development Conference, The Hague, January 2006.

Marshall, H.I., 1997 [1922]. *The Karen People of Burma: A Study in Anthropology and Ethnology*. Bangkok: White Lotus.

MST (Movimento dos Trabalhadores Rurais Sem Terra), 2001. *Strong Roots (Raiz Forte)*. Video disk. San Francisco, CA: Global Exchange.

Piven, F.F. and R.A. Cloward, 1977. *Poor People's Movements: Why They Succeed, How They Fail*. New York: Pantheon.

Scott, J.C., 1985. *Weapons of the Weak: Everyday Forms of Peasant Resistance*. New Haven, CT: Yale University Press.

Scott, J.C., 1998. *Seeing Like a State: How Certain Schemes to Improve the Human Condition Have Failed*. New Haven, CT: Yale University Press.

Selth, A., 2002. *Burma's Armed Forces: Power Without Glory*. Norwalk: Eastbridge.

Smith, M., 1991. *Burma: Insurgency and the Politics of Ethnicity*. London: Zed Books.

Smith, M., 2002. *Burma/Myanmar: The Time for Change*. London: Minority Rights Group International.

South, A., 2006. *Forced Migration in Myanmar: Patterns, Impacts and Responses*. (unpublished). Report commissioned for the Office of the Resident Coordinator of the United Nations System in Myanmar.

Steinberg, D., 1999. 'A Void in Myanmar: Civil Society in Burma'. In *Strengthening Civil Society in Burma: Possibilities and Dilemmas for International NGOs*, eds Burma Center Netherlands and Transnational Institute, 1–14. Chiang Mai: Silkworm Books.

Stewart, C.J., C.A. Smith and R.E. Denton, Jr., 2001. *Persuasion and Social Movements*. Fourth Edition. Prospect Heights, IL: Waveland Press.

Tarrow, S., 1998. *Power in Movement: Social Movements and Contentious Politics*, 2nd edn. Cambridge: Cambridge University Press.

TBBC (Thailand Burma Border Consortium), 2005. *Internal Displacement and Protection in Eastern Burma*. Bangkok: TBBC.

TBBC (Thailand Burma Border Consortium), 2007. *Internal Displacement in Eastern Burma: 2007 Survey*. Bangkok: TBBC.

Thawnghmung, A.M., 2004. *Behind the Teak Curtain: Authoritarianism, Agricultural Policies and Political Legitimacy in Rural Burma/Myanmar*. London: Kegan Paul.

Vía Campesina, 2006. 'Annual Report: Violations of Peasants' Human Rights: A Report on Cases and Patterns of Violances [sic] 2006'. Jakarta Selatan: Vía Campesina.

Index

Note: 'n' after a page reference indicates a note on that page.

Acevedo Vogl, A. 74
Action Aid 151, 155, 168, 183, 189, 197
Adams, M. 133
Adas, M. 330, 337
Aditjondro, G. 213, 217
Adiwibowo, S. 228
ADM 184, 186
Advisory and Research Centre (ESPLAR) 183, 189, 197
Afiff, S. 22, 23, 25, 29, 209–33
AfricaBio 154
African National Congress (ANC) 138, 165, 166
Agarwala, R. 25n
Agrarian Orange Guard 5–6
Agrawal, A. 210
Agrevo 195
agrifood supply chains 239–46
Agroceres 184
agrofuels 30, 41n, 43, 53, 328, 340
Ainger, K. 42n, 43
Akram Lodhi, H. 1, 93
Alavi, H. 39
Alegria, R. 96, 313
Alexander, A. 133, 136–7, 138, 140
Alvarez, S. 28, 81, 334
AMAN see National Indigenous Peoples' Association
AME Foundation 168
American Federation of Labor Congress of Industrial Organizations (AFLCIO) 16n, 68
Amin, S. 297
Amorim, Jaime 128
Andalucía 109
Andhra Pradesh Farmers' Association 151
Andrade-Eekhoff, K. 78, 80
Andrioli, A. 191
Anheier, H. 62n
Anti-Big-Dam campaign 216
Apartheid 125, 132, 133, 138, 147, 169
Appadurai, A. 91, 149
Araghi, F. 240
Arana, A. 79
Argentina 178, 180, 181, 185, 186, 190, 197, 202, 203, 204
Arias, Salvador 71, 72
Ark of Taste 248
Armitage, C. 303

Arquilla, J. 81n
AruPA see Volunteer Alliance to Save the Environment
Ashton, Glen 152
Asian Peasant Coalition (APC) 2, 111
Asociación Nacional de Empresas Comercializadoras de Productores del Campo (ANEC) 111, 115
Asociación de Trabajadores del Campo (ATC) 68, 96, 108, 115
ASOCODE see Association of Central American Peasant Organizations for Cooperation and Development
ASPTA see Services and Advice to Alternative Agriculture Projects
Assadi, M. 107, 170
Associated Country Women of the World (ACWW) 9–10
Association of Central American Peasant Organizations for Cooperation and Development (ASOCODE) 15, 29, 63–70, 81–2, 85, 96, 108, 115, 199
Association for Rural Advancement (AFRA) 134
auditing and certification, lineage of 241
Austria 5, 6
avocados 244, 250–2
Awang, S. 229

Bachriadi, D. 210
Backpack Health Worker Teams 333n
Bada, Xochitl 16, 17, 19, 21, 277, 280, 283, 284
Bain, C. 244
Baletti, B. 14, 19, 23, 123–43
Banco Cuscatlán 76n
Bangalore 67
Bangladesh 109
Bangladesh Krishok Federation (BKF) 109, 115
Bantu Consolidation Act 1946 132
Baber, C. 213
Baranyi, S. 102
Barchiesi, F. 140
Barham, B. 71
Barling, D. 241, 244
Barquero, M. 74
Barr, C. 213, 215
Barry, T. 254

Barta, R. 38
Bartra, Armando 69, 70, 286
Basic Agrarian Law no.5 (BAL) 219
Basic Christian Base Communities 190
Batliwala, S. 93, 100
Bebbington, A. 76
Becker, J. 302
Belize 61n, 67, 69
Bell, J. D. 6
Bellisario, A. 179
Bello, W. 92n
Belsky, J. 226
Benbrook, C. 186
Benson, P. 79
Berdegué, J. A. 239
Bernstein, H. 1, 25n, 40, 51, 56, 92, 93, 133, 240, 243n, 246, 247n, 300, 301, 302
Besserer, F. 278
Beverley, J. 44
Bevington, D. 63
Bharatiya Kisan Union 108
Biekart, K. 63n, 64, 231, 232
bi-national citizenship 267
Bioagricert 253
Biosafety National Association (ANBIO) 195
Biosecurity law 160
Biowatch 153, 154, 161, 167
Bo Xilai 299
Bob, C. 99, 100, 105
Bodnar, J. 51
Boff, L. 190
Bolivia 177, 185, 186, 202, 204
Bolsheviks 6
Bond, P. 169
boomerang effect 97, 203, 283n, 341, 343, 344, 345
Borlaug, N. 55, 151
Borras, S. 1–31, 41n, 62, 91–116, 135, 187, 198, 210, 212, 223, 231, 311, 313
Bové, J. 51, 52, 106, 155, 158, 191, 198, 257, 341, 342
box schemes 239
Boy, A. 186
Brandford, S. 126, 130
Brass, T. 4, 108
Brazil 14, 15, 17, 20, 23, 26, 27, 54, 55, 76n, 95, 96, 103, 104, 106, 107, 109, 111, 112, 123–43, 178, 179, 180, 183, 184, 185, 186, 190, 195, 197, 202, 203
    Biosafety Law 1995 147, 156, 204
    GM crops 154–6
    Workers' Party (PT) 28
Brenner, R. 296
Bretton Woods framework 30

Britain 9, 17, 41, 69, 244, 261n
Broad, R. 91
Brockett, C. 63
Brooks, D. 268n
Brown, D. 93
Bryceson, D. 1, 16
Brysk, A. 25n
Buckley, C. 303
Buechler, S. 211
Bukharin, Nikolai 7
Bulgaria 5, 6
Bulgarian Agrarian National Union (BANU) 5, 6
Bukharin, N. 7
Bunge 184, 186
Burbach, R. 193
Burch, D. 243
Burkett, P. 296, 298
Burnham, P. 193
Burma 323–46
    armed resistance 326
    Buddhist monk protests 339–40
Busch, L. 242, 244
Bush, George 68n
Bush, R. 97
Byres, T. J. 1, 93

Cai, Y. 305
Caldeira, R. 39, 54
Callon, M. 244
Cameron, A. 48
Campbell, H. 241, 246
Campesino a Campesino 4, 38, 65n
Campling, L. 246
Campos, W. 64
Canada 9, 64, 150, 186
Cancún 3, 40, 51
Cape Town 134
Carazo Vargas, E. 74
Cardoso, President Fernando Henrique 126, 155, 196
Cargill 24n, 158n, 184, 186
Carlsen, L. 53
Carr, E. 7
Cartagena Protocol on Biosafety 195
Castells, M. 170
Castro, C. 80
Catholic Church 126, 130, 139
Catholic Relief Services 68
CECCAM *see* Centro de Estudios para el Cambio el Campo Mexicano
Central America Solidarity Coordinating Group 65
Central American Association of Peasant

Organizations for Cooperation and
Development   15
Central American Indigenous Council
(CICA)   65
Central American Rural Coordinating
Group   65
Central Workers' Union (CUT)   130
Centro de Estudios para el Cambio en Campo
Mexicano (CECCAM)   70
Chan, J.   297, 299, 309
Chapela, I.   253
Chapin, M.   228
Cheeseman, N.   325
Chen, G.   301, 302, 306
Chen, J.   301
Chile   4, 28
China   7, 14, 17, 18–19, 41n, 113, 162
  Communist Party membership   297
  corrupt local power   299–303
  disparity in incomes   298
  entry into WTO   299
  foreign investment in   297
  land seizures   303–10, 311
  market socialism in   296–9
  migrant populations   308–9
  NGOs in   314–15
  organized crime in   305–6
  regional inequalities   298
  tax burden   301, 302
  trade unions   300
Chinese Communist Party (CCP)   7
  Peasant Movement Training Institute
  (PMTI)   7
Chomthongdi, J-C.   44
CICAFOC see Indigenous and Peasant
  Community Agroforestry Coordinating
  Group
Circle for Village and Agrarian Reform
  (KARSA)   221
Citi Group   76n
citizens' juries   151, 159, 197
Civil Initiative for Central American
  Integration (ICIC)   65
Clapp, J.   185
class   24–7, 106–9
CLOC see Latin American Coordination of
  Peasant Organizations
Cloward, R.   334, 336
CNCR see Conseil National de Concertation et de
  Coopération des Ruraux
Coalition Paysanne de Madagaskar
  (CPM)   107, 116
COCOCH see Consejo Coordinador de
  Organizaciones Campesinas de Honduras

Cody, E.   297, 304, 307, 308
coffee   63, 65, 71n, 73–6, 83–4, 243, 247,
  272, 285
Cohen, J.   334
Cohen, R.   11
Cohen, S.   8
Colby, F.   6
Cold War   40
Collor, President Fernando   184
Colombia   95
Comaroff, J. and J.   81n
Comintern   6–7
  Fifth Congress   7
Communists   5, 6, 8
  Chinese   7
  Yugoslav   7
Community Shared Agriculture   239
Comprehensive Agrarian Reform Program
  (CARP)   110
Confederation of Labour Unions (CUT)   197
CONIC see Coordinadora Nacional Indígena y
  Campesina
*Conseil National de Concertation et de Coopération
  des Ruraux* (CNCR)   20, 27, 107, 109,
  112, 115
Consejo Coordinador de Organizaciones
  Campesinas de Honduras
  (COCOCH)   67, 68, 69
Consultative Council of the Central American
  Integration System (SICA)   64, 67
CONTAG see National Confederation of
  Agricultural Workers
Conway, G.   179
Coordinadora Nacional Indígena y Campesina
  (CONIC)   66n, 68, 69
Coordinadora Nacional de Pequeños y
  Medianos Productores de Guatemala
  (CONAMPRO)   66
Coordinadora Nacional Plan de Ayala
  (CNPA)   111, 116
Coordination National de Organisations
  Paysannes (CNOP)   107, 116
Coop Italia   249
Copenhagen Initiative for Central America   64
Cordero-Guzmán, H.   275
Cornelius, W.   269n
Cornell University   9
Costa Rica   17, 61n, 63, 72, 73, 74, 75, 76n,
  77, 78, 79, 80, 83
Cousins, B.   94, 113, 132, 133
Coyote Rogo   253–5, 260, 261
CPE see European Peasant Coordination
CPT see Pastoral Land Commission
Cranshaw, M.   66

Crib, R. 210, 212
Croat Peasant Party 7
Croatia 7
Cruz, Santiago Rafael 16n
CTNBio (National Biosafety Technical Commission) 147, 154, 155, 156, 190, 195, 196
Cuba 62n
Czechoslovakia 5, 6

D'Andrea, C. 228
Da Silveir, J. 147
Dauvergne, P. 213
Davies, C. 9
Davis, M. 37, 40, 43, 44
Davis, S. 76n
De Angelis, M. 297
De Bremond, A. 61n, 96
De Carvalho Borges, I. 147
De Carvalho Filho, J. 105
Deere, C. 106, 126
deforestation 253
De Janvry, A. 126
De la Rosa, R. 111
Della Porta, D. 20, 105, 334, 335, 336
Democracia sin pobreza 71
Democratic Karen Buddhist Army (DKBA) 326, 330, 333
Democratic Peasant Movement of the Philippines (DKMP) 15, 27, 110–11, 116
Deng Xiaoping 296
Denmark 65
Denomination of Origin (DOC) 242, 247
Desai, A. 136
Desmarais, A. 2, 11, 18, 37, 39, 42n, 46, 52, 54, 62, 65, 69, 93, 105, 123
De Soto, H. 1, 94, 113
De Souza Martins, J. 130
De Ste. Croix 193
development, migration and 284–7
de Vries, F. 55
Diani, M. 334, 335, 336
Díaz Porras, R. 72
Divjak, C. 302
Dixon, C. 63
Dixon, J. 244
Djeweng, S. 215, 219
DKMP *see* Democratic Peasant Movement of the Philippines
Domínguez Santos, R. 286
Dominican Republic-US-Central America Free Trade Agreement (DR-CAFTA) 68n, 72, 84
Dove, Michael 230n

Drage, D. 9
Dufour, F. 51, 52
Duncan, C. 47, 51
Du Pont 195
Durantt, W. 6
Durban Social Forum 134

Earth Women 153
Earthlife Africa 153
Eber, C. E. 54
ECM *see* Mesoamerican Peasant Platform or Meeting
Ecuador 185, 204
Ecumenical Service for socio-economic transformation 153
Edelman, M. 1–31, 61–84, 96, 97n, 123, 149, 187, 189, 193, 194, 198, 201, 211, 212, 221, 223, 311
Edwards, M. 13, 95
Egypt 40, 97
Egziabher, Tewolde Berhan 152
Ekogaia 153
Ellis, F. 1, 16
Ellison, N. 172
El Salvador 61, 62, 64, 67, 72n, 73, 75, 79, 96
Embrapa 156
England 9, 37, 310
Enríquez, E. 70
environmental justice 212, 214–8
Environmental Justice Networking Forum 153
Environmental Monitoring Group 153, 160
Erosion, Technology and Concentration Group (ETC Group) 2, 150n, 186, 188
Escobar, A. 28, 334
ESPLAR *see* Advisory and Research Centre
Ethiopia 152
ethnicity 24–5
  in Burma 325–8
European Peasant Coordination (CPE) 18, 43n, 48
European Union (EU) 43, 48, 68, 76n, 84, 103, 104, 116, 154, 180
  food standards 242, 243, 245, 247
Euro Retailer Producer Working Group Good Agricultural Practices (EurepGAP) 245, 251
exit-voice dichotomy 21, 62, 77, 267

fair trade 240, 244, 285
Farid, H. 209, 210
Farm Labor Organizing Committee (FLOC) 16n
Farmers' Institutes 9

farmers' markets 239
FASE *see* Social Assistance and Education Federation
Fauzi, N. 22, 23, 25, 29, 209–33
Featherstone, D. 123
Federação dos Trabalhadores na Agricultura Familiar (FETRAF) 96, 116
Ferguson, J. 125
Fernandes, B. M. 126, 129
Fernández Buey, F. 81
Fernando, T. 325
Fiesta Farms 256
Fifth Comintern Congress 7
Filho, J. 336
Finland 8
Fischer, E. 79
Fitting, E. 185, 203
Fitzgerald, D. 276, 278n
Flavio de Almeida, L. 46
Florini, A. 11
Fonte, M. 239
Food Alliance 256
Food and Agriculture Organization of the United Nations (FAO) 3, 4, 8n, 10, 19, 47, 56, 102, 103, 104, 105, 116, 215, 220, 345
Food and Allied Workers Union 153
food certification 240–57
  benchmarking 242, 246
  Third Party (TCP) 241
food sovereignty 21, 39n, 42–57, 82, 83–4, 92, 98–9, 123, 157, 165, 172, 177, 184, 186, 188, 189, 192, 240, 260, 312
  IPC for 2, 19, 20, 27, 29, 95
FoodFirst Information and Action Network (FIAN) 2, 28, 82–3, 84, 97–8, 103, 105, 106, 111, 116, 311, 312, 314
FoodShare Toronto 257
Forman, S. 126
Forum for Biological and Cultural Diversity 188
Forum for Biotechnology and Food Security 168
Forum on the Free Use of Knowledge 184
Fox, J. 2n, 11, 12n, 13, 16, 19, 21, 81, 91, 100, 340
Fox, R. 346
Fox, Vicente 67
France 3, 8
Franco, J. 105, 111, 212
Free Trade Area of the Americas (FTAA) 177, 186, 187, 189, 194
Friedberg, S. 152
Friedman, H. 1, 21, 29, 40, 41, 43, 92, 243, 245, 255, 258, 259
Friends of the Earth 2, 179
FSPI *see* Indonesian Federation of Peasant Unions
Fujimori, Alberto 72n
Fujita, K. 327
Fúnez, F. 61n, 76

G-8 summit (2001) 3
Gacitúa, E. 76
Gaete, M. 63, 80
Galián, C. 74
Gallagher, A. 203
Ganie-Rochman, M. 214
GATT 41
Gauster, S. 61n, 96
Gaventa, J. 13, 20, 95
Gene Campaign 168
Genetic Resources Action International (GRAIN) 2, 150, 153
Genetic Use Restriction Technologies (GURT) 185, 188, 192
Genetic Snowball 159, 193
genetically modified (GM) crops 17, 19, 21, 24, 25, 30, 147–72
Geographical Indications (GI) 239, 247
Gereffi, G. 244
Germany 6
Gerschenkron, A. 124, 143
Gher Der 331–2, 333, 337
Ghimire, K. 93
Gianaris, N. 6
Gilboy, G. 315
Gioanetto, F. 253
Gledhill, J. 277
Glick Schiller, N. 12n
Global Campaign for Agrarian Reform (GCAR) 21–2, 29, 91–115, 116, 311–12, 313
global-local processes 1–2
Global People's Action Group 150n
GLOBALGAP (GG) 240, 245–6, 251, 259
Glover, D. 177, 191, 195, 196
GM crops 147–72
Gohn, M. d. G. 126
Goldring, L. 276
Goodman, D. 249
Goodman, G. 271
Goodman, M. K. 249
Goodspeed, P. 298, 302
Gordon, Ishbel, Countess of Aberdeen 9
Green Foundation 168
Green International 5–8
Greenberg, S. 97, 133, 134, 137, 139, 140

Greenfield, G. 298
Greenpeace 152, 154, 155, 159, 160, 162, 168, 181, 183, 186, 189, 190, 192, 193, 198, 202
Greenstein, R. 133, 140
Griffin, K. 93
Grupo Reflexión Rural 186, 201
Gu, Q. 307
Guatemala 61, 62, 64, 65, 67, 68, 71, 73, 75, 77, 79, 96
Guidry, J. 93
Guo, X. 305
Gupta, A. 41
Guthman, J. 241, 248
Gwynne, R. 1, 92

Hadiz, V. 210
Hadley, K. 253
Hajer, M. 192, 209
Hall, A. L. 126
Hall, R. 97
Hamm, G. 255
Hart, G. 132, 133, 143
Hart-Landsberg, M. 296, 298
Harvey, D. 44, 296, 297
Harvey, M. 245
Harvey, N. 4
Havana 62n
He, Q. 304, 306
Hecht, S. 61, 79
Heginbotham, E. 315
Heim, Dr. Georg 6
Helleiner, E. 30
Heller, P. 28, 163
Hemingway, Ernest 5
Hemispheric Social Alliance (HSA) 187, 189, 194
Herring, R. 18, 25n, 28, 157, 164
Hertel, S. 101
Heruputri, A. 219
Hewitt, C. 111
Hikam, M. 214
Hinton, P. 325
Hirschman, A. 62n, 267n, 271, 287
Hiskey, J. 271
HIV/AIDS 166, 167
HIVOS 64, 168
Houtzager, P. 127
Howell, J. 133
Ho Chi Minh 7
Ho, P. 305
Hochstetler, K. 195
Holland, L. 300, 302
Holt-Giménez, E. 4, 11, 38, 55, 65n, 268
Homeland System 132

hometown associations 267, 274–6
Honduran Coordinating Council of Peasant Organizations (COCOCH) 67, 68, 69, 96, 116
Honduras 61, 63, 67, 69, 71, 73, 79, 96, 221
Honey Bee Network 168
Hong Kong 53
Horowitz, L. 152
Horton, G. 339n, J. 162
Houtzager, P. 127
Howell, J. 133
Hu, Ping, J. 162
Huang, J. 162
Hudson-Rodd, N. 327
human rights
 Burma 323, 333, 340–6
 China 300, 311, 312–13, 314
 globalization of them 52
Hungary 5, 6
Hurricane Mitch 65
Hurst, W. 302
Husken, F. 209, 213

Icaza, R. 194
ICIC see Civil Initiative for Central American Integration
IDEC see Institute for the Defence of Consumers
IFAD see International Fund for Agricultural Development
IFAP see International Federation of Agricultural Producers
illegal planting 147, 164, 204
India 17, 18, 24, 26–7, 37, 41, 42, 96, 106, 108
 GM crops 150–2
Indigenous and Peasant Community Agroforestry Coordinating Group (CICAFOC) 65
Indonesia 3, 22, 25–6, 41, 53, 74, 107, 113n, 209–33, 313
Indonesian Federation of Peasant Unions (FSPI) 23, 221, 223–4, 226, 230–1, 223, 224, 230, 231
Indonesian Forum for the Environment (WALHI) 215, 216, 218, 229
Indonesian Legal Aid Foundation (YLBHI) 216, 218, 219, 221
INESC see Socio-Economic Studies Institute
Institute for the Defence of Consumers (IDEC) 154, 155, 181, 183, 196
intellectual property 247
intellectual property rights (IPRs) 178, 184, 185, 187, 188
Inter-American Development Bank 67

Inter-Church Organization for Development
    Cooperation (ICCO) 29
International Agrarian Bureau 6, 8
International Agrarian Institute 8
International Coffee Agreement 74
International Collective in Support of Fish
    Workers (ICSFW) 2
International Commission of Agriculture
    (ICA) 8, 10
International Conference on Agrarian Reform
    and Rural Development (ICARRD) 4,
    102, 104, 116, 342, 343
International Coordinating Committee – of Vía
    Campesina (ICC) 16, 46, 106, 110, 116
International Council of Women (ICW) 9
International Federation of Adult Catholic
    Farmers' Movements (FIMARC) 2
International Federation of Agricultural
    Producers (IFAP) 2, 3, 4, 8, 10, 19–20,
    25, 27, 92–3, 107, 116
International Fund for Agricultural
    Development (IFAD) 3, 4, 19, 103–4,
    116
International Institute of Agriculture (IIA) 8,
    10
International Labour Organization 76n, 341
International Land Coalition (ILC) 20, 27, 223
International Monetary Fund (IMF) 3, 19, 42,
    82, 242, 311
International Movement of Catholic
    Agricultural and Rural Youth
    (MIJARC) 2–3
International Organization for Aid to
    Revolutionaries (MOPR) 7
International Planning Committee (IPC) for
    Food Sovereignty 2, 19, 20, 27, 29, 46,
    47, 95, 99, 116, 188
International Union of Food, Agricultural,
    Hotel, Restaurant, Catering, Tobacco and
    Allied Workers' Associations (IUF) 2
interstitial social transformation 258, 260–61
IPC see International Planning Committee
    (IPC) for Food Sovereignty
Isakson, S. 61n, 96
Isaula, R. 61

Jacka, T. 302
Jackson, G. 5, 8
Jacobs, R. 134–6, 138–9
Jacovetti, Chantal 50
Jaffee, D. 253
Jansen, K. 177
Japan 106
Jelsma, M. 64n

Jepson, W. 158, 228
Johnson, Tamara 14, 20, 23, 123–43
Johnston, H. 11, 102
Jonakin, J. 61
Jordan, L. 104n
Jornal Sem Terra 128
Joshi, Sharad 151

Kahn, J. 297, 298, 303, 308
Karen people 324–46
Karen National Union/Karen National
    Liberation Army (KNU/KNLA) 326,
    331–2, 333, 330, 331
Karen Office of Relief and Development
    (KORD) 333
Karen Rivers Watch 333
Karnataka Coalition Against GM Crops 152
Karnataka State Farmers' Association
    (KRRS) 18, 24, 25, 26n, 27, 42n, 96,
    107–8, 111, 112, 116, 150–1, 158–9, 161,
    165, 168, 169
KARSA see Circle for Village and Agrarian
    Reform
Kartodihardjo, H. 225
Katayama, Sen 7
Katzenstein, M. F. 169
Kautsky, K. 39
Kay, C. 1, 28, 51, 69n, 104n, 179
Kearney, M. 278
Keck, M. 11, 20, 21, 22, 82, 93, 97, 99, 102,
    105, 124, 203, 217, 341
Kedung Ombo 216
Kenfield, I. 193
Kentucky Fried Chicken 3, 158n
Keohane, R. 92
Kerkvliet, B. 18, 113, 114
Keyes, C. 325
Khagram, S. 2n, 92n
Khubeka, Mangaliso 134
Kiwi Fruit 243
Klooster, D. 253
KMP see Peasant Movement of the Philippines
Konefal, J. 244, 245
Korzeniewicz, M. 244
Kossoy, A. 185
Kowalchuk, L. 212
Krestintern see Peasant International
KRRS see Karnataka State Farmers' Association
Kuomintang (KMT) 7
Kurtz, M. 127

labelling 239–61
    bioregional 250
    for Muslims 241

Lahiff, E. 1, 97, 133
land enclosure 303–4
land occupations 129–30, 136–8
Land Research and Action Network (LRAN) 2, 28–9, 98, 99, 111, 116
Landless People's Charter 2001 137
Landless People's Movement (LPM) 14–15, 23, 97, 107, 116, 124–5, 131, 133–42, 136, 155, 166
Landsberger, H. 111
Lang, T. 241, 244
Lara, F. 110
Lass, C. 9
Latin American Alliance for Progress 40
Latin American Coordination of Peasant Organizations (CLOC) 3, 30, 62n, 65, 68, 69, 76n, 82, 188, 191
Laur, Dr, Ernst 8
Lawrence, G. 243
Leach, M. 172
League of Nations 8, 9, 10
Lee, C. 300, 302
Lee Kyang Hae 3
LeHeron, R. 241, 246
Lenin, V. I. 39
Leong, A. 298
Levenson, D. 79
Levido, L. 196
Lewis, W. A. 37
Leyva Reus, J. 270
Li, C. 298, 301, 303
Li, F. 306
Li, L. 299, 300, 301
Li, T. 211, 228
Li, Z. 302
Links of Friendship 9
Lipton, M. 179
Loblaws 244
Local Food Plus (LFP) 255–7, 260, 261
Lodge, T. 165
Lofredo, Gino 82
López Obrador, Andrés Manuel 70
Lowe, C. 212, 214, 228
Lozano Ascencio, F. 270
Lubin, David 8n
Lucas, A. 210, 214, 220, 223, 226
Lukmanudin, I. 229
Lula administration 103, 105, 159–60, 168, 184, 196, 190, 196, 203
Lyon, M. 212
Lyons, K. 241

Madagascar 107
Madeley, J. 43n
Madrid 67
Mainwaring, S. 124
Malaysia 17, 153
Maldonado, C. 278
Malseed, Kevin 14, 17, 19, 113, 323–46
Managua 66, 96, 108
Mandela, N. 133
Manthorpe, J. 303
Manwaring, M. 79
Mao Tse-tung 7
Mao Zedong 302
market-led agrarian reform (MLAR) 22, 75, 93, 116, 135
market liberalism 10
Marsden, T. 78, 241, 248
Marshall, H. 325, 328
Martin, P. 269, 270
Martinez, M. 108n, 113n, 114n
Martinez-Alier, J. 45, 55
Martorell, J. 135
Marx, K. 50, 258, 310
Mas'oet, M. 214
Masset, E. 61
Maybury-Lewis, B. 126
Mayhco seed firm 150
Mayo, M. 11, 91
Mazoyer, M. 11, 37
McAdam, D. 170, 202
McCalla, A. F. 48n
McCarthy, J. 53, 222, 226, 228
McDonalds 3, 51, 186
McKeon, N. 20, 211
McMichael, P. 1, 21, 37–57, 92, 124, 143, 211, 240n, 245, 312
McNabb, M. 9
McNair, A. 21, 29
McVey, R. 212
Medeiros, L. 96
Mečiř, Karel 8
Medeiros, L. 126
Meier, M. 9
Melin, Jules 8n
Mendonça, M. 105, 336
Mercosur 182, 194–5
Mesoamerican Initiative for Trade, Integration and Sustainable Development (Iniciativa CID) 68
Mesoamerican Peasant Platform or Meeting (ECM) 68
Meszaros, G. 111
Mexico 7, 16, 17, 23, 28, 38, 53, 61n, 65n, 68, 74, 106, 123, 135, 186
  bioregional labelling 250–5
  hometown associations 274–6

migration 267–88
  population decrease 269–70
  poverty rates 272–4
  remittances 276–84
  Three-for-One programme 279–84
México Calidad Suprema 246, 251, 253, 259
Miao, Y. 304
micro-banks 286
Mijangos-Cortés, J. 252
Ming dynasty 310
Ministry of Environment 216
Mitchell, T. 40, 51
Mngxitama, A. 97, 138, 140
MOICAM *see* Movimiento Indígena y Campesino Mesoamericano
Moniaga, S. 211, 213, 219
Monsalve, S. 106
Monsanto 55, 147–68, 179, 183, 184, 191, 193, 195, 204
Montreal 67
Moore, B. 37
Morales, A. 66, 80, 110
Moreno, R. 74
Morgan, K. 255
Morocco 17
Mortimer, R. 212, 214
Moss, J. 9
Movement of the Landless Rural Workers (MST) 15, 18, 20, 23, 26, 27, 28, 46, 55, 95n, 112, 116, 124, 162, 165, 166, 169, 177, 183, 184, 190, 191–2, 196, 197, 199–200, 335, 338–9
Movimiento Indígena y Campesino Mesoamericano (MOICAM) 68, 69, 70, 81, 82
Moyo, S. 51, 211
Mozambique 107
MST *see* Movement of the Landless Rural Workers
Mugabe, Robert 137
Müller, B. 18, 165
Mumbai World Social Forum (WSF) 24n
Murray, D. 240
Murray, W. 179
Mutersbaugh, T. 272, 278
Myhre, D. 269n

Nadal, A. 252
Nairobi 67, 113n
Namibia 95, 97, 113n
Nanjundaswamy, Prof. M. D. 24n, 107, 108, 150, 151, 158–9, 161, 168, 170
Narmada project 216
National Biosafety Framework 197

National Confederation of Agricultural Workers (CONTAG) 96, 104, 105, 116, 182
National Consumer Forum 153
National Coordination of Autonomous Local Rural People's Organizations (UNORKA) 23, 24, 27, 42n, 110–11, 116
National Farmers' Union (NFU) 10
  of Canada 18, 42
National Forum for Agrarian Reform 95
National Indigenous Peoples' Association (AMAN) 222, 225, 230
National Institute for Colonization and Agrarian Reform (INCRA) 131
National Land Agency (NLA) 211
National Land Committee (NLC) 97, 116, 135, 139–41
National Union of Autonomous Regional Peasant Organizations (UNORCA) 23, 42n, 111, 116
National Union of Peasants (UNAC) 18, 107, 113, 116
Natives Land Act 1913 131, 133
Native Laws Amendment Act 1957 132
Navarro, Z. 95, 130
Neabel, L. 9
Nelson, Paul 82
neoliberalism 1, 13, 29, 296, 310
Nepal 95
Netherlands 5, 6
Networks and Netwars 81n
Network of Peasants' and Producers' Organizations of West Africa (ROPPA) 3, 30, 114, 116
New Economic Policy (NEP) 6
New Order Regime 210, 216
New Zealand 243
Newell, Peter 17, 19, 20, 21, 29, 177–204
Newsweek 3
Nguyen Ai-quoc *see* Ho Chi Minh
Nhampossa, Diamantino 113
Nicaragua 4, 17, 61, 62, 63, 64, 67, 71n, 73, 75, 78, 79, 83, 108
niche markets 243, 249
Nicholson, Paul 26, 29, 46, 115
nobility, women from 9–10
nongovernmental organizations (NGOs)
  in China 301, 314–15
  overlap with popular organizations 81–2
  peasant movement relations 28–9
non-traditional crops 243
North America Free Trade Agreement (NAFTA) 74, 194, 199, 252, 254–5, 268n, 269, 270, 271

Novaes, R. 126
Novartis 195
Ntsebeza, L. 97, 132, 134
Nyamu-Musembi, C. 94
Nye, S. 92
Nyunt, M. 327

O'Brien, K. 300, 301, 303
O'Brien, R. 11, 18, 82, 113
Observatorio Social del Agro Mesoamericano 69n
Okamoto, I. 327
Olivera Lozano, F. 270
Onorati, Antonio 99n
Ontario 259
  Local Food Plus (LFP) 255–7
  Presidum for Red Fife Wheat 257
Organic Agriculture Association of South Africa 153
Organic Agriculture Network 168
organic food certification 21
Otero, G. 56
Owen, R. 47
Oxfam 68, 299
Oxfam-Novib 29
  Dutch 29
  Great Britain 69
Oya, C. 109n, 179

Paarlberg, R. 149, 157, 179, 180, 190
Padayachee, V. 163
Palan, R. 48
Pan Wei 315
Panama 61n, 64, 69
Paraguay 180, 195, 203
Parmigiano Cheese 247
Pastoral Land Commission 128 (CPT) 128, 130, 190
Patel, R. 21, 46, 49, 52, 92, 165
Peasant International 5–8
  founding congress 7
Peasant Movement of the Philippines (KMP) 24, 27, 96, 110–11, 116
Peasant Movement Training Institute (PMTI) 7
Peleaz, V. 177, 178, 181, 182, 184, 195, 203
Pelupessy, W. 72
Peluso, N. 22, 23, 25, 29, 107, 209–33
Pengue, W. 177
People's Food Sovereignty 41
Pereira, J. 96
Perry, E. 302, 306
Pertriaux, J-B. 49
Peru 177, 185, 202, 204

Peters, P. 113n
Petras, J. 26, 56, 127, 201
Petrini, C. 240, 247, 249, 259
Philippines 4, 15, 17, 24, 27, 95, 96, 105, 107, 110–11, 123, 222
Pianta, M. 11
Pick 'n Pay 153
Pietermaritzburg Agency for Christian Awareness 153
Pils, E. 305, 308
Pimbert, M. 49, 50, 151
Pithouse, R. 136
Piven, F. 334, 336
Plan Puebla-Panamá (PPP) 67, 68
Poland 5, 6, 15, 16, 17
Polanyi, K. 40, 47, 48, 241, 261
Polaski, S. 269
Pollan, M. 246
Poole, T. 298
Posas, M. 61
Powelson, J. 221
Prague International Agrarian Bureau 8
Prakash, C. S. 151
President's Choice brand 244
Projeto Cedula da Terra (PCT) 95, 96, 116
Pundeff, M. 5, 6
Putzel, J. 110

'quality' foods 243
Quinn-Judge, S. 7
Quiróz-Becerra, V. 275
Quist, D. 253

Rabadán, L. 278
Radič, Stjepan 7
  Croat Peasant Party 7
Rai, S. 11
Ray, R. 169
Raynolds, L. 240, 248
Razavi, S. 106
Reardon, T. 239
Red International of Trade Unions 7
Red Interamericana de Agriculturas y Democracia (RIAD) 188
Red Latinoamericana contra los Monocultivos de Arboles 188
Red por una América Latina Libre de Transgénicos (RALLT) 187, 188, 201
Reddy, Chengal 151
Reis, E. 126
remittances
  Chinese peasants 299
  Mexican migrant 270, 276–84
Requião, Roberto 193

Index 359

Research Foundation for Science Technology and Ecology (RFSTE)   151, 153, 168
Resosudarmo, B.   222, 228
Ribeiro, S.   253
Riechmann, J.   81
Rigg, J.   1, 16
Riles, A.   65n, 81
Rivera-Salgado, G.   268n, 275, 278
Robinson, W.   72, 76
Robison, R.   210, 213, 214, 215
Rocha, J.   126, 130
Rodrigues, Roberto   156, 203
Rodriguez, J.   189
Romania   5, 6, 8, 9
Ronfeldt, D.   81n
ROPPA *see* Network of Peasants' and Producers' Organizations of West Africa
Rosa, M.   127, 139, 140
Rosset, P.   1, 21, 41, 92, 93, 108n, 113n, 114n, 158
Rostow, W.   37
Roudart, L.   11, 37
Routledge, P.   123, 211
Ruaysoongnern, S.   55
Ruben, Raúl   76
Ruben, Ruerd   61, 76
Rubin, J. W.   15, 30
Rudolph, I. and S.   164
Ruggie, J. G.   54
Ruhl, J. M.   61
Rumansara, A.   216
Rupp, L.   9
Rural Advancement Foundation International (RAFI)   150, 153, 186n
Rural Coalition   16n
Russia   14 *see also* Soviet Union
Ruwiastuti, M.   218

Saachi, S.   61, 79
Sachs, Jeffrey   37
Safe Food Coalition   153
Sahai, Suman   168
Saiget, R.   307
Salamini, H.   28
Salim, Dr Emil   216
San Antonio Pajonal   80
Sanchez, F.   46
Sandinistas   61, 63
Sangkoyo, H.   225
Santos, B.   62n
Santos, M.   184, 196
Saragih, Henry   3, 23, 221, 224
Sassen, S.   92
Sauer, S.   95, 104

Scherr, S.   250
Schmeiser, Percy   158, 191, 198
Schmidt, W.   177, 179, 181, 182, 184, 195, 203
Scholte, J. A.   92
Schwartz, S.   125
Scoones, Ian   17, 19, 24, 25, 29, 107, 147–72, 181, 182, 183, 190, 195, 197
Scott, James   17, 18, 295, 300, 326, 331, 333, 336, 337
Seeing Like a State   17
Segovia, A.   71n, 72, 75, 76, 80
Seligmann, R.   126
Selth, A.   327
Senegal   27, 107, 109, 112
Seoane, J.   11
Serbia   7
Services and Advice to Alternative Agriculture Projects (ASPTA)   183
Shah, E.   157
Shanghai   7
Sharma, D.   41, 151n, 168
Shetkari Sanghatana   108
Shiva, Vandana   151, 152, 157, 158, 168, 191, 249
SICA *see* Consultative Council of the Central American Integration System
Sihlongonyane, M.   132, 140
Sikkink, K.   11, 20, 21, 22, 82, 93, 97, 99, 102, 105, 124, 203, 211, 217, 341
Silva, F.   11, 28
Silva Ávalos, C.   78, 80
Simon, H.   229
Sindicato Obrero del Campo (SOC)   109, 116
Sivaramakrishnan, K.   210
Slow Food   240, 246–9, 254
Slow Food Foundation for Biodiversity (SFBD)   248, 258, 259
  Presidium   246, 248, 259
Smith, J.   11, 102
Smith, M.   325
Smith, N.   123
Smith, R.   275
Social Assistance and Education Federation (FASE)   183, 190, 197
Social Movements Indaba   135
Social Origins of Dictatorship and Democracy   37
Socio-Economic Studies Institute (INESC)   183, 190
Solidarnosc-Rural   15–16
Solinger, D.   299
Somoza, Debayle, Anastasio   63
South Africa   4, 14–15, 17, 23, 95, 97, 106, 113n, 123–43
  GM crops   152–4

South Africa GMO Act 1997 147
South African Communist Party 136, 165
South African Council of Churches 153
South African Freeze Alliance on Genetic Engineering (SAFeAGE) 152, 153, 167
South Korea 106
Soviet Union 6, 11, 17, 18, 113
  New Economic Policy (NEP) 6
  see also Russia
soybeans 243
Spain 17, 26, 109, 115, 150
Stahlbrand, L. 256, 257
Stalin, Joseph 6–8
Stamboliski, Alexander 5–6
Stanford, L. 250
StarLink corn 185
Starn, O. 341
Starr, A. 47, 56
Stédile, João Pedro 15, 26, 112, 130, 184, 106, 112, 130, 184, 228, 231
Stein, Eduardo 71
Steinberg, D. 332
Stephen, L. 15
Stewart, C. 334
Stock, R. 221
Stockholm 65
Stone, G. 157
Suharto regime 213, 214, 216, 219, 220, 222
supermarkets 240, 241, 242–6
Surplus People's Project (SPP) 135, 221–2, 223, 224, 225, 226, 230, 231
Swiss Peasant Union 8
Switzerland 6
Swords, A. 54
Syngenta 153, 193

Taddei, E. 11
Tadem, E. 100
Tanaka, K. 244
Tang, H. 305
Tang, Y. 301
Tangermann, K-D. 64n
Tanner, M. 300, 301
Tarrow, S. 11, 21, 62n, 64, 91, 92, 95, 99, 102, 105, 114, 191, 309, 314, 334
Tatmadaw 324–5, 327, 329, 330–2, 337
Taylor, R. 11
Tegucigalpa 62, 67, 68, 69, 84
Tegucigalpa Protocol 64
Téllez, L. 269
Tenure Security Coordinating Committee (TSCC) 134
terminator genes 148, 149

Teubal M. 189
Thailand 44, 55
Thawnghmung, A. 327
The Countrywoman 9
The Great Transformation 261
Themudo, N. 62n
Thompson, E. 309
Thornton, P. 302, 308
Thrupp, L. 179
Thwala, W. 132
Tilly, C. 62n, 92, 126
Tiney, J. 69
Tomich, D. W. 40
Toni, A. 155
Toronto, University of 255–6
traceability 241–2
trade unions 300, 332
Transnational Agrarian Movements (TAMs) 2–31
  alliance building 27–30
  arenas of interaction 22–4
  China and 310–16
  early and late mobilizers 23
  historical antecedents 4–10
  ideological and political differences 26–7
  personality differences 27
  powerful actors within 20–1
  representation within 13
  similarities and differences 18–21
Transnational Peasant Network (TPN) 123–43
transnational social movements, gap in literature 10–12
Trees for Africa 153
TRIPS agreement 157
Trouillot, M. 51, 217
Tussie, D. 178, 189, 199

União Nacional de Camponesess (UNAC) 18, 107, 113, 116
Unión Nacional de Agricultores y Ganaderos (UNAG) 68, 108–9, 116, 68
Unión Nacional de Pequeños y Medianos Productores Agropecuarios (UPANACIONAL) 68
Unión Nacional de Productores Asociados (UNAPA) 68
Union Seed Act 150
United Nations (UN) 3, 9n, 101, 134, 211, 343
  Declaration of Independence (1960) 41
  World Conference Against Racism and Discrimination 134
  World Food Programme 27

United Nations Commission on Human Rights (UNCHR) 3
United States 9, 16, 17, 21, 38, 53, 72, 73, 78, 83, 135, 177, 179, 185, 186, 204, 243
 corn production in 252
UNORCA *see* National Coordination of Autonomous Local Rural People's Organizations
UPANACIONAL *see* Unión Nacional de Pequeños y Medianos Productores Agropecuarios
Uruguay 203

Valdés, I. 64n
van Dong, J. K. 97
van Tuijl, P. 104
Vandergeest, P. 213, 227
Vara, A. 180
Vasconcellos, Ronaldo 195
Vellema, S. 177
Veltmeyer, H. 26n, 106n, 201
Verhagen, Nico 114
Vía Campesina
 members of 3
 move to Indonesia 221
 multiple tactics and strategies 20
 origins 91–3
 silence on China 311, 313–14
 solidarity with non-members 341–6
Viales Hurtado, R. 78
Vietnam 74
Viotti da Costa, E. 125
Volunteer Alliance to Save the Environment (ARuPA) 229
von Braun, J. 155

Wakeford, T. 151
Waldinger, R. 275
WALHI *see* Indonesian Forum for the Environment
Walker, C. 97
Walker, Kathy Le Mons 15, 18, 113, 300, 309, 310
Wal Mart 251, 254
Walton, J. 37
Wargadipura, N. 221
Warren, C. 214, 220, 223, 226
Warwick, H. 179
Waterman, P. 11
Watt, Madge 9
Weapons of the Weak: Everyday Forms of Peasant Resistance 295
websites, Chinese 314
Webster, N. 211

Weil, R. 302
Wen Jiabao 308
Wen, T. 305
What the Countrywomen of the World Are Doing 9
Whitaker, C. 62n
Whittaker, R. 228
White, B. 209, 213
White International 6
Whole Foods 243
Wiebe, Nettie 53
Wilkinson, J. 244
Wittman, H. 46, 48
Wolf, Eric vii, 37
Wolford, Wendy 14, 20, 23, 39, 47, 52, 54, 55, 95n, 112, 123–43
Women's Institute (WI) 9
Wood, A. 231, 232
Woolworths 153
Works, M. 253
World Bank 3, 19, 22, 27, 42, 82, 84, 96, 99, 100, 101, 103, 104, 105, 112, 113, 215, 223, 242, 311, 312, 313, 323
World Conservation Union 153
World Forum of Fish Harvesters and Fishworkers (WFF) 2
World Forum of Fisher Peoples (WFFP) 2
World Network of Agricultures, Peasants, Food and Globalization (APM Mondial) 315
World Rainforest Movement (WRM) 186
World Rural Women's Day 10
World Social Forum (WSF) 62n, 155, 158, 162, 191, 199, 201, 313
World Summit on Sustainable Development 135, 158
World Trade Organization (WTO) 3, 19, 30, 39n, 41, 42, 43, 44, 47, 52, 56, 82, 166, 170, 180, 184, 239, 242, 312, 345
 China's entry into 299
World Wildlife Fund 257
Wright, A. 39, 47, 52, 95n, 126, 127
Wright, E. 241, 258
Wu, C. 301, 302, 306
Wynberg, R. 152

Xue, D. 162

Yao, R. 305
Yardley, J. 297, 298, 307, 315
Yashar, D. 11, 12n, 106n
Year of the World's Indigenous People 1993 211

Yeros, P. 211
YLBHI *see* Indonesian Legal Aid Foundation
Yu, D. 299
Yue, S. 302
Yugoslavia 5, 6

Zakaria, R. 225
Zapatista movement 270, 272
Zerner, C. 213
Zhu, S. 305
Zimbabwe 17, 113n, 125, 137, 138

Printed and bound by CPI Group (UK) Ltd, Croydon, CR0 4YY
09/06/2025

14685997-0002